Proofs

Definitions

Axioms

(An Alternative to Discrete Mathematics)

G. Viglino

Ramapo College of New Jersey

April 2018

CONTENTS

Chapter 1 A Logical Beginning

1.1	Propositions	1
1.2	Quantifiers	14
1.3	Methods of Proof	23
1.4	Principle of Mathematical Induction	33
1.5	The Fundamental Theorem of Arithmetic	43

Chapter 2 A Touch of Set Theory

2.1	Basic Definitions and Examples	53
2.2	Functions	63
2.3	Infinite Counting	77
2.4	Equivalence Relations	88
2.5	What is a Set?	99

Chapter 3 A Touch of Analysis

3.1	The Real Number System	111
3.2	Sequences	123
3.3	Metric Space Structure of $\Re$	135
3.4	Continuity	147

Chapter 4 A Touch of Topology

4.1	Metric Spaces	159
4.2	Topological Spaces	171
4.3	Continuous Functions and Homeomorphisms	185
4.4	Product and Quotient Spaces	196

Chapter 5 A TOUCH OF GROUP THEORY

5.1	Definition and Examples	207
5.2	Elementary Properties of Groups	219
5.3	Subgroups	227
5.4	Homomorphisms and Isomorphisms	236

Appendix A: Solutions to CYU-boxes

Appendix B: Answers to Selected Exercises

PREFACE

The underlying goal of this text is to help you develop an appreciation for the essential roles axioms and definitions play throughout mathematics.

We offer a number of paths toward that goal: a ***Logic*** path (Chapter 1); a ***Set Theory*** path (Chapter 2); an ***Analysis*** path (Chapter 3); a ***Topology*** path (Chapter 4); and an ***Algebra*** path (Chapter 5). You can embark on any of the last three independent paths upon completion of the first two:

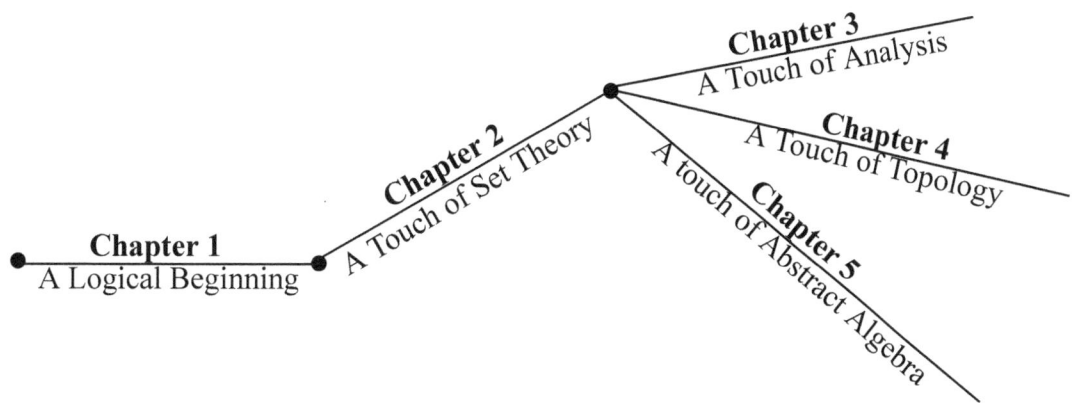

A one-semester course may include one, two, or even all three of the last three chapters, in total or in part, depending on the preparation of the students, and the ambition of the instructor.

For our part, we have made every effort to assist you in the journey you are about to take. We did our very best to write a readable book, without compromising mathematical integrity. Along the way, you will encounter numerous ***Check Your Understanding*** boxes designed to challenge your understanding of each newly-introduced concept. Detailed solutions to each of the Check Your Understanding problems appear in Appendix A, but you should only turn to that appendix after making a valiant effort to solve the given problem on your own, or with others. In the words of Desecrates:

> ***We never understand a thing so well, and make it our own, when we
> learn it from another, as when we have discovered it for ourselves.***

There are also plenty of exercises at the end of each section, covering the gamut of difficulty, providing you with ample opportunities to further hone your mathematical skills.

I wish to thank my colleague, Professor Marion Berger, for her numerous comments and suggestions throughout the development of the text. Her generous participation is deeply appreciated. I am also grateful to Professor Maxim Goldberg-Rugalev for his invaluable input.

CHAPTER 1
A Logical Beginning

Mathematics strives, as best it can, to transpose our physical universe into a non-physical form—a thought-universe, as it were, where definitions are the physical objects, and where the reasoning process rules. At some point, one has to ask whether we humans are creating mathematics or whether, through some kind of wonderful mental telescope, we are capable of discovering a meager portion of the mathematical universe which, in all of its glory, may very well be totally independent of us the spectators. Whichever; but one thing appears to be clear: the nature of mathematics, as we observe it or create it, appears to be inspired by what we perceive to be intuitive logic.

§1. PROPOSITIONS

We begin by admitting that "True" and "False" are undefined concepts, and that the word "sentence" is also undefined. Nonetheless:

DEFINITION 1.1 A **proposition** (or **statement**) is a sentence
PROPOSITION   that is True or False, but not both.

Fine, we have a definition, but its meaning rests on undefined words! We feel your frustration. Still, mathematics may very well be the closest thing to perfection we have. So, let's swallow our pride and proceed.

We can all agree that $5 + 3 = 8$ is a True proposition, and that $5 + 3 = 9$ is a False proposition. But how about this sentence:

THE SUM OF ANY TWO ODD INTEGERS IS EVEN.

First off, in order to understand the sentence one has to know all of its words, including: sum, integer, odd, and even. Only then can one hope to determine if the sentence is a proposition; and, if it is, to establish whether it is True or False. All in good time. For now, let us consider two other sentences:

HE IS AN AMERICAN CITIZEN. This sentence cannot be evaluated to be True or False without first knowing exactly who "*He*" is. It is a **variable proposition** (more formally called a **predicate**), involving the variable "*He.*" Once *He* is specified, then we have a proposition which is either True or False.

THE TWO OF US ARE RELATED. Even if we know precisely who the "*two of us*" are, we still may not have a proposition. Why not? Because the word "*related*" is just too vague.

COMPOUND PROPOSITIONS

The two numbers 3 and 5 can be put together in several ways to arrive at other numbers: $3+5$, $3-5$, $3 \cdot 5$, and so on. Similarly, two propositions, p and q, can be put together to arrive at other propositions, called compound propositions. One way, is to "and-them:"

DEFINITION 1.2
p and q
$p \wedge q$

Let p, q be propositions. The **conjunction of p and q** (or simply: **p and q**), written $p \wedge q$, is that proposition which is True if both p and q are True, and is False otherwise.

In truth table form:

p	q	$p \wedge q$
T	T	T
T	F	F
F	T	F
F	F	F

For example, the proposition:
$7 > 5$ **and** $3 + 5 = 8$ is True.
(since **both** $7 > 5$ and $3 + 5 = 8$ are True)

The proposition:
$7 > 5$ **and** $3 + 5 = 9$ is False.
(since $7 > 5$ and $3 + 5 = 9$ are not **both** True)

Another way to put two propositions together is to "or-them:"

DEFINITION 1.3
p or q
$p \vee q$

Let p, q be propositions. The **disjunction of p and q** (or simply: **p or q**), written $p \vee q$, is that proposition which is False when both p and q are False, and is True otherwise.

In truth table form:

p	q	$p \vee q$
T	T	T
T	F	T
F	T	T
F	F	F

Note that p or q is True when both p and q are True. As such, it is said to be the **inclusive-or**. The **exclusive-or** is typically used in conversation, as in: *Do you want coffee or tea* (one or the other, but not both)

For example, the proposition: $7 > 5$ **or** $3 + 5 = 8$ is True, as is the proposition $7 > 5$ **or** $3 + 5 = 9$. The proposition $7 < 5$ **or** $3 + 5 = 9$ is False, since neither $7 < 5$ nor $3 + 5 = 9$ is True.

CHECK YOUR UNDERSTANDING 1.1

Let p be the proposition $7 = 5$, and q be the proposition $3 + 5 = 8$.
(a) Is $p \vee q$ True or False? (b) Is $p \wedge q$ True or False?

(a) True (b) False

The above *and/or* operators, which act on **two** given propositions, are said to be **binary** operators. The following operator is a **unary** operator in that it deals with only **one** given proposition:

DEFINITION 1.4
NOT P

Let p be a proposition. The **negation of p**, read "**not p**," and written $\sim p$, is that proposition which is False if p is True, and True if p is False.

In truth table form:

p	$\sim p$
T	F
F	T

For example, the proposition:
$\sim(7 < 5)$ is True, since the proposition $7 < 5$ is False.

The proposition:
$\sim(3 + 5 = 8)$ is False, since the proposition $3 + 5 = 8$ is True.

CHECK YOUR UNDERSTANDING 1.2

Indicate if the proposition $\sim p$ is True of False, given that

(a) p is the proposition $5 > 3$.

(b) p is the proposition $\sim q$, where q is the proposition $3 = 5$.

(a) False (b) False

The negation operator "$\sim$" takes precedence (is performed prior) to both the "$\wedge$" and the "$\vee$" operators.

For example, the proposition:

$$\sim (3 > 5) \vee (3 + 5 = 9) \text{ is True.}$$
$$[\text{since } \sim (3 > 5) \text{ is True}]$$

As is the case with algebraic expressions, the order of operations in a logical expression can be overridden by means of parentheses.

For example, the proposition:

$$\sim [(5 > 3) \wedge (3 + 5 = 8)] \text{ is False}$$
$$[\text{since } (5 > 3) \wedge (3 + 5 = 8) \text{ is True }]$$

Brackets "[ ]" play the same role as parentheses, and are used to bracket an expression which itself contains parentheses.

CHECK YOUR UNDERSTANDING 1.3

Assume that p and q are True propositions, and that s is a False proposition. Determine if the given compound proposition is True or False.

(a) $\sim (p \wedge q)$ (b) $\sim (p \vee q)$ (c) $\sim (s \vee q)$ (d) $\sim (s \wedge q)$

(e) $\sim p \wedge q$ (f) $\sim p \vee q$ (g) $\sim s \vee q$ (h) $\sim s \wedge q$

(i) $(p \vee s) \wedge (q \wedge s)$ (j) $(p \vee s) \vee (q \wedge s)$ (k) $\sim (p \vee s) \wedge (q \vee s)$

True: d, f, g, h, j
False: a, b, c, e, i, k

CONDITIONAL STATEMENT

Consider the sentence:

IF IT IS MONDAY, THEN JOHN WILL CALL HIS MOTHER.

For it to become a proposition, we must attribute logical values to it (True or False); and so we shall:

DEFINITION 1.5

if p then q

$p \rightarrow q$

Let p, q be propositions. The **conditional of q by p**, written $p \rightarrow q$, and read *if p then q*, is False if p is True and q is False, and is True otherwise.

p is called the **hypothesis** and q is called the **conclusion** of $p \rightarrow q$.

In truth table form:

p	q	$p \rightarrow q$
T	T	T
T	F	F
F	T	T
F	F	T

You probably feel comfortable with the first two rows in the adjacent truth table but maybe not so much with the last two. Why is it that for p False, the proposition $p \rightarrow q$ is specified to be True for every proposition q, whether q is True or False?

In an attempt to put your mind a bit at ease, let's reconsider the conditional proposition:

IF IT IS MONDAY, THEN JOHN WILL CALL HIS MOTHER.

Suppose it is not Monday. Then John may or may not call his mother. So, if p is False (it is not Monday), then $p \to q$ should not be assigned a value of False.

> Yes, but can't the same "argument" be given to support the assertion that if p is False, then $p \to q$ should not be assigned a value of True? It could, but that option would lead us up an intuitively illogical path (see Exercise 85).

Does the statement:

IF IT IS MONDAY, THEN I WILL GO TO SCHOOL

allow you to go to school on Tuesday? Sure, but the following statement does not:

I WILL GO TO SCHOOL IF AND ONLY IF IT IS MONDAY

Or, if you prefer:

IT IS MONDAY IF AND ONLY IF I GO TO SCHOOL

Formally:

p	q	$p \leftrightarrow q$
T	T	T
T	F	F
F	T	F
F	F	T

DEFINITION 1.6

p if and only if q

p iff q

$p \leftrightarrow q$

Let p, q be propositions. The **biconditional of** p **and** q, written $p \leftrightarrow q$, and read p *if and only if* q (abbreviated "p *iff* q"), is True if p and q have the same truth values and is False if they have different truth values.

ORDER OF OPERATIONS

First "$\sim$", then "$\wedge, \vee$", and finally "$\to, \leftrightarrow$"
Use parentheses to establish precedence
between $\wedge, \vee$ and between $\to, \leftrightarrow$

LOGICAL IMPLICATION AND EQUIVALENCE

A **tautology** is a proposition which is necessarily always True; as is the case with the proposition $[(p \to q) \wedge p] \to q$:

The third column follows from Definition 1.5; the fourth from Definition 1.2, and the fifth from Definition 1.5

p	q	$p \to q$	$(p \to q) \wedge p$	$[(p \to q) \wedge p] \to q$
T	T	T	T	T
T	F	F	F	T
F	T	T	F	T
F	F	T	F	T

— Definition 1.5 —

> Subjected to the logical implication " $\rightarrow$ " the proposition:
> *If pigs can fly then Donald Duck is the president of the United States*
> is True.
>
> It is a well-known fact that most mathematicians ignore flying pigs. Typically, they start with a True proposition in the hope of being able to establishing the validity of another.

We use the symbol $p \Rightarrow q$, read **p implies q**, to mean that if the proposition p is True, than q must also be True; equivalently, that $p \rightarrow q$ is a tautology:

> For given propositions p and q, the only thing that can keep $p \rightarrow q$ from being a tautology is if p is True and q is False (Definition 1.5).

In particular, as we have just observed: $[(p \rightarrow q) \wedge p] \Rightarrow q$.

> Note that $p \Rightarrow q$ is not a proposition for it does not assume values of True or False. It is an assertion (sometimes called a *meta-proposition*).

> **It is only logical**:
> If p or q is True and p is False, then q must be True.

EXAMPLE 1.1 (a) Show that $[(p \vee q) \wedge \sim p] \Rightarrow q$.
(b) Show that $[(p \rightarrow q) \wedge (q \rightarrow s)] \Rightarrow (p \rightarrow s)$.

SOLUTION: (a) We verify that $(p \vee q) \wedge \sim p \rightarrow q$ is a tautology:

> The Truth-Values in the fifth column are a consequence of the Truth-Values in columns 3 and 4 (Definition 1.2). The tautology (column 6) follows from Definition 1.5.

p	q	$p \vee q$	$\sim p$	$(p \vee q) \wedge \sim p$	$[(p \vee q) \wedge \sim p] \rightarrow q$
T	T	T	F	F	T
T	F	T	F	F	T
F	T	T	T	T	T
F	F	F	T	F	T

(b) We verify that $[(p \rightarrow q) \wedge (q \rightarrow s)] \rightarrow (p \rightarrow s)$ is a tautology:

p	q	s	$p \rightarrow q$	$q \rightarrow s$	$(p \rightarrow q) \wedge (q \rightarrow s)$	$p \rightarrow s$	$[(p \rightarrow q) \wedge (q \rightarrow s)] \rightarrow (p \rightarrow s)$
T	T	T	T	T	T	T	**T**
T	T	F	T	F	F	F	**T**
T	F	T	F	T	F	T	**T**
T	F	F	F	T	F	F	**T**
F	T	T	T	T	T	T	**T**
F	T	F	T	F	F	T	**T**
F	F	T	T	T	T	T	**T**
F	F	F	T	T	T	T	**T**

> **CHECK YOUR UNDERSTANDING 1.4**
>
> Verify:
> $$[(p \rightarrow q) \wedge \sim q] \Rightarrow \sim p$$

Answer: See page A-1.

DEFINITION 1.7
LOGICALLY EQUIVALENT

Two propositions p and q are **logically equivalent**, written $p \Leftrightarrow q$, if $p \Rightarrow q$ and $q \Rightarrow p$.

In other words:
$p \Leftrightarrow q$ if the proposition $p \leftrightarrow q$ is a tautology; or, equivalently:
p is True if and only if q is True (Definition 1.6).

Augustus DeMorgan (1806-1871).

THEOREM 1.1
DEMORGAN'S LAWS

(a) $\sim(p \wedge q) \Leftrightarrow \sim p \vee \sim q$

(b) $\sim(p \vee q) \Leftrightarrow \sim p \wedge \sim q$

PROOF:

(a)

The driving force in the table was to arrive at the truth-values of the compound propositions: $\sim(p \wedge q)$ and $\sim p \vee \sim q$

p	q	$\sim p$	$\sim q$	$p \wedge q$	$\sim(p \wedge q)$	$\sim p \vee \sim q$
T	T	F	F	T	F	F
T	F	F	T	F	T	T
F	T	T	F	F	T	T
F	F	T	T	F	T	T

— same —

Observing that the values of $\sim(p \wedge q)$ and $\sim p \vee \sim q$ coincide, we conclude that $\sim(p \wedge q) \Leftrightarrow \sim p \vee \sim q$.

(b) $\sim(p \vee q) \Leftrightarrow \sim p \wedge \sim q$:

p	q	$\sim p$	$\sim q$	$p \vee q$	$\sim(p \vee q)$	$\sim p \wedge \sim q$
T	T	F	F	T	F	F
T	F	F	T	T	F	F
F	T	T	F	T	F	F
F	F	T	T	F	T	T

— same —

EXAMPLE 1.2 Use DeMorgan's laws to negate the given proposition:

(a) Mary is a math major and lives on campus.
(b) Bill will go running or swimming.
(c) $5 < x \leq 19$.

SOLUTION:

(a) $\sim$ (Mary is a math major $\wedge$ Mary lives on campus)
$\Leftrightarrow \sim$(Mary is a math major) $\vee \sim$(Mary lives on campus):

Mary is not a math major **or** she does not live on campus

(Either Mary is not a math major or she does not live on campus.)

(b)   ~ [Bill will go running ∨ Bill will go swimming]
⇔ ~(Bill will go running) ∧ ~(Bill will go swimming):

Bill will not go running **and** he will not go swimming

(Bill will neither go running nor swimming)

(c) The negation of the proposition $5 < x \leq 19$:
$$5 < x \text{ AND } x \leq 19$$
is the proposition
$$5 \not< x \text{ OR } x \not\leq 19$$
equivalently: $5 \geq x \text{ OR } x > 19$

CHECK YOUR UNDERSTANDING 1.5

Use DeMorgan's laws to negate the given proposition:
(a) Mary is going to a movie or she is going shopping.
(b) Bill weighs more than 200 pounds and is less than 6 feet tall.
(c) $x > 0$ or $x \leq -5$.

Answer: See page A-2.

There are several distributive properties involving propositions. Here is one of them:

THEOREM 1.2 $p \vee (q \wedge s) \Leftrightarrow (p \vee q) \wedge (p \vee s)$

PROOF: The eight possible value-combinations of p, q, and s, are listed in the first three columns of the table below. The remaining columns speak for themselves:

p	q	s	$q \wedge s$	$p \vee (q \wedge s)$	$p \vee q$	$p \vee s$	$(p \vee q) \wedge (p \vee s)$
T	T	T	T	**T**	T	T	**T**
T	T	F	F	**T**	T	T	**T**
T	F	T	F	**T**	T	T	**T**
T	F	F	F	**T**	T	T	**T**
F	T	T	T	**T**	T	T	**T**
F	T	F	F	**F**	T	F	**F**
F	F	T	F	**F**	F	T	**F**
F	F	F	F	**F**	F	F	**F**

↑——— same ———↑

In general, a truth table involving n propositions will contain 2^n rows. That being the case, they quickly become "unmanageably tall": A four-proposition-table calls for $2^4 = 16$ rows, while a five-proposition-table already has 32 rows.

EXAMPLE 1.3   Are $(p \wedge q) \rightarrow \sim p$ and $p \wedge (q \rightarrow \sim p)$ logically equivalent?

SOLUTION: No:

p	q	$p \wedge q$	$\sim p$	$(p \wedge q) \to \sim p$	$q \to \sim p$	$p \wedge (q \to \sim p)$
T	T	T	F	F	F	F
T	F	F	F	T	T	T
F	T	F	T	T	T	F
F	F	F	T	T	T	F

↑ not the same truth values, so ↑
not logically equivalent

CHECK YOUR UNDERSTANDING 1.6
Verify: (a) $\sim(\sim p) \Leftrightarrow p$ (b) $p \to q \Leftrightarrow \sim p \vee q$ (c) $\sim(p \to q) \Leftrightarrow \sim(\sim p \vee q)$ (d) $p \leftrightarrow q \Leftrightarrow (p \to q) \wedge (q \to p)$

Answer: See page A-2.

To say that $p \to q$ is **Not True** is to say that p is True **and** q is False. To put it another way:

THEOREM 1.3   NEGATION OF A CONDITIONAL PROPOSITION:
$$\sim(p \to q) \Leftrightarrow p \wedge \sim q$$

PROOF: We offer two proofs for your consideration:

Truth Table Method:

p	q	$p \to q$	$\sim(p \to q)$	$\sim q$	$p \wedge \sim q$
T	T	T	F	F	F
T	F	F	T	T	T
F	T	T	F	F	F
F	F	T	F	T	F

Using Established Results

             ↙ CYU 1.6(b)
$\sim(p \to q) \Leftrightarrow \sim(\sim p \vee q)$
1.1: $\Leftrightarrow \sim(\sim p) \wedge \sim q$
CYU 1.6(a) $\Leftrightarrow p \wedge \sim q$

Formal proof aside, we suggest that this result is intuitively appealing. After all, to say that it is not true that p implies q, is to say that p is true while q is false.

While the above truth table method is the more straight-forward approach, the development on the right is a better representation of what you will be doing in subsequent math courses. Indeed, truth be told:

Once you leave this course, you may never again encounter a truth table.

CONTRAPOSITIVE

Are the following two statements saying the same thing?

IF IT IS MONDAY, THEN I WILL GO TO SCHOOL.

IF I DO NOT GO TO SCHOOL, THEN IT IS NOT MONDAY.

Yes:

This is a very useful result. It tells you that:

    as $p \to q$ goes
    so does $\sim q \to \sim p$

THEOREM 1.4 $p \to q \Leftrightarrow \sim q \to \sim p$

PROOF:

p	q	$p \to q$	$\sim p$	$\sim q$	$\sim q \to \sim p$
T	T	T	F	F	T
T	F	F	F	T	F
F	T	T	T	F	T
F	F	T	T	T	T

↑——— same ———↑

> We note that the proposition $\sim q \to \sim p$ is said to be the **contrapositive** of the proposition $p \to q$.

Getting a bit ahead of the game, we point out that by virtue of 1.4, one can establish the validity of the proposition:

If n^2 is even then n is even

by verifying that:

If n is odd, then n^2 is not even

CHECK YOUR UNDERSTANDING 1.7

Prove that $[(p \vee q) \to \sim s] \Leftrightarrow [s \to (\sim p \wedge \sim q)]$ using:
(a) A truth table.   (b) Previously established results.

Answer: See page A-3.

CONVERSE AND INVERSE OF A CONDITIONAL STATEMENT

DEFINITION 1.8 The **converse** of $p \to q$ is $q \to p$.

The **inverse** of $p \to q$ is $\sim p \to \sim q$.

For example:

Proposition	$p \to q$	If it is Monday, then I will go to school.
Converse	$q \to p$	If I go to school, then it is Monday.
Inverse	$\sim p \to \sim q$	If it is not Monday, then I will not go to school.

It is important to note that neither the converse nor the inverse of a conditional proposition is logically equivalent to the given proposition:

p	q	$p \to q$	$q \to p$	$\sim p$	$\sim q$	$\sim p \to \sim q$
T	T	T	T	F	F	T
T	F	F	T	F	T	T
F	T	T	F	T	F	F
F	F	T	T	T	T	T

↑ not the same ↑   ↑ not the same ↑

Truth table aside, consider the True statement:
  If n is divisible by 4, then n is even.
Its converse is False:
  If n is even, then n is divisible by 4.
  (6 is even but is not divisible by 4)
Its inverse is also False:
  If n is not divisible by 4, then n is not even.
  (6 is not divisible by 4 but is even)

So, the inverse is the contrapositive of the converse.

Answer: See page A-3.

CHECK YOUR UNDERSTANDING 1.8

Prove that the converse and the inverse of $p \to q$ are logically equivalent.

SET NOTATION

A bit of set notation is needed for the development of the remainder of this chapter. Roughly speaking, a **set** is a collection of objects, or elements. A "small" set can be specified by simply listing all of its elements inside brackets, as is done with the set A below:

$$A = \{-2, 5, 11, 99\}$$

The order in which elements appear in a set is of no consequence. The sets $\{1, 2, 3\}$ and $\{2, 3, 1\}$ for example, are considered to be one and the same, or equal.

The above so-called **roster method** can also be used to denote a set whose elements follow a discernible pattern, as is done with the set O of odd positive integers below:

$$O = \{1, 3, 5, 7, 9, \ldots\}$$

The **descriptive method** can also be used to define a set. In this method, a statement or condition is used to specify the elements of the set, as is done with the set O below:

$$O = \{x | x \text{ is an odd positive integer}\}$$

Read: O is the set of all x such that x is an odd positive integer

For a given set A, $x \in A$, is read: x **is an element of** A (or x is contained in A), and $x \notin A$ is read: x is **not** an element of A. For example:

$$5 \in \{-2, 5, 11, 99\} \quad \text{while} \quad 9 \notin \{-2, 5, 11, 99\}$$

Finally, we note that throughout the text:

$\Re$ denotes the set of **real numbers**.

$\Re^+$ denotes the set of **positive real numbers**.

Z denotes the set of **integers**: $Z = \{\ldots, -3, -2, -1, 0, 1, 2, 3, \ldots\}$.

Z^+ denotes the set of **positive integers**: $Z^+ = \{1, 2, 3, 4, \ldots\}$.

Q denotes the set of rational numbers ("fractions").

Q^+, the set of positive rational numbers.

	EXERCISES	

Exercises 1-10. State whether the given proposition is True or False.

1. 10 is an even number.
2. 15 is an even number.
3. 10 or 15 is an even number.
4. 10 and 15 are even numbers.
5. $3 < 2$ or $(5 > 7$ or 9 is odd$)$
6. $3 < 2$ and $(5 > 7$ or 9 is odd$)$
7. $3 < 2$ or $(5 > 7$ and 9 is odd$)$
8. $(3 < 2$ and $5 > 7)$ or 9 is odd
9. $(3 < 2$ or $5 > 7)$ and 9 is odd
10. $(3 < 2$ and $5 > 7)$ or 9 is not odd

Exercises 11-31. Assume that p and q are True propositions, and that r and s are False propositions. Determine if the given compound proposition is True or False.

11. $p \wedge s$
12. $\sim(p \wedge s)$
13. $\sim(p \wedge \sim s)$
14. $r \wedge s$
15. $\sim(r \wedge s)$
16. $\sim(r \wedge \sim s)$
17. $r \vee s$
18. $\sim(r \vee s)$
19. $\sim(r \vee \sim s)$
20. $(p \vee s) \wedge (r \wedge s)$
21. $(p \vee s) \vee (r \wedge s)$
22. $(p \wedge q) \wedge (\sim r \wedge \sim s)$
23. $(p \wedge q) \vee (r \wedge s)$
24. $(p \wedge q) \wedge (r \vee s)$
25. $(p \wedge \sim r) \wedge (q \vee s)$
26. $\sim(p \wedge q) \vee \sim(r \wedge s)$
27. $\sim[(p \wedge q) \vee (r \wedge s)]$
28. $\sim(p \wedge q) \wedge (p \wedge s)$
29. $(p \wedge q) \wedge \sim(r \wedge s)$
30. $\sim[(p \vee r) \wedge \sim(p \wedge s)]$
31. $\sim[(p \wedge q) \wedge \sim(r \wedge \sim s)]$

Exercises 32-39. Determine if the given statement is a tautology.

32. $p \vee \sim p$
33. $p \vee (\sim p \vee q)$
34. $p \wedge (\sim p \vee q)$
35. $p \vee (\sim p \wedge q)$
36. $(\sim p \vee q) \vee (p \wedge \sim q)$
37. $(p \wedge q) \vee [\sim p \vee (p \wedge \sim q)]$
38. $[(p \wedge q) \to s] \to [p \to (q \to s)]$
39. $[(p \vee q) \to s] \to [\sim p \to (q \to s)]$

Exercises 40-43. Proceed as in Example 1.1 to establish the given implication.

40. $p \wedge q \Rightarrow p$
41. $p \wedge q \Rightarrow p \vee \sim q$
42. $(p \vee q) \wedge (\sim p \wedge q) \Rightarrow q$
43. $p \wedge (\sim p \vee q) \Rightarrow p \vee q$

Exercises 44-59. Determine if the given pair of statements are logically equivalent.

44. $p \vee (p \wedge q), p$
45. $p \vee (p \wedge q), q$
46. $(p \wedge \sim q) \vee p, p$
47. $(p \wedge \sim q) \vee p, q$

48. $\sim(p \wedge \sim q), q \vee \sim p$

49. $\sim[(p \vee \sim q) \vee (\sim p \wedge \sim q)], \sim q$

50. $\sim[p \vee (\sim p \wedge q)], \sim p \wedge \sim q$

51. $\sim[p \wedge (\sim p \vee q)], \sim p \vee \sim q$

52. $\sim(p \to q), p \wedge (\sim q)$

53. $\sim(p \to q), \sim q \to \sim p$

54. $(p \wedge q) \vee (p \wedge s), p \vee (q \wedge s)$

55. $(p \vee q) \vee (p \wedge s), (p \vee q) \wedge s$

56. $(p \to r) \wedge (q \to s), (p \vee q) \to s$

57. $(p \to s) \wedge (q \to s), p \vee (q \to s)$

58. $(p \to q) \wedge (q \to s), p \to s$

59. $(p \to q) \leftrightarrow (q \to s), p \leftrightarrow s$

Exercises 60-62. Establish the given logical equivalence.

60. **Commutative Laws:**
    (a) $p \wedge q \Leftrightarrow q \wedge p$
    (b) $p \vee q \Leftrightarrow q \vee p$

61. **Associative Laws:**
    (a) $(p \wedge q) \wedge r \Leftrightarrow p \wedge (q \wedge r)$
    (b) $(p \vee q) \vee r \Leftrightarrow p \vee (q \vee r)$

62. **Distributive Laws:**
    (a) $p \wedge (q \vee r) \Leftrightarrow (p \wedge q) \vee (p \wedge r)$
    (b) $p \vee (q \wedge r) \Leftrightarrow (p \vee q) \wedge (p \vee r)$
    Already proved (Theorem 1.2)

Exercises 63-65. Indicate True or False. For any proposition p and any tautology t:

63. $p \wedge t \leftrightarrow p$

64. $p \vee t \leftrightarrow p$

65. $p \vee t \leftrightarrow t$

66. **Contradiction.** A contradiction is a proposition which is False for all possible values of the statements from which it is constructed; as is the case with $p \wedge \sim p$:

p	$\sim p$	$p \wedge \sim p$
T	F	F
F	T	F

(a) Verify that $(\sim p \vee q) \wedge (p \wedge \sim q)$ is a contradiction.

(b) Verify that $(p \wedge q) \wedge [\sim p \wedge (p \vee \sim q)]$ is a contradiction.

(c) Indicate True or False. For any proposition p and any contradiction c:
   (i) $p \wedge c \Leftrightarrow p$ (ii) $p \wedge c \Leftrightarrow c$ (iii) $p \vee c \Leftrightarrow p$

(d) Indicate True or False. For any tautology t and any contradiction c:
   (i) $t \wedge c$ is a tautology.
   (ii) $t \vee c$ is a tautology.
   (iii) $t \wedge c$ is a contradiction.
   (iv) $t \vee c$ is a contradiction.

(e) Indicate True or False: $(\sim t \vee c) \wedge p$ is a contradiction for any tautology t, contradiction c, and proposition p.

Exercises 67-74. Negate the given statement.

67. Joe is a math major and Mary is a biology major.

68. Joe is a math major or Mary is a biology major.

69. Joe is neither a math major nor a biology major.

70. Mary does not live in the dorms but does eat lunch in the cafeteria.

71. $3x + 5 = y$ and both x and y are integers.

72. x is divisible by both 2 and 3, but is not divisible by 7.

73. x is not divisible by either 2 or 3, but is divisible by 7.

74. x is greater than y and z, but is less than $y - z$.

Exercises 75-83. Formulate the contrapositive, converse, and inverse of the given conditional statement.

75. If it rains, then I will stay home.

76. If it does not rain, then I will go to the game.

77. If Nina feels better, then she will either go to the library or go shopping.

78. If either Jared or Tommy gets paid, then the two of them will go to the concert.

79. If $X = Z$ then $M > N$.

80. If $X \neq Z$ then M is a solution of the given equation.

81. If $X \neq Z$ then M or N is a solution of the given equation.

82. If $X > Z$ then M and N are solutions of the given equation.

83. If $X < Z$ then the given equation has no solution.

84. **Exclusive-Or.** Let p and q be propositions. Define, via a Truth Table, the exclusive-or proposition: p or q but not both p and q.

85. **In defense of Definition 1.5.**

(a) Prove that If we chose to define:

p	q	$p \to q$
T	T	T
T	F	F
F	T	F
F	F	T

Then $p \to q \Leftrightarrow q \to p$ (which runs contrary to our logical instincts)

(b) Prove that If we chose to define:

p	q	$p \to q$
T	T	T
T	F	F
F	T	T
F	F	F

Then $p \to q \not\Leftrightarrow \sim q \to \sim p$ (which runs contrary to our logical instincts)

(c) What if we chose to define:

p	q	$p \to q$
T	T	T
T	F	F
F	T	F
F	F	F

Anything "bad" happens?

> Chances are you already know most of the material of this section. Not necessarily because you formally saw it before, but because it is, well, "logical." You know, for example, that to disprove the claim that *every dog has fleas* one needs to exhibit a specific dog which has no fleas. On the other hand, finding a dog with fleas will not prove that all dogs have fleas.

§2. QUANTIFIERS

In the previous section, we considered the variable proposition: *He is an American citizen.* As it stands, the sentence is not a (determined) proposition, for its truth value depends on the variable "*He*." Here is another variable proposition, $p(x)$:

$$p(x): \quad x + 5 \geq 15$$

Once we substitute a number for the variable x, we arrive at a proposition which can be evaluated to be True or False. For example:

$$p(7) \text{ is False, since } 7 + 5 \not\geq 15$$

while $p(25)$ is True, since $25 + 5 \geq 15$

Now consider the sentences:

For all real numbers x, $x + 5 \geq 15$ (*)

For some real numbers x, $x + 5 \geq 15$ (**)

Are they propositions? Yes: (*) is a False proposition since it fails to hold for the number 7 (among other numbers), while (**) is a True proposition since it holds for the number 25 (among others). In both cases a variable x is present, but in each case it is quantified by a condition: "**For all**" in (*), and "**For some**" in (**). Lets begin by focusing on the "**For all**" situation:

THE UNIVERSAL QUANTIFIER: $\forall$

The symbol $\forall$, called the **universal quantifier**, is read "for all."

The proposition $\forall x \in \Re, x + 5 \geq 15$ is False, since it does **not** hold **for every** real number x.

The proposition $\forall x \in \Re, x^2 \geq 0$ is true, since it **does** hold **for every** real number x.

> We remind you that $\Re$ denotes the set of real numbers, and that $x \in \Re$ is read: x is an element of $\Re$; which is to say, x is a real number.

In general:

DEFINITION 1.9

UNIVERSAL PROPOSITION

Let $p(x)$ be a variable proposition, and let X denote the set of values which the variable x can assume (called the **domain** of x).

If $p(x)$ is True $\forall x \in X$, then the universal proposition $\boxed{\forall x \in X, p(x)}$ is **True**.

If $p(x)$ is false for at least one $x \in X$, then the universal proposition $\boxed{\forall x \in X, p(x)}$ is **False** (any such $p(x)$ is said to be a **counterexample**).

The above is a long definition, but it's really quite straightforward. Consider the following examples.

We remind you that Z denotes the set of all integers, and that Z^+ denotes the set of all positive integers.

The expression $\forall n, m \in Z^+$ in (c) is shorthand for:

$\forall n \in Z^+$ and $m \in Z^+$

EXAMPLE 1.4 Indicate if the given proposition is True or False. Justify your claim.

(a) $\forall n \in Z^+, 2n > n$

(b) $\forall n \in Z, 2n \geq n$

(c) $\forall n, m \in Z^+, n + m < nm$

SOLUTION: (a) We show $\forall n \in Z^+, 2n > n$ is True:

For any $n \in Z^+$: $2n = n + n > n + 0 = n$

(b) We show that the proposition $\forall n \in Z, 2n \geq n$ is False by exhibiting a **specific counterexample**:

For $n = -5$, $2n = 2(-5) = -10 < -5$

Any negative integer whatsoever can be used to show that the given proposition is False.

(c) The proposition $\forall n, m \in Z^+, n + m < nm$ is also False. We need to display a **counterexample** — a <u>SPECIFIC</u> pair of integers, n and m, for which $n + m \not< nm$. Here is one such pair: $n = 1$ and $m = 2$:

$n + m = 1 + 2 = 3$ while $nm = 1 \cdot 2 = 2$

CHECK YOUR UNDERSTANDING 1.9

Indicate if the given proposition is True or False. If False, exhibit a counterexample.

(a) All months have at least thirty days.

(b) Every month contains (at least) three Sundays.

(c) $\forall n, m \in Z^+, n + m \in Z^+$.

(d) $\forall n \in Z^+$ and $\forall m \in Z, n + m \in Z^+$

(b) and (d) are True
(a) and (d) are False

THE EXISTENTIAL QUANTIFIER: $\exists$

The symbol "$\exists$," called the **existential quantifier**, is read "there exists."

The proposition $\exists n \in Z^+$ such that $n + 3 = 5$ is True, since $2 \in Z^+$ and $2 + 3 = 5$.

The proposition $\exists n \in Z^+$ such that $n + 5 = 3$ is False, for there does not exist a positive integer which when added to 5 yields 3.

We point out that the symbol "$\ni$" is read "**such that**." For example:

$\exists n \in Z^+ \ni n + 5 = 3$ translates to: $\exists n \in Z^+$ such that $n + 5 = 3$

and: $\exists x \in X \ni p(x)$ translates to $\exists x \in X$ such that $p(x)$.

In general:

The proposition:
$\exists x \in X \ni p(x)$
can be further abbreviated:
$\exists x \in X, p(x)$

DEFINITION 1.10

EXISTENTIAL PROPOSITION

Let $p(x)$ be a variable proposition with domain X.

If $p(x)$ is True for at least one $x \in X$, then the existential proposition $\boxed{\exists x \in X \ni p(x)}$ is **True**.

If $p(x)$ is false for every $x \in X$, then the existential proposition $\boxed{\exists x \in X \ni p(x)}$ is **False**.

EXAMPLE 1.5 Indicate if the given proposition is True or False.
(a) $\exists n \in Z \ni 2n < -n$
(a) $\exists x \in \Re \ni 2x = x$
(b) $\exists n \in Z^+ \ni 2n = n$

SOLUTION:
(a) $\exists n \in Z \ni 2n < -n$ is True: for $n = -5$, $2n = -10 < 5 = -n$.
_{Any negative integer whatsoever can be used to establish the validity of this proposition}

(b) $\exists x \in \Re \ni 2x = x$ is True: for $x = 0$, $2x = 2 \cdot 0 = 0 = x$.
<sub>Only the number 0 can be used to establish the validity of this proposition.
One is all we needed!</sub>

(c) $\exists n \in Z^+ \ni 2n = n$ is False: for any positive integer n, $2n > n$.

CHECK YOUR UNDERSTANDING 1.10

Indicate if the given proposition is True or False. If True, verify.
(a) There exists a month with more than thirty days.
(b) There exists a week with more than seven days.
(c) $\exists n, m \in Z^+ \ni n + m = 100$
(d) $\exists n, m \in Z^+ \ni nm = n + m$

(a), (c), and (d) are True.
(b) is False.

PROPOSITIONS CONTAINING MULTIPLE QUANTIFIERS

When a statement contains more than one quantifier, the left-most quantifier takes precedence. For example, the proposition:
$$\forall n \in Z, \exists m \in Z \ni n + m = 5$$
is True, since:
> For **every** integer n, there does exists an integer m such that $n + m = 5$; namely, $m = 5 - n$.

On the other hand, the proposition:
$$\exists n \in Z \ni \forall m \in Z, n + m = 5$$
is False, since:
> For **any** given integer n we can find some m for which $n + m \neq 5$. One such m is $6 - n$: $n + (6 - n) \neq 5$.

The four mathematical sentences of this example are nice and compact. Their interpretation may call for a less compact consideration. Please try to arrive at each answer before looking at our solution.

EXAMPLE 1.6 Indicate if the given proposition is True or False.

(a) $\forall n, m \in Z^+ \exists s \in Z^+ \ni ns = m$.

(b) $\forall n, m \in Z^+ \exists s \in \Re \ni ns = m$.

(c) $\exists n \in Z \ni \forall m \in Z^+ \; n < m$.

(d) $\exists n \in Z \ni \forall m \in Z \; n \leq m$.

SOLUTION:

(a) Here is what the proposition "$\forall n, m \in Z^+ \exists s \in Z^+ \ni ns = m$" is saying:

> *For all positive integers, n and m, there exists a positive integer, s, such that ns = m.*

Is it True? No:

> For $n = 2$ and $m = 1$, there does not exist a positive integer s such that $ns = m$; which is to say, there is no positive integer s such that $2s = 1$

(b) $\forall n, m \in Z^+, \exists s \in \Re \ni ns = m$ is True:

> For given $n, m \in Z^+$ let $s = \dfrac{m}{n} \in \Re$. Then: $ns = n \cdot \dfrac{m}{n} = m$.

(c) Here is what the proposition $\exists n \in Z \ni \forall m \in Z^+, n < m$ is saying:

> *There is an integer that is smaller than every positive integer.*

This proposition is True:

We had to **exhibit a particular n**, and went with $n = -7$. Any non-positive integer would do as well.

> The integer $n = -7$ is smaller than every positive integer.

(d) The proposition $\exists n \in Z \ni \forall m \in Z, n \leq m$ is False:

> For any given $n \in Z$, the integer $m = n - 1$ is less than n.

CHECK YOUR UNDERSTANDING 1.11

Indicate if the given proposition is True or False. Justify your claim.

(a) $\forall n, m \in Z^+, \exists s \in Z^+ \ni s > nm$.

(b) $\forall s \in Z^+, \exists n, m \in Z^+ \ni s > nm$.

(c) $\exists x \in \Re \ni \forall n \in Z^+, x^n > x$.

(d) $\exists n \in Z^+ \ni \forall m \in Z^+, n^m = n$.

(a), and (d) are True.
(b) and (c) are False

NEGATION OF QUANTIFIED PROPOSITIONS

The negation of the proposition: *All dogs have fleas.*
is **NOT**: ~~*No dog has fleas*~~.

It **IS** the proposition: *At least one dog does not have fleas.*

In general:

> The negation of the universal proposition:
> $$\forall x \in X, p(x)$$
> is the existential proposition:
> $$\exists x \in X \ni \sim p(x)$$

EXAMPLE 1.7 Negate the given proposition.

(a) All things go bang in the night.

(b) $\forall x \in X, x$ is this and x is that.

(c) $\forall x \in X, x$ is this or x is that.

SOLUTION:

(a) The negation of: *All things go bang in the night.*
   is: Some thing does not go bang in the night.

(b) The negation of: $\forall x \in X, x$ is this and x is that.
   is: $\exists x \in X$ such that x is not this or x is not that.
   (possibly neither)

(c) The negation of: $\forall x \in X, x$ is this or x is that.
   is: $\exists x \in X$ such that x is not this and x is not that.

CHECK YOUR UNDERSTANDING 1.12

Write an existential negation of the given universal proposition.

(a) All college students study hard.

(b) Everyone takes a bath at least once a week.

(c) $\forall x \in X, p(x) \wedge q(x)$

(d) $\forall x \in X, p(x) \vee q(x)$

Answer: See page A-4.

The negation of the proposition: *Some music rocks.*
Is the proposition: *No music rocks*
(or: *all music does not rock*)

In general:

> The negation of the existential proposition:
> $$\exists x \in X \ni p(x)$$
> is the universal proposition:
> $$\forall x \in X, \sim p(x)$$

EXAMPLE 1.8 Negate the given proposition.
   (a) There is a perfect person.
   (b) $\exists x \in X$ such that x is this and x is that.
   (c) $\exists x \in X$ such that x is this or x is that.

SOLUTION:
(a) The negation of: *There is a perfect person.*
   is: *Nobody is perfect.*
(b) The negation of: $\exists x \in X$ such that x is this and x is that.
   is: $\forall x \in X$, x is not this or x is not that.
   (possibly neither)
(c) The negation of: $\exists x \in X$ such that x is this or x is that.
   is: $\forall x \in X$, x is not this and x is not that.

CHECK YOUR UNDERSTANDING 1.13

Write a universal negation of the given existential proposition.
   (a) There are days when I don't want to get up.
   (b) $\exists x \in X \ni [p(x) \wedge q(x)]$
   (c) $\exists x \in X \ni [p(x) \vee q(x)]$

Answer: See page A-4.

The negation of the universal/existential proposition:
   In every day there is a beautiful moment.
Is the existential/universal proposition:
   Some day contains no beautiful moment.
In general:

> The negation of the universal/existential proposition:
> $$\forall x \in X, \exists y \in Y \ni p(x, y)$$
> is the existential/universal proposition:
> $$\exists x \in X \ni \forall y \in Y, \sim p(x, y)$$

The notation $p(x, y)$ is used to indicate that the truth value of the proposition, p, is a function of two variables, x and y.

EXAMPLE 1.9 Negate the given proposition.
   (a) To every problem there is a solution.
   (b) $\forall x \in \Re, \exists n \in Z^+ \ni n > x$

SOLUTION:
(a) The negation of: *To every problem there is a solution.*
   is: *There is a problem that has no solution.*
(b) The negation of: $\forall x \in \Re, \exists n \in Z^+ \ni n > x$
   is: $\exists x \in \Re \ni \forall n \in Z^+, n \leq x$

CHECK YOUR UNDERSTANDING 1.14

Write an existential/universal negation of the given universal/existential proposition.

(a) For every $x \in X$ there exists a $y \in Y$ such that y blips at x.

(b) $\forall x \in \Re, \exists n \in Z^+ \ni x = 2n$

Answer: See page A-4.

The negation of the existential/universal proposition:
  There is a girl at school who can jump higher than every boy.
Is the universal/existential proposition:
  For any girl in school, there is a boy who can jump as high or higher than that girl.
  Or: *No girl in school can jump higher than every boy.*

In general:

> The negation of the existential/universal proposition:
> $\exists x \in X \ni \forall y \in Y, p(x, y)$
> is the universal/existential proposition:
> $\forall x \in X, \exists y \in Y \ni \sim p(x, y)$

EXAMPLE 1.10 Negate the given proposition.

(a) Some cheetahs run faster than every gazelle.

(b) $\exists x \in \Re \ni \forall n \in Z^+, 0 < xn \leq 100$

(c) $\exists x \in \Re \ni \forall n \in Z^+, xn \leq 0$ or $xn > 100$

SOLUTION:

(a) The negation of: *Some cheetahs run faster than every gazelle.*
  is: *For any cheetah there is a gazelle that runs as fast or faster than that cheetah.*

(b) The negation of: $\exists x \in \Re \ni \forall n \in Z^+, 0 < xn \leq 100$
  is: $\forall x \in \Re, \exists n \in Z^+ \ni xn \leq 0$ or $xn > 100$.

(c) The negation of: $\exists x \in \Re \ni \forall n \in Z^+, xn \leq 0$ or $xn > 100$
  is: $\forall x \in \Re, \exists n \in Z^+ \ni 0 < xn \leq 100$.

CHECK YOUR UNDERSTANDING 1.15

Write a universal/existential negation of the given existential/universal proposition.

(a) There is a motorcycle that gets better mileage than any car.

(b) $\exists n \in Z \ni \forall m \in Z^+ \, n < m$

Answer: See page A-4.

EXERCISES

Exercises 1-51. Indicate if the given statement is True or False.
If you indicate False, then justify your claim by means of a specific counterexample. To illustrate:
$$\forall a, b \in \Re: (a+b)^2 = a^2 + b^2 \text{ is False: } (3+2)^2 = 25 \text{ while } 3^2 + 2^2 = 13.$$

1. $\forall n \in Z, -n \leq n+1$
2. $\forall r \in \Re, -r \neq r$
3. $\exists n \in Z \ni -n \leq n+1$
4. $\exists r \in \Re \ni -r \neq r$
5. $\forall n, m \in Z, nm > n+m$
6. $\forall n, m \in Z^+, nm \in Z$
7. $\exists n, m \in Z \ni nm \in Z^+$
8. $\forall n, m \in Z: n > m \text{ or } m > n$
9. $\forall a, b \in \Re, |a+b| = |a|+|b|$
10. $\forall a, b \in \Re, |a+b| \leq |a|+|b|$
11. $\exists a, b \in \Re \ni |a+b| = |a|+|b|$
12. $\forall n, m \in Z^+, |n+m| = |n|+|m|$
13. $\forall n, m \in Z, \exists s \in Z \ni n+s = m$
14. $\forall n, m \in Z, \exists s \in Z^+ \ni n+s = m$
15. $\forall x, y \in \Re, \exists n \in Z \ni x+n = y$
16. $\forall x, y \in \Re, \exists r \in \Re \ni x+r = y$
17. $\forall n \in Z: \sqrt{n^2} = n$
18. $\forall n \in Z^+: (\sqrt{n})^2 = n$
19. $\forall x, y \in \Re, \exists r \in \Re \ni xr = y$
20. $\forall n \in Z^+, \exists a, b \in Z^+ \ni a+n = b$
21. $\forall n \in Z, \exists a, b \in Z^+ \ni a+n = b$
22. $\forall n \in Z^+, \exists a, b \in Z \ni a+n = b$
23. $\forall n \in Z^+, n < 2n$
24. $\forall n \in Z, n < 2n$
25. $\exists n \in Z^+ \ni n < 2n \text{ and } n \geq 2n$
26. $\exists n \in Z^+ \ni n < 2n \text{ or } n \geq 2n$
27. $\forall n \in Z^+ \exists r \in \Re \ni 5r = n$
28. $\forall r \in \Re \exists n \in Z^+ \ni 5n = r$
29. $\exists n \in Z \ni n > n^2$
30. $\exists n \in Z^+ \ni n < n^2$
31. $\exists n \in Z^+ \ni n \leq n^2$
32. $\exists x, y \in \Re \ni x = x+y$
33. $\forall n \in Z^+, n \leq n^2$
34. $\forall x, y \in \Re \exists s \in \Re \ni x+s = x+y$
35. $\exists x, y \in \Re \ni (x+y)^2 < 2xy$
36. $\forall x, y \in \Re: (x+y)^2 < 2xy$
37. $\exists n \in Z \ni \forall m \in Z^+, nm < n$
38. $\exists n \in Z^+ \ni \forall m \in Z, n \neq m$
39. $\forall n \in Z, \exists m \in Z^+ \ni nm > n$
40. $\forall n \in Z^+, \exists m \in Z \ni nm > n$
41. $\exists n \in Z \ni \forall m \in Z^+, n \neq m$
42. $\exists n, m \in Z \ni \forall s \in Z, n+m \neq s$
43. $\forall n \in Z, \exists m \in Z \ni n = 2m$
44. $\forall a \in \Re, \exists b \in \Re \ni a = 2b$
45. $\exists n, m \in Z^+ \ni \forall s \in Z, n+m \neq s$
46. $\exists n, m \in Z^+ \ni \forall s \in Z^+, n+m \neq s$

47. $\exists n \in Z$ and $x \in \Re \ni \forall m \in Z^+, x + m > n$
48. $\exists n \in Z^+$ and $x \in \Re \ni \forall m \in Z^+, n + m < x$
49. $\exists n \in Z^+$ and $x \in \Re \ni \forall m \in Z^+, n + m > x$
50. $\exists n \in Z^+ \ni \forall m, s \in Z, n + m > n + s$
51. $\exists n \in Z^+ \ni \forall m, s \in Z^+, nm = ns$
52. $\exists n \in Z \ni \forall m, s \in Z^+, nm = ns$

Exercises 53-64. Write an existential negation of the given universal proposition.

53. All roads lead to Rome.
54. Every cloud has a silver lining.
55. Every time it rains, it rains pennies from heaven.
56. All good things must come to an end.
57. $\forall n \in Z, n < a$ for $a \in \Re$
58. $\forall x \in \Re, x \geq a$ for $a \in \Re$
59. $\forall n \in Z, n < a$ and $n < b$ for $a, b \in \Re$
60. $\forall x \in \Re, x < a$ or $x > b$
61. $\forall x \in X, p(x) \wedge q(x) \wedge s(x)$
62. $\forall x \in X, p(x) \vee q(x) \vee s(x)$
63. $\forall x \in X, p(x) \wedge [q(x) \vee s(x)]$
64. $\forall x \in X, [p(x) \vee q(x)] \wedge s(x)$

Exercises 65-74. Write a universal negation of the given existential proposition.

65. There is a reason for everything.
66. There is no room for error.
67. $\exists n \in Z \ni n < a$ for $a \in \Re$
68. $\exists n \in Z \ni n < a$ for $a \in \Re$
69. $\exists n \in Z \ni n < a$ and $n \geq b$ for $a, b \in \Re$
70. $\exists x \in \Re \ni x < a$ or $x > b$ for $a, b \in \Re$
71. $\exists x \in X \ni p(x) \wedge q(x) \wedge s(x)$
72. $\exists x \in X \ni p(x) \vee q(x) \vee s(x)$
73. $\exists x \in X \ni p(x) \wedge [q(x) \vee s(x)]$
74. $\exists x \in X \ni [p(x) \vee q(x)] \wedge s(x)$

Exercises 75-84. Write an existential/universal negation of the given universal/existential proposition.

75. Everybody loves somebody.
76. All dogs go to heaven.
77. Every day contains a special moment.
78. Every good deed has a reward.
79. $\forall x \in Z, \exists y \in Z \ni x + y = 0$
80. $\forall x \in Z, \exists y \in \Re \ni x \neq y$
81. $\forall a, b \in \Re, \exists m, n \in Z \ni a + b = mn$
82. $\forall a, b \in Z, \exists x \in \Re \ni a + b = x$
83. $\forall x \in Z, \exists a, b \in R \ni a = x + b$ or $b = a + x$
84. $\forall x \in \Re$ and $y \in \Re, \exists z \in Z \ni z = x$ or $z = y$

Exercises 85-94. Write a universal/existential negation of the given existential/universal proposition.

85. Some operas are longer than every symphony.
86. There is a solution to every problem.
87. Someone is greater than everyone else.
88. At some point in time, everything will be fine.
89. $\exists x \in \Re \ni \forall y \in \Re, x + y = 0$
90. $\exists x \in \Re \ni \forall y \in \Re, x \neq y$
91. $\exists a, b \in \Re \ni \forall m, n \in Z, a + b = mn$
92. $\exists a, b \in Z \ni \forall x \in \Re, a + b = x$
93. $\exists x \in \Re \ni \forall a, b \in \Re, a = x + b$ or $b = a + x$
94. $\exists x \in \Re$ and $y \in Z \ni \forall z \in Z, z = x$ or $z = y$

§3. METHODS OF PROOF

THROUGHOUT THIS SECTION WE WILL BE DEALING EXCLUSIVELY WITH INTEGERS.
(A TOUCH OF NUMBER THEORY)

DIRECT PROOF

The **direct method** of proving $p \Rightarrow q$ is to assume that p is True and then apply mathematical reasoning to deduce that q is True.

Let us accept the fact that: n is odd if and only if it is not even.

DEFINITION 1.11
EVEN AND ODD

$n \in Z$ is **even** if $\exists k \in Z \ni n = 2k$.

$n \in Z$ is **odd** if $\exists k \in Z \ni n = 2k+1$.

For example, $n = 14$ is even since:

$$14 = 2 \cdot \overset{k}{\underset{\downarrow}{7}}$$

While $n = -23$ is odd since:

$$-23 = 2(\overset{k}{\underset{\downarrow}{-12}}) + 1$$

EXAMPLE 1.11 Prove that the sum of any two even integers is even.

*A proof is like a journey, which begins at a given point (the **hypothesis**), and ends at a given point (the **conclusion**). You must be mindful of both the beginning and the end of the journey. For if you know where you want to go (the conclusion), but do not know where you are (the hypothesis), then chances are that you aren't going to reach your destination. Moreover, if you know where you are, but do not know where you are going, then you probably won't get there; and even if you do, you won't know it.*

SOLUTION: The hypothesis and conclusion of the journey establishes a frame within which you are to paint a logical path:.

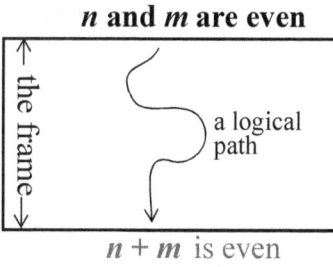

Let n and m be even integers.
By Definition 1.11, $\exists k, h \in Z$ such that:
$$n = 2k \text{ and } m = 2h$$
Then:
$$n + m = 2k + 2h = 2(k+h)$$
Thus: $n + m$ is even (Definition 1.11).

Note how Definition 1.11 is used in both directions in the above proof. It was used in one direction to accommodate the given information that n and m are even integers, and was then used in the other direction to conclude that $n + m$ is even.

CHECK YOUR UNDERSTANDING 1.16

(a) Prove that the sum of any two odd integers is even.

(b) Formulate a conjecture concerning the sum of an even integer with an odd integer. Establish the validity of your conjecture.

Answer: See page A-4.

EXAMPLE 1.12 Prove that $2m + n$ is odd **if and only if** n is odd.

SOLUTION: We need to establish validity in both directions; namely:

(1) If $2m + n$ is odd, then n is odd.

(2) If n is odd, then $2m + n$ is odd.

Lets do it, beginning with (1):

If $2m + n$ is odd, then $2m + n = 2k + 1$ for some k.

Solving for n we have: $n = 2k + 1 - 2m = 2(k - m) + 1$.
Conclusion: n is odd.

Now for (2):

If n is odd, then $n = 2k + 1$ for some k.

Consequently: $2m + n = 2m + 2k + 1 = 2(m + k) + 1$.

Conclusion: $2m + n$ is odd.

Sometimes, one may be able establish both directions of an "if and only if" proposition simultaneously; as we are able to do here:

$$\begin{aligned} 2m + n \text{ is odd} &\overset{\text{Definition 1.11}}{\Leftrightarrow} 2m + n = 2k + 1 \\ &\Leftrightarrow n = 2k - 2m + 1 \\ &\Leftrightarrow n = 2(k - m) + 1 \underset{\text{Definition 1.11}}{\Leftrightarrow} n \text{ is odd} \end{aligned}$$

CHECK YOUR UNDERSTANDING 1.17

Answer: See page A-4.

Prove that $2m + n$ is even **if and only if** n is even.

CONTRAPOSITIVE PROOF

$p \to q \Leftrightarrow \sim q \to \sim p$

Theorem 1.4, page 8 (see margin) tells us that if you can show that $\sim q \Rightarrow \sim p$, then $p \Rightarrow q$ will follow. Formality aside:

Suppose you know that $\sim q \Rightarrow \sim p$. Can p be True and q False? No, for if p were True and q False, then $\sim p$ would be False and $\sim q$ True — contradicting $\sim q \Rightarrow \sim p$.

Another argument:

Suppose (i): $\sim q \Rightarrow \sim p$. Can it be that (ii): $p \Rightarrow \sim q$? No, for if so:

$$p \underset{\text{(ii)}}{\Rightarrow} \sim q \underset{\text{(i)}}{\Rightarrow} \sim p \quad \text{Tisk!}$$

EXAMPLE 1.13 Show that for all integers n:
(a) n is even if and only if n^2 is even.
(b) n is odd if and only if n^2 is odd.

PROOF: (a) We use a direct argument to show that

$$n \text{ even} \Rightarrow n^2 \text{ even}:$$

$$\mathbf{n \text{ even}} \Rightarrow n = 2k \text{ (for some } k\text{)}$$

$$\Rightarrow n^2 = 4k^2 = 2(2k^2)$$

$$\Rightarrow \mathbf{n^2 \text{ is even}}$$

We use a contrapositive argument (margin) and show that:

$$n^2 \text{ even} \Rightarrow n \text{ even}$$

Specifically, we show that: n not even n not even $\Rightarrow n^2$ not even i.e :

$$n \text{ odd} \Rightarrow n^2 \text{ odd}:$$

$$\mathbf{n \text{ odd}} \Rightarrow n = 2k + 1 \text{ (for some } k\text{)}$$

$$\Rightarrow n^2 = (2k+1)^2$$

$$= 4k^2 + 4k + 1$$

$$= 2(2k^2 + 2k) + 1 \Rightarrow \mathbf{n^2 \text{ odd}}$$

(b) n is odd $\Leftrightarrow n$ is not even

Part (a): $\Leftrightarrow n^2$ is not even

$\Leftrightarrow n^2$ is odd

> A proof which established the validity of $p \Rightarrow q$ by showing that $\sim q \Rightarrow \sim p$ is said to be a **Contrapositive Proof**.

CHECK YOUR UNDERSTANDING 1.18

Prove that:
(a) $3n$ is odd if and only if n is odd.
(b) n^3 is odd if and only if n is odd.

Answer: See page A-5.

PROOF BY CONTRADICTION

One can establish that a proposition p is True by demonstrating that the assumption that p is False leads, via a logical argument, to a False conclusion. Invoking the "logical commandment" that from *Truth only Truth can follow*, one can then conclude that the assumption that p is False must itself be False, and that therefore p has to be True.

> This method of proof is called: **proof by contradiction** or, if you prefer Latin: *reductio ad absurdum*.

Whatever you can do by the contrapositive method can also be done by the contradiction method. To illustrate, in Figure 1.1(b) we recall the contrapositive proof of n^2 even $\Rightarrow n$ even given in Example 1.13(a), and offer a proof by contradiction in Figure 1.1(b). As you can see there is precious little difference between the two methods. In (b), rather than showing that n not even $\Rightarrow n^2$ not even, we start off accepting he given condition that n^2 is even, and then go on to show that the assumption that n is not even contradicts that given condition.

$$n^2 \text{ is even} \Rightarrow n \text{ is even}$$

Contrapositive Proof:
We show n odd $\Rightarrow n^2$ odd :

n **odd** $\Rightarrow n = 2k + 1$ (for some k)

$\Rightarrow n^2 = (2k+1)^2$
$\quad\quad\, = 4k^2 + 4k + 1$
$\quad\quad\, = 2(2k^2 + 2k) + 1 \Rightarrow n^2$ **odd**

(a)

Proof by Contradiction:

Suppose n^2 is even (given condition), and **assume** that n is odd, say $n = 2k + 1$ for some k. Then:
$n^2 = 4k^2 + 4k + 1$
$\quad\, = 2(2k+2) + 1 \Rightarrow n^2$ is odd

— contradicting our stated assumption that n^2 **is even**.

(b)

Figure 1.1

We have exhibited three methods of proof:

| Direct Proof | Contrapositive Proof | Proof by Contradiction |

In attempting to prove something, which method should you use? No general answer can be provided, other than if one does not work then try another. We do note, however, that a direct proof is considered to be more aesthetically pleasing than either a contrapositive proof or a proof by contradiction, but a direct proof may not always be a viable option.

EXAMPLE 1.14 Prove that if $3n + 2$ is even, then n is even.

SOLUTION: Let's try to prove that $3n + 2$ even $\Rightarrow n$ even using a direct approach:

$3n + 2 = 2k \Rightarrow 3n = 2k - 2$ — Now what? If you try something like $n = \frac{2k-2}{3}$, where does it take you (nowhere, other than to an expression that may be outside the realm of integers)?

Okay the direct approach does not appear to work, should we try the contrapositive approach or the proof by contradiction approach? That is basically a matter of taste, and so we offer both methods for your consideration:

$3n + 2$ even $\Rightarrow n$ even

Contrapositive	**Proof by Contradiction:**
Proof: n not even $\Rightarrow 3n + 2$ not even i.e: n odd $\Rightarrow 3n + 2$ odd	Let $3n + 2$ be **even** (given condition)
n odd $\Rightarrow n = 2k + 1$ (for some k)	**Assume** that n is odd, say $n = 2k + 1$.
$\Rightarrow 3n + 2 = 3(2k + 1) + 2$ $= 6k + 5$ $= 6k + 4 + 1$ $= 2(3k + 2) + 1$ $\Rightarrow 3n + 2$ odd	Then: $3n + 2 = 3(2k + 1) + 2$ $= 6k + 5 = 6k + 4 + 1$ $= 2(3k + 2) + 1$ $\Rightarrow 3n + 2$ **odd** — contradicting the stated condition that **$3n + 2$ is even**.

CHECK YOUR UNDERSTANDING 1.19

Answer: See page A-5.

Prove that $3n + 2$ is odd if and only if n is odd.

Note that while $\frac{15}{5}$ is the number 3. $5 \mid 15$ is **not** a number; it is the statement that there exists an integer k (in this case 3) such that $15 = 5k$.

DEFINITION 1.12
DIVISIBILITY

We say that a nonzero integer a **divides** an integer b, written $a \mid b$, if $b = ak$ for some integer k.

In the event that $a \mid b$, we say that b is **divisible** by a and that b is a **multiple** of a.

Here are two statements for your consideration:

(a) If $a \mid b$ and $b \mid c$, then $a \mid c$.

(b) If $a \mid bc$, then $a \mid b$ or $a \mid c$.

Let's Challenge (a) with some specific integers:

Challenge 1: 3 divides 9 and 9 divides 18; does 3 divide 18? Yes.

Challenge 2: 2 divides 4 and 4 divides 24; does 2 divide 24? Yes.

Challenge 3: 5 divides 55 and 55 divides 110; does 5 divide 110? Yes.

The above "Yeses" may certainly suggest that (a) does indeed hold for all $a, b, c \in Z$ — suggest, yes, but **NOT PROVE**. A proof is provided in Theorem 1.5(a), below.

Statement (b): If $a \mid bc$, then $a \mid b$ or $a \mid c$ is False.

A **counterexample**: $6 \mid (2 \cdot 3)$ but $6 \nmid 2$ and $6 \nmid 3$.

TYPICALLY:

A GENERAL ARGUMENT is needed to establish the validity of a statement.
A (<u>specific</u>) COUNTEREXAMPLE is needed to establish that a statement is False.

THEOREM 1.5 Let b and c be nonzero integers. Then:

(a) If $a|b$ and $b|c$, then $a|c$.

(b) If $a|b$ and $a|c$, then $a|(b+c)$.

(c) If $a|b$, then $a|bc$ for every c.

PROOF: (a) If $a|b$ and $b|c$, then, by Definition 1.12:
$$b = ak \text{ and } c = bh \text{ for some } h \text{ and } k.$$
Consequently:
$$c = bh = (ak)h = a(kh) = at \text{ (where } t = kh).$$
It follow, from Definition 1.12, that $a|c$.

> Note how Definition 1.12 is used in both directions in the above proof.

(b) If $a|b$ and $a|c$, then $b = ah$ and $c = ak$ for some h and k. Consequently:
$$b+c = ah+ak = a(h+k) = at \text{ (where } t = h+k).$$
It follows that $a|(b+c)$.

(c) If $a|b$, then $b = ak$ for some k. Consequently, for any c:
$$bc = (ak)c = a(kc) = at \text{ (where } t = kc).$$
It follows that $a|bc$.

EXAMPLE 1.15 Prove or give a counterexample.

(a) If $a|(b+c)$, then $a|b$ or $a|c$.

(b) If $a|b$ and $a|(b+c)$, then $a|c$.

SOLUTION: (a) Unless you are fairly convinced that a given statement is True, you may want to start off by challenging it:

Challenge 1. $4|(8+16)$, and 4 certainly divides 8 or 16 (in fact, it divides both 8 and 16). Inconclusive.

Challenge 2. $4|(3+1)$, and 4 divides neither 3 nor 1 — a **counterexample**! The statement is False.

(b) Challenging the statement "If $a|b$ and $a|(b+c)$, then $a|c$" will not yield a counterexample. It can't, since the statement is True. To establish its validity, a **general argument** is called for.

One proof:

Since $a|b$, there exists h such that: (1) $b = ah$.

Since $a|(b+c)$, there exists k such that: (2) $b+c = ak$.

(Now we have to go ahead and show that $c = at$ for some t)

From (2): $c = ak - b$.

From (1): $c = ak - ah = a(k-h)$.

Since $c = at$ (where $t = k-h$): $a|c$.

$a|b$ and $a|(b+c)$
↓
$a|c$

An alternate proof: Observing that:
$$c = (b+c) - b = (b+c) + (-1)b$$

we employ Theorem 1.6(b) and (c) to conclude that $a|c$.

(Recall that we are given that $a|(b+c)$ and that $a|b$)

CHECK YOUR UNDERSTANDING 1.20

(a) Prove:
- (a-i) If $a|n$ and $a|m$ then $a|(n+m)$.
- (a-ii) If $n|a$ then $n|ca$ for every $c \in Z$.

(b) Prove or give a counterexample:
- (b-i) If $a|b$ or $a|c$, then $a|(b+c)$.
- (b-ii) If a and b are even and if $a|(b+c)$, then c must be even.
- (b-iii) If $(a+b)|(c+d)$, then there exist k such that:
$$ak + bk + c + d = 0.$$

Answer: See page A-6.

As you worked your way through this section, you must have observed how:

DEFINITIONS RULE!

They are the physical objects in the mathematical universe.

30 Chapter 1 A Logical Beginning

EXERCISES

Exercises 1-25. Establish the validity of the given statement.

1. 0 is even and 1 is odd.
2. The product of any two even integers is even.
3. The product of any two even integers is divisible by 4.
4. The sum of any two odd integers is even.
5. If $a|b$ then $a^2|b^2$.
6. $3|n$ if and only if $3|(n+9)$.
7. If $a|b$, and $a|c$ then $a|(bn+cm)$ for every n and m.
8. If $a|c$, and $b|d$ then $ab|cd$.
9. If $5n-7$ is even then n is odd.
10. $11n-7$ is even if and only if n is odd.
11. If $4|(n^2-m^2)$ then either n and m are both even or they are both odd.
12. n^2+3n+5 is odd for all n.
13. $3n+1$ is even if and only if n is odd.
14. n^3 is even if and only if n is even.
15. If n^4 is even then so is $3n$.
16. $3n^3$ is even if and only if $5n^2$ is even.
17. $5n-11$ is even if and only if $3n-11$ is even.
18. If $3n+5m$ is even, then either n and m are both even or they are both odd.
19. The square of every odd integer is of the form $4n+1$ for some $n \in Z$.
20. If $4|(n^2-1)$ then $2|(n-1)$.
21. Let $b = aq+r$. Prove that if $c|a$ and $c|b$, then $c|r$.
22. If bc is not a multiple of a, then neither b nor c can be a multiple of a.
23. If $b+c$ is a multiple of a, and b is a multiple of a, then c is a multiple of a.
24. If a and b are odd positive integers and if $c|(a+b)$, then c is even.
25. If a is odd and b is even and if $c|(a+b)$, then c is odd.

Exercises 26-33. Disprove the given statement. .

26. The sum of any two even integers is divisible by 4.
27. The sum of any three odd integers is divisible by 3.
28. The product of any two even integers is divisible by 6.
29. If $4|(n^2-1)$ then $4|(n-1)$.
30. If $2m+n$ is odd then both m and n are odd.
31. If $2m+n$ is even then both m and n are even.
32. If $a|b$ or $a|c$ then $a|(b+c)$.
33. If $a|b$ and $b|a$ then $a=b$.

PROVE OR GIVE A COUNTEREXAMPLE

34. If n is odd and m is even then $n-m$ is odd.
35. If $n+m$ is odd then neither n nor m can be even.
36. If $n+m$ is even then neither n nor m can be odd.
37. If $n+m$ is odd then n or m must be odd.
38. If $n+m$ is even then n or m must be even.
39. If $n+1$ is even then so is n^3-1.
40. If $n+1$ is even then so is n^2-1.
41. If $n+1$ is even then so is n^3+1.
42. If $n+1$ is even then so is n^2+1.
43. If $n+1$ is odd then so is n^3-1.
44. If $n+1$ is odd then so is n^2-1.
45. If $n+1$ is odd then so is n^3+1.
46. If $n+1$ is odd then so is n^2+1.
47. If a is even and b is odd then a^2+2b is even.
48. If a is even and b is odd then a^2+3b is odd.
49. If a is odd then so is a^2+2a.
50. If a is odd then so is a^2+3a.

51. If a is even and b is odd then $(a+2)^2 + (b-1)^2$ is even.

52. If $9|(n+3)$ then $3|n$

53. If $(a+2)^2 + (b-1)^2$ is even then a or b has to be even.

54. If $a|b, b|c$, and $c|a$, then $a = b = c$.

55. If $a|b, b|c$, and $c|a$, then $a = b$ or $a = c$ or $b = c$.

§4. Principle of Mathematical Induction

A form of the Principle of Mathematical Induction is actually one of Peano's axioms, which serve to define the positive integers.
[Giuseppe Peano (1858-1932).]

This section introduces a most powerful mathematical tool, the Principle of Mathematical Induction (*PMI*). Here is how it works:

PMI		
Let $P(n)$ denote a proposition that is either true or false, depending on the value of the integer n.		
If:	I.	$P(1)$ is True.
And if, from the **assumption** that:	II.	$P(k)$ is True
one can show that:	III.	$P(k+1)$ is also True.
then the proposition $P(n)$ is valid **for all integers** $n \geq 1$		

Step II of the induction procedure may strike you as being a bit strange. After all, if one can assume that the proposition is valid at $n = k$, why not just assume that it is valid at $n = k+1$ and save a step! Well, you can assume whatever you want in Step II, but if the proposition is not valid for all n you simply are not going to be able to demonstrate, in Step III, that the proposition holds at the next value of n. It's sort of like the domino theory. Just imagine that the propositions $P(1), P(2), P(3), \ldots, P(k), P(k+1), \ldots$ are lined up, as if they were an infinite set of dominoes:

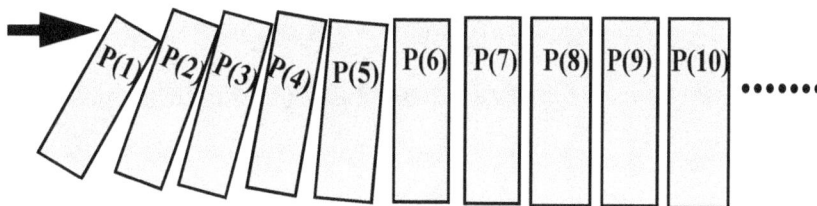

If you knock over the first domino (Step I), and if when a domino falls (Step II) it knocks down the next one (Step III), then all of the dominoes will surely fall. But if the falling k^{th} domino fails to knock over the next one, then all the dominoes need not fall.

The *Principle of Mathematical Induction* might have been better labeled the *Principle of Mathematical Deduction*, for inductive reasoning is used to formulate a hypothesis or conjecture, while deductive reasoning is used to rigorously establish whether or not the conjecture is valid.

To illustrate how the process works, we ask you to consider the sum of the first n odd integers, for $n = 1$ through $n = 5$:

Sum of the first n odd integers	Sum
1	1
1 + 3	4
1 + 3 + 5	9
1 + 3 + 5 + 7	16
1 + 3 + 5 + 7 + 9	25

n	Sum
1	1
2	4
3	9
4	16
5	25
6	?

Figure 1.2

34 Chapter 1 A Logical Beginning

Looking at the pattern of the table on the right in Figure 1.1, you can probably anticipate that the sum of the first 6 odd integers will turn out to be $6^2 = 36$, which is indeed the case. Indeed, the pattern suggests that: The sum of the first n odd integers is n^2

Using the Principle of Mathematical Induction, we now establish the validity of the above conjecture:

Let $P(n)$ be the proposition that the sum of the first n odd integers equals n^2.

I. Since the sum of the first 1 odd integers is 1^2, $P(1)$ is true.

II. **Assume** $P(k)$ is true; that is:
$$1 + 3 + 5 + \cdots + (2k-1) = k^2$$
see margin

III. We show that $P(k+1)$ is true, thereby completing the proof:

the sum of the first $k+1$ odd integers
$$[1 + 3 + 5 + \cdots + (2k-1)] + (2k+1) = k^2 + (2k+1) = (k+1)^2$$
induction hypothesis: Step II

> The sum of the first **3** odd integers is:
> $1 + 3 + 5 \leftarrow \boxed{2\cdot 3 - 1}$
> The sum of the first **4** odd integers is:
> $1 + 3 + 5 + 7 \leftarrow \boxed{2\cdot 4 - 1}$
> Suggesting that the sum of the first **k** odd integers is:
> $1 + 3 + \ldots + \boxed{(2k-1)}$
> (see Exercise 1).

EXAMPLE 1.16 Use the Principle of Mathematical Induction to establish the following formula for the sum of the first n integers:
$$1 + 2 + 3 + \ldots + n = \frac{n(n+1)}{2}$$

SOLUTION: Let $P(n)$ be the proposition:
$$1 + 2 + 3 + \ldots + n = \frac{n(n+1)}{2} \quad (*)$$

I. $P(1)$ is true: $1 = \frac{1(1+1)}{2}$ Check!

II. **Assume** $P(k)$ is true: $1 + 2 + 3 + \ldots + k = \frac{k(k+1)}{2}$

III. We are to show that $P(k+1)$ is true; which is to say, that $(*)$ holds when $n = k+1$:
$$1 + 2 + 3 + \ldots + k + (k+1) = \frac{(k+1)[(k+1)+1]}{2} = \frac{(k+1)(k+2)}{2}$$

Let's do it:
$$1 + 2 + 3 + \ldots + k + (k+1) = [1 + 2 + 3 + \cdots + k] + (k+1)$$
$$\text{induction hypothesis:} = \frac{k(k+1)}{2} + (k+1)$$
$$= \frac{k(k+1) + 2(k+1)}{2} = \frac{(k+1)(k+2)}{2}$$

1.4 Principle of Mathematical Induction   35

CHECK YOUR UNDERSTANDING 1.21

(a) Use the formula for the sum of the first n odd integers, along with that for the sum of the first n integers, to derive a formula for the sum of the first n even integers.

(b) Use the Principle of Mathematical Induction directly to establish the formula you obtained in (a).

Answer: See page A-6.

The "domino effect" of the Principle of Mathematical Induction need not start by knocking down the first domino $P(1)$. Consider the following example where domino $P(0)$ is the first to fall.

EXAMPLE 1.17 Use the Principle of Mathematical Induction to establish the inequality $n < 2^n$ for all $n \geq 0$.

SOLUTION: Let $P(n)$ be the proposition $n < 2^n$.

I. $P(0)$ is true: $0 < 2^0$, since $2^0 = 1$.

II. **Assume** $P(k)$ is true: $k < 2^k$.

III. We show $P(k+1)$ is true; namely that $k + 1 < 2^{k+1}$:

$$k + 1 < 2^k + 1 \leq 2^k + 2^k = 2(2^k) = 2^{k+1}$$

II ↑   ↑ $1 \leq 2^k$

EXAMPLE 1.18 Use the Principle of Mathematical Induction to show that $4 | (5^n - 1)$ for all integers $n \geq 0$.

SOLUTION: Let $P(n)$ be the proposition $4 | (5^n - 1)$.

I. $P(0)$ is true: $4 | (5^0 - 1)$, since $5^0 - 1 = 1 - 1 = 0$.

II. **Assume** $P(k)$ is true: $4 | (5^k - 1)$.

III. We show $P(k+1)$ is true; namely, that $4 | (5^{k+1} - 1)$:

$$5^{k+1} - 1 = 5(5^k) - 1 = 5(5^k) - 5 + 4 \text{ (see margin)}$$
$$= 5(5^k - 1) + 4$$

The desired conclusion now follows from CYU 1.17, page 24:

CYU 1.20(a), page 29: $4 | (5^k - 1) \Rightarrow 4 | 5(5^k - 1)$ and then:

CYU 1.20(b): $4 | 5(5^k - 1)$ and $4 | 4 \Rightarrow 4 | [5(5^k - 1) + 4]$

What motivated us to write −1 in the form −5 + 4? Necessity did:

We had to do something to get "$5^k - 1$" into the picture (see II).

Clever, to be sure; but such a clever move stems from stubbornly focusing on what is given and on what has to be established.

EXAMPLE 1.19 Use the Principle of Mathematical Induction to show that $3 | (2^{2n} - 1)$ for all integers $n \geq 1$.

SOLUTION: Let $P(n)$ be the proposition: $3|(2^{2n}-1)$.

I. $P(1)$ is true: $2^{2 \cdot 1} - 1 = 3$ Check!

II. **Assume** $P(k)$ is true: $3|(2^{2k}-1)$.

III. We show that $P(k+1)$ is true; which is to say: $3|(2^{2(k+1)}-1)$:

$$2^{2(k+1)} - 1 = 2^{2k+2} - 1$$
$$\boxed{a^{n+m} = a^n a^m:} = 2^2 \cdot 2^{2k} - 1$$
$$= 4 \cdot 2^{2k} - 1$$
$$\boxed{\text{wanting to get a 3 into the picture:}} = (3+1)2^{2k} - 1$$
$$\boxed{\text{regrouping:}} = 3 \cdot 2^{2k} + (2^{2k} - 1)$$

Clearly 3 divides $3 \cdot 2^{2k}$, and, by the induction hypothesis, $3|(2^{2k}-1)$. The desired result now follows from CYU 1.18(a), page 25.

CHECK YOUR UNDERSTANDING 1.22

Use the Principle of Mathematical Induction to show that $6|(n^3 + 5n)$ for all integers $n \geq 1$.

Answer: See page A-7.

Recall that:
$n! = 1 \cdot 2 \cdot \ldots \cdot n$

EXAMPLE 1.20 Use the Principle of Mathematical Induction to show that $n! > n^2$ for all integers $n \geq 4$.

SOLUTION: Let $P(n)$ be the proposition $n! > n^2$:

I. $P(4)$ is true: $4! = 1 \cdot 2 \cdot 3 \cdot 4 = 24 \geq 4^2$.

II. **Assume** $P(k)$ is true: $k! > k^2$ (for $k \geq 4$)

III. We show $P(k+1)$ is true; namely, that $(k+1)! > (k+1)^2$:

$$(k+1)! = k!(k+1) > k^2(k+1)$$
$$\text{II} \uparrow$$

Now what? Well, if we can show that $k^2(k+1) > (k+1)^2$, then we will be done. Let's do it:

Since $k \geq 4$, $k \geq 2$, and therefore

$$k^2 = k \cdot k > 2k > k+1$$

Multiplying both sides by the positive number $(k+1)$:

$$k^2(k+1) > (k+1)^2.$$

CHECK YOUR UNDERSTANDING 1.23

Use the Principle of Mathematical Induction to show that $2^n < (n+2)!$ for all integers $n \geq 0$.

Answer: See page A-7.

1.4 Principle of Mathematical Induction

Our next application of the Principle of Mathematical Induction involves the following **Tower of Hanoi** puzzle:

Start with a number of washers of differing sizes on spindle A, as is depicted below:

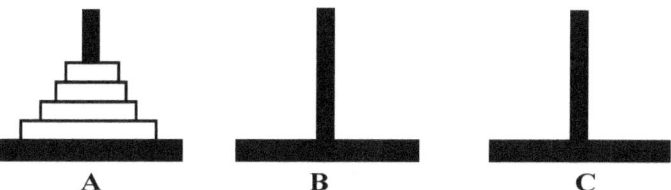

Edouard Lucas formalized the puzzle in 1883, basing it on the following legend: *In a temple at Benares, there are 64 golden disks mounted on one of three diamond needles. At the beginning of the world, all the disks were stacked on the first needle. The priests attending the temple have the sacred obligation to move all the disks to the last needle without ever placing a larger disk on top of a smaller one. The priests work day and night at this task. If and when they finally complete the job, the world will end.*

The objective of the game is to transfer the arrangement currently on spindle A to one of the other two spindles. The rules are that you may only move one washer at a time, without ever placing a larger one on top of a smaller one.

EXAMPLE 1.21 Show that the tower of Hanoi game is winnable for any number n of washers.

SOLUTION: If spindle A contains one washer, then simply move that washer to spindle B to win the game (Step I).

Assume that the game can be won if spindle A contains k washers (Step II — the induction hypothesis).

We now show that the game can be won if spindle A contains $k + 1$ washers (Step III):

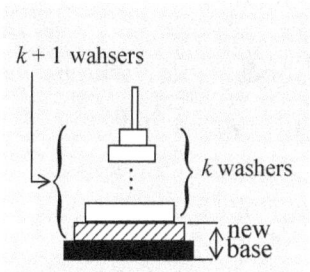

Just imagine that the largest bottom washer is part of the base of spindle A. With this sleight of hand, we are looking at a situation consisting of k washers on a modified spindle A (see margin). By the induction hypothesis, we can move those k washers onto spindle B. We now take the only washer remaining on spindle A (the largest of the original $k + 1$ washers), and move it to spindle C, and then think of it as being part of the base of that spindle. Applying the induction hypothesis one more time, we move the k washers from spindle B onto the modified spindle C, thereby winning the game.

CHECK YOUR UNDERSTANDING 1.24
(Challenging)

Use the Principle of Mathematical Induction to show that any n lines, no two of which are parallel and no three of which pass through a common point, will separate the plane into $\dfrac{n^2 + n + 2}{2}$ regions.

Answer: See page A-7.

TWO ALTERNATE FORMS OF THE PRINCIPLE OF MATHEMATICAL INDUCTION

We complete this section by introducing two equivalent forms of the Principle of Mathematical Induction — equivalent in that any one of them can be used to establish the remaining two.

> API is often called the Strong Principle of Induction. A bit of a misnomer, since it is, in fact, equivalent to PMI.

One version, which we will call the Alternate Principle of Induction (API), is displayed in Figure 1.3(b). As you can see, the only difference between PMI and API surfaces in (*) and (**). Specifically, the proposition "$P(k)$ True" in (a) is replaced, in (b), with the proposition "$P(m)$ True for all integers m **up to and including** k".

Let $P(n)$ denote a proposition that is either true or false, depending on the value of the integer n.

PMI	API
If $P(1)$ is True, and if:	If $P(1)$ is True, and if
(*) $P(k)$ **True** $\Rightarrow P(k+1)$ **True**	(**): $P(m)$ **True for** $1 \leq m \leq k \Rightarrow P(k+1)$ **True**
then $P(n)$ is True for all integers $n \geq 1$	then $P(n)$ is True for all integers $n \geq 1$
(a)	(b)

Figure 1.3

We establish the equivalence of PMI and API by showing that (*) holds if and only (**) holds. Clearly, if (*) holds then (**) must also hold. As for the other way around:

Assume that (**) holds and that (*) does not.
(we will arrive at a contradiction)

If (*) does not hold, then there must exist some k_0 for which $P(k_0)$ is True and $P(k_0 + 1)$ is False. Since $P(k_0 + 1)$ is False, and since (**) holds, we know that $P(k_1)$ is False for some $1 \leq k_1 \leq k_0$. But we are assuming that $P(k_0)$ is True. Hence $P(m)$ is False for some $1 \leq m < k_0$.

Let $m = k_1$ be the smallest positive integer less than k_0 for which $P(m)$ is false. Repeating the above procedure with k_1 playing the role of k_0 we arrive at $P(k_2)$ is False for some $1 \leq k_2 < k_1$.

Continuing in this fashion we shall, after at most $k_0 - 1$ steps, be forced to conclude that $P(1)$ is False

— contradicting the assumption that $P(1)$ is True.

EXAMPLE 1.22 Use API to show that for any given integer $n \geq 12$ there exist integers $a > 0, b \geq 0$ such that $n = 3a + 7b$.

SOLUTION:

I. Claim holds for $n = 12$: $12 = 3 \cdot 4 + 7 \cdot 0$

II. Assume claim holds for all m such that $12 \leq m \leq k$.

III. To show that the claim holds for $n = k+1$ we first show, directly, that it does indeed hold if $k + 1 = 13$ or if $k + 1 = 14$:
$$13 = 3 \cdot 2 + 7 \cdot 1 \text{ and } 14 = 3 \cdot 0 + 7 \cdot 2$$
Now consider any $k + 1 \geq 15$.

If $k + 1 \geq 15$, then $12 \leq (k+1) - 3 \leq k$. Appealing to the induction hypothesis, we choose $a > 0, b \geq 0$ such that:
$$(k+1) - 3 = 3a + 7b$$
It follows that $k + 1 = 3(a + 1) + 7b$, and the proof is complete.

Here is another important property which turns out to be equivalent to the Principle of Mathematical Induction:

> **THE WELL-ORDERING PRINCIPLE FOR Z^+**
>
> Every nonempty subset of Z^+ has a smallest (or least, or first) element.

Note that subsets of Z need not have first elements. A case in point
$$\{..., -4, -2, 0, 2, 4, ...\}$$
Nor does the bounded set:
$$\{x \in \Re | 5 < x < 9\}$$
contain a smallest element (note that 5 is not in the above set).

We show that the Alternate Principle of Mathematical Induction implies the Well-Ordering Principle:

Let S be a NONEMPTY subset of Z^+.

If $1 \in S$, then it is certainly the smallest element in S, and we are done.

Assume $1 \notin S$, and suppose that S **does not** have a smallest element (we will arrive at a contradiction; namely, that S would have to be empty):

Let $P(n)$ be the proposition that $n \notin S$ for $n \in Z^+$. Since, $1 \notin S$, $P(1)$ is True. Suppose that $P(m)$ is True for all $1 \leq m \leq k$, can $P(k+1)$ be False? No:

> To say that $P(k+1)$ is False is to say that $k + 1 \in S$. But that would make $k+1$ the smallest element in S, since none of its predecessors are in S. This cannot be, since S was assumed not to have a smallest element.

Since $P(1)$ is True ($1 \notin S$) and since the validity of $P(m)$ for all $1 \leq m \leq k$ implies the validity of $P(k+1)$, $P(n)$ must be True for all $n \in Z^+$; which is the same as saying that no element of Z^+ is in S — contradicting the assumption that S is NONEMPTY.

> **CHECK YOUR UNDERSTANDING 1.25**
>
> Show that the Well-Ordering Principle implies the Principle of Mathematical Induction.

Answer: See page A-8.

EXERCISES

Exercises 1-33. Establish the validity of the given statement.

1. The nth odd integer is $2n-1$.

2. For every integer $n \geq 1$, $1 + 4 + 7 + \cdots + (3n-2) = \dfrac{3n^2 - n}{2}$.

3. For every integer $n \geq 1$, $1^2 + 3^2 + 5^2 + \cdots + (2n-1)^2 = \dfrac{n(2n-1)(2n+1)}{3}$.

4. For every integer $n \geq 1$, $1^2 + 2^2 + 3^2 + \cdots + n^2 = \dfrac{n(n+1)(2n+1)}{6}$.

5. For every integer $n \geq 1$, $4 + 4^2 + 4^3 + \cdots + 4^n = \dfrac{4(4^n - 1)}{3}$.

6. For every integer $n \geq 1$, $\dfrac{1}{2} + \dfrac{1}{4} + \dfrac{1}{8} + \cdots + \dfrac{1}{2^n} = 1 - \dfrac{1}{2^n}$.

7. For every integer $n \geq 1$, $\dfrac{1}{1 \cdot 2} + \dfrac{1}{2 \cdot 3} + \dfrac{1}{3 \cdot 4} + \cdots + \dfrac{1}{n(n+1)} = \dfrac{n}{n+1}$.

8. For every integer $n \geq 1$, $\dfrac{1}{2 \cdot 3} + \dfrac{1}{3 \cdot 4} + \cdots + \dfrac{1}{(n+1)(n+2)} = \dfrac{n}{2n+4}$.

9. For every integer $n \geq 1$, $\left(1 + \dfrac{1}{1}\right)\left(1 + \dfrac{1}{2}\right)\left(1 + \dfrac{1}{3}\right) \cdots \left(1 + \dfrac{1}{n}\right) = n+1$.

10. For every integer $n \geq 1$ and any real number $x \neq 1$, $x^0 + x^1 + x^2 + \cdots + x^n = \dfrac{1 - x^{n+1}}{1 - x}$.

11. For every integer $n \geq 1$, and any real number $r \neq 1$, $\displaystyle\sum_{i=0}^{n} ar^i = \dfrac{a(1 - r^{n+1})}{1 - r}$.

12. For every integer $n \geq 0$: $5 \mid (2^{4n+2} + 1)$.

13. For every integer $n \geq 1$: $9 \mid (4^{3n} - 1)$.

14. For every integer $n \geq 1$: $3 \mid (5^n - 2^n)$.

15. For every integer $n \geq 1$: $7 \mid (3^{2n} - 2^n)$.

16. For every integer $n \geq 1$: $6 \mid (n^3 + 5n)$.

17. For every integer $n \geq 1$, $5^{2n} + 7$ is divisible by 8.

18. For every integer $n \geq 1$, $3^{3n+1} + 2^{n+1}$ is divisible by 5.

19. For every integer $n \geq 1$, $4^{n+1} + 5^{2n-1}$ is divisible by 21.

20. For every integer $n \geq 1$, $3^{2n+2} - 8n - 9$ is divisible by 64.

21. For every integer $n \geq 0$, $2^n > n$.

22. For every integer $n \geq 5$, $2n - 4 > n$.

23. For every integer $n \geq 5$, $2^n > n^2$.

24. For every integer $n \geq 4$, $3^n > 2^n + 10$.

25. For every integer $n \geq 1$, $\dfrac{(2n)!}{2^n n!}$ is an odd integer.

26. For every integer $n \geq 4$, $2n < n!$.

27. (Calculus Dependent) Show that the sum of n differentiable functions is again differentiable.

28. (Calculus Dependent) Show that for every integer $n \geq 1$, $\dfrac{d}{dx} x^n = n x^{n-1}$.

    Suggestion: Use the product Theorem: If f and g are differentiable functions, then so is $f \cdot g$ differentiable, and $\dfrac{d}{dx}[f(x)g(x)] = f(x)\dfrac{d}{dx}g(x) + g(x)\dfrac{d}{dx}f(x)$.

29. Let $a_1 = 1$ and $a_{n+1} = 3 - \dfrac{1}{a_n}$. Show that $a_{n+1} > a_n$.

30. Let $a_1 = 2$ and $a_{n+1} = \dfrac{1}{3 - a_n}$. Show that $a_{n+1} < a_n$.

31. For every integer $n \geq 1$, $1 + \dfrac{1}{\sqrt{2}} + \dfrac{1}{\sqrt{3}} + \cdots + \dfrac{1}{\sqrt{n}} > 2(\sqrt{n+1} - 1)$.

32. For any positive number x, $(1 + x)^n \geq 1 + nx$ for every $n \geq 1$.

33. For every integer $n \geq 8$, there exist integers $a > 0$, $b > 0$ such that $n = 3a + 5b$.

34. Let m be any nonnegative integer. Use the Well-Ordering Principle to show that every nonempty subset of the set $\{n \in Z | n \geq -m\}$ contains a smallest element.

35. Use the Principle of Mathematical Induction to show that there are $n!$ different ways of ordering n objects, where $n! = 1 \cdot 2 \cdot 3 \cdot \ldots \cdot n$.

36. What is wrong with the following "Proof" that any two positive integers are equal:

    Let $P(n)$ be the proposition: *If a and b are any two positive integers such that* $\max(a, b) = n$, *then* $a = b$.

    I. $P(1)$ is true: If $\max(a, b) = 1$, then both a and b must equal 1.
    II. Assume $P(k)$ is true: If $\max(a, b) = k$, then $a = b$.
    III. We show $P(k+1)$ is true:
        If $\max(a, b) = k+1$ then $\max(a-1, b-1) = k$.
        By II, $a - 1 = b - 1 \Rightarrow a = b$.

§5. The Division Algorithm and Beyond

ALL LETTERS IN THIS SECTION WILL BE UNDERSTOOD TO REPRESENT INTEGERS.

In elementary school you learned how to divide one integer into another to arrive at a quotient and a remainder, and could then check your answer (see margin). That checking process reveals an important result:

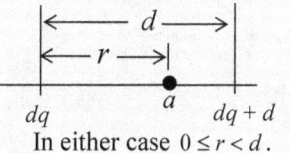

Check: $17 = 3 \cdot 5 + 2$

$a = dq + r$

THEOREM 1.6 For any given $a \in Z$ and $d \in Z^+$, there exist unique integers q and r, with $0 \leq r < d$, such that:
THE DIVISION ALGORITHM
$$a = dq + r$$

PROOF: We begin by establishing the existence of q and r such that:
$$a = dq + r \text{ with } 0 \leq r < d$$

Here is a "convincing argument" for your consideration:

Mark off multiples of d on the number line:

```
—+——+——+——+——+—
-2d  -d  0   d   2d
```

Case 1. If $a = dq$, then let $r = 0$.

Case 2. If a is not a multiple of d, then let dq be such that $dq < a < (d+1)q$. We then have $a = dq + r$, where:

```
|←——— d ———→|
|←— r —→|
              •
              a
dq         dq + d
```

In either case $0 \leq r < d$.

Consider the set:
$$S = \{a - dn | n \in Z \text{ and } a - dn \geq 0\} \quad (*)$$

We first show that S is not empty:

If $a \geq 0$, then $a = a - d \cdot 0 \geq 0$, and therefore $a \in S$.
[0 is playing the role of n in (*)]

If $a < 0$, then $a - da \geq 0$, and therefore $a - da \in S$.
[a is playing the role of n in (*) and remember that $d \in Z^+$]

Since S is a nonempty subset of $\{0\} \cup Z^+$, it has a least element (Exercise 34, page 42); let's call it r. Since r is in S, there exists $q \in Z$ such that:
$$r = a - dq \quad (**)$$

To complete the existence part of the proof, we show that $r < d$.

Assume, to the contrary, that $r \geq d$. From:
$$r - d \underset{(**)}{=} (a - dq) - d = a - d(q + 1)$$

we see that $r - d$ is of the form $a - dn$ (with $n = q + 1$). Moreover, our assumption that $r \geq d$ implies that $r - d \geq 0$. It follows that $r - d \in S$, **contradicting** the minimality of r.

To establish uniqueness, assume that:
$$a = dq + r \text{ with } 0 \leq r < d \text{ and } a = dq' + r' \text{ with } 0 \leq r' < d$$
[We will show that $q = q'$ and $r = r'$ (see margin)]

This is a common mathematical theme:
To establish that something is unique, consider two such "somethings" and then go on to show that the two "somethings" are, in fact, one and the same.

Since $r \geq 0$ and $r' < d$ (or $-r' > -d$): $r - r' \geq 0 - r' > 0 - d = -d$.

Since $r < d$ and $r' \geq 0$ (or $-r' \leq 0$): $r - r' < d - r' \leq d - 0 = d$

Thus: $-d < r - r' < d$, or $|r - r'| < d$

From $dq + r = dq' + r'$ we have: $r - r' = d(q' - q)$

(a multiple of d)

But if $|r - r'| < d$ and if $r - r'$ is a **multiple of d**, then $r - r' = 0$ (or $r = r'$). Returning to $dq + r = dq' + r'$ we now have:
$$dq + r = dq' + r' \Rightarrow dq = dq' \Rightarrow d(q - q') = 0 \underset{d \neq 0}{\Rightarrow} q = q'$$

EXAMPLE 1.23 Show that for any **odd** integer n, $8 \mid (n^2 - 1)$.

SOLUTION: There are, at times, more than one way to stroke a cat:

Using Induction

We show that the proposition:
$$8 \mid [(2m+1)^2 - 1]$$
holds for all $m \geq 0$ (**thereby covering all odd integers n**).

I. Valid at $m = 0$: $(2 \cdot 0 + 1)^2 - 1 = 0$.

II. Assume valid at $m = k$; that is:
$$(2k+1)^2 - 1 = 8t \text{ or } \mathbf{4k^2 + 4k = 8t}$$
for some integer t.

III. We are to establish validity at $m = k+1$; that is, that:
$$[2(k+1)+1]^2 - 1 = 8s$$
for some integer s. Let's do it:

$$[2(k+1)+1]^2 - 1$$
$$= (2k+3)^2 - 1$$
$$= 4k^2 + 12k + 8$$
$$= (4k^2 + 4k) + (8k+8)$$
$$= \mathbf{8t} + 8(k+1) = 8(t+k+1) = 8s$$
$\quad \uparrow$
$\quad$ II

Using the Division Algorithm

We know that for any n there exists q such that:
$n = 2q$ or $n = 2q+1$ $\quad$ (*)
$n = 3q$ or $n = 3q+1$ or $n = 3q+2$ $\quad$ (**)
$\mathbf{n = 4q \text{ or } n = 4q+1 \text{ or } n = 4q+2 \text{ or } n = 4q+3}$

While (*) and (**) may not lead us to a fruitful conclusion, the bottom line does. Specifically:

For any n:
$$n = 4q \text{ or } n = 4q+1 \text{ or } n = 4q+2 \text{ or } n = 4q+3$$
If n is **odd**, then there are but the two possibilities:
$$n = 4q+1 \text{ or } n = 4q+3$$
We now show that, in either case $8 \mid (n^2 - 1)$.

If $n = 4q+1$, then:
$$n^2 - 1 = (4q+1)^2 - 1 = 16q^2 + 8q + 1 - 1 = 8k$$
$$\text{(with } k = 2q^2 + q\text{)}$$

If $n = 4q+3$, then:
$$n^2 - 1 = (4q+3)^2 - 1 = 16q^2 + 24q + 9 - 1 = 8h$$
$$\text{(with } h = 2q^2 + 3q + 1\text{)}$$

CHECK YOUR UNDERSTANDING 1.26

Prove that for any integer n, $n^2 = 3q$ or $n^2 = 3q + 1$ for some integer q.

Answer: See page A-8.

DEFINITION 1.13
GREATEST COMMON DIVISOR

For given a and b not both zero, the **greatest common divisor** of a and b, denoted by $\gcd(a,b)$, is the largest positive integer that divides both a and b.

If the magnitude of either a or b is "small," then one can easily find their greatest common divisor. Consider, for example, the two numbers 245 and 21. The only divisors of 21 are 1, 3, 7 and 21. Since 21 does not divide 245, it cannot be the greatest common divisor of 245 and 21. The next contender is 7, and 7 does divide 245 ($245 = 7 \cdot 35$). Consequently: $\gcd(245, 21) = 7$.

The following example illustrating a procedure that can be used to find the greatest common divisor of any two numbers — a procedure that will surface within the proof of Theorem 1.7 below.

EXAMPLE 1.24   Determine $gcd(4942, 1680)$

SOLUTION: Employing the Division Algorithm:

Divide 1680 into 4942 to arrive at:	$4942 = 2 \cdot 1680 + 1582$ (1) $\quad\uparrow\quad r_1$
Divide $r_1 = 1582$ into 1680:	$1680 = 1 \cdot 1582 + 98$ (2) $\quad\uparrow\quad\uparrow\quad r_1\quad r_2$
Divide $r_2 = 98$ into $r_1 = 1582$:	$1582 = 16 \cdot 98 + 14$ (3) $\quad\uparrow\quad\uparrow\quad\uparrow\quad r_1\quad r_2\quad r_3$
Divide $r_3 = 14$ into $r_2 = 98$:	$98 = 7 \cdot 14 + 0$ (4)

Figure 1.4

Looking at equation (4), we see that 14 divides 98. Moving up to equation (3) we can conclude that 14 must also divide 1582 [see Theorem 1.5(b), page 28]. Equation (2) then tells us that 14 divides 1680, and moving up one more time we find that 14 divides 4942. At this point, we know that 14 divides both 1680 and 4942.

> Since each resulting remainder is strictly smaller than its predecessor, the algorithm must eventually terminate, as it did in step (4), with a zero remainder.

To see that 14 is, in fact, the greatest common divisor of 1680 and 4942, consider any divisor d of 1680 and 4942. From equation (1), rewritten in the form $1582 = 4942 - 2 \cdot 1680$, we see that d must also divide 1582. From Equation (2), rewritten as $1680 - 1582 = 98$, we see that d must also divide 98. Equation (3) then tells us that d divides 14.

Having observed that any divisor of 1680 and 4942 also divides 14, we conclude that $gcd(1680, 4942) = 14$.

Working the algorithm of Figure 1.4 in reverse, we show that the greatest common divisor of 4942 and 1680, namely **14**, can be expressed as a multiple of **4942** plus a multiple of **1680**:

$$\begin{aligned}
\text{From (3):} \mathbf{14} &= 1582 - 16 \cdot 98 \\
\text{From (2):} &= 1582 - 16(1680 - 1582) \\
\text{From (1):} &= 4942 - 2 \cdot 1680 - 16[1680 - (4942 - 2 \cdot 1680)] \\
\text{regrouping:} &= 17 \cdot \mathbf{4942} + (-50)\mathbf{1680}
\end{aligned}$$

In general:

> In the above illustration:
> $a = 4942, b = 1680$
> $s = 17$, and $t = -50$

THEOREM 1.7
If a and b are not both 0, then there exist s and t such that:
$$gcd(a, b) = sa + tb$$

Proof: Let
$$G = \{x > 0 | x = ma + nb \text{ for some } m \text{ and } n\}$$
Assume, without loss of generality that $a \neq 0$. Since both a and $-a$ are of the form $ma + nb$: $a = 1a + 0b$ while $-a = (-1)a + 0b$; and since either a or $-a$ is positive: $G \neq \emptyset$. That being the case the Well Ordering Principle (page 39) assures us that G has a **smallest element** $g = sa + tb$. We show that $g = gcd(a, b)$ by showing that (1): g divides both a and b, and that (2): every divisor of a and b also divides g.

(1) Applying the Division Algorithm we have:
$$a = qg + r \text{ with } 0 \leq r < g.$$
$$\text{(*)} \qquad \qquad \text{(**)}$$

Substituting $g = sa + tb$ in (*) brings us to:
$$a = q(sa + tb) + r$$
$$r = (1 - qs)a - qtb$$

Since r is of the form $ma + nb$ with $r < g$, it cannot be in G, and must therefore be 0 [see (**)]. Consequently $a = qg$, and $g|a$. The same argument can be used to show that $g|b$.

(2) If $d|a$ and $d|b$, then, by Theorem 1.5(b) and (c), page 28: $d|g$.

CHECK YOUR UNDERSTANDING 1.27

(a) Determine $gcd(1870, 5605)$.

(b) Find integers s and t such that $gcd(1870, 5605) = sa + tb$.

(c) Show that for any a and b not both zero:
$$gcd(a, b) = gcd(|a|, |b|).$$

(a) 5 (b) $s = 3, t = -1$
(c) See page A-8

DEFINITION 1.14
RELATIVELY PRIME

Two integers a and b, not both zero, are **relatively prime** if:
$$gcd(a, b) = 1$$

For example:
Since $gcd(15, 8) = 1$, 15 and 8 are relatively prime.
Since $gcd(15, 9) = 3 \neq 1$, 15 and 9 are not relatively prime.

THEOREM 1.8 Two integers, a and b, are relatively prime if and only if there exist $s, t \in Z$ such that $1 = sa + tb$

Proof: To say that a and b are relatively prime is to say that $gcd(a, b) = 1$. The existence of integers s and t such that $1 = sa + tb$ follows from Theorem 1.7.

For the converse, assume that there exist integers s and t such that $1 = sa + tb$. Since $gcd(a, b)$ divides both a and b, it divides 1 [Theorem 1.6(b) and (c), page 28]; and, being positive, must equal 1 [CYU 1.20(a-1), page 29].

1.5 The Division Algorithm and Beyond

THEOREM 1.9 Let $a, b, c \in Z$. If $a|bc$, and if $gcd(a, b) = 1$, then $a|c$.

PROOF: Let s and t be such that:
$$1 = sa + tb$$
Multiplying both sides of the above equation by c:
$$c = sac + tbc$$

Clearly $a|sac$. Moreover, since $a|bc: a|tbc$. The result now follows from Theorem 1.5(b), page 28.

CHECK YOUR UNDERSTANDING 1.28

Answer: See page A-8.

Let $a, b, c \in Z$. Show that if $a|bc$ and $a \nmid b$, then a and c can not be relatively prime.

PRIME NUMBERS

Chances are that you are already familiar with the important concept of a prime number; but just in case:

DEFINITION 1.15
PRIME
An integer $p > 1$ is **prime** if 1 and p are its only divisors.

For example: 2, 5, 7, and 11 are all prime, while 9 and 25 are not. Moreover, since any even number is divisible by 2, no even number greater than 2 is prime.

So, 2 is the oddest prime (sorry).

THEOREM 1.10 If p is prime and if $p|ab$, then $p|a$ or $p|b$.

PROOF: If $p|a$, we are done. We complete the proof by showing that if $p \nmid a$, then $p|b$:

Since the greatest common divisor of p and a divides p, it is either 1 or p. As it must also divide a, and since we are assuming $p \nmid a$, it must be that $gcd(p, a) = 1$. The result now follows from Theorem 1.9.

CHECK YOUR UNDERSTANDING 1.29

Answer: See page A-8.

Let p be prime. Use the Principle of Mathematical Induction to show that if $p|a_1 a_2 \cdots a_n$, then $p|a_i$ for some $1 \leq i \leq n$.

The following result is important enough to be called the Fundamental Theorem of Arithmetic.

THEOREM 1.11 Every integer n greater than 1 can be expressed uniquely (up to order) as a product of primes.

PROOF: We use *API* of page 38 (starting at $n = 2$) to establish the existence part of the theorem:

I. Being prime, 2 itself is already expressed as a product of primes.

II. Suppose a prime factorization exists for all m with $2 \leq m \leq k$.

III. We complete the proof by showing that $k+1$ can be expressed as a product of primes:

If $k+1$ is prime, then we are done.

If $k+1$ is not prime, then $k+1 = ab$, with $2 \leq a \leq b \leq k$. By our induction hypothesis, both a and b can be expressed as a product of primes. But then, so can $k+1 = ab$.

For uniqueness, consider the set:

$$S = \{n \in Z^+ | n \text{ has two different prime decompositions}\}$$

Assume that $S \neq \varnothing$ (we will arrive at a contradiction).

The Well-Ordering Principle of page 39 assures us that S has a **smallest element**, let's call it m. Being in S, m has two distinct prime factorizations, say:

$$m = p_1 p_2 \cdots p_s = q_1 q_2 \cdots q_t$$

Since $p_1 | p_1 p_2 \cdots p_s$ and since $p_1 p_2 \cdots p_s = q_1 q_2 \cdots q_t$ we have $p_1 | q_1 q_2 \cdots q_t$. By CYU 1.28, $p_1 | q_j$ for some $1 \leq j \leq t$.

Without loss of generality, let us assume that $p_1 | q_1$. Since q_1 is prime, its only divisors are 1 and itself. It follows, since $p_1 \neq 1$, that $p_1 = q_1$. Consequently:

$$p_1 p_2 \cdots p_s = q_1 q_2 \cdots q_t \Rightarrow p_1 p_2 \cdots p_s = p_1 q_2 \cdots q_t$$

$$\Rightarrow p_1 p_2 \cdots p_s - p_1 q_2 \cdots q_t = 0$$

$$\Rightarrow p_1 (p_2 \cdots p_s - q_2 \cdots q_t) = 0$$

$$p_1 \neq 0: \Rightarrow p_2 \cdots p_s - q_2 \cdots q_t = 0$$

$$\Rightarrow p_2 \cdots p_s = q_2 \cdots q_t$$
↑
two distinct prime decompositions for an integer smaller than m — contradicting the minimality on m in S

> A pairs of prime number, such as $(3, 5)$, $(5, 7)$, and $(11, 13)$, that differ by 2 are said to be **twin primes**. Whether or not there exist infinitely many twin primes remains an open question.

THEOREM 1.12 There are infinitely many primes.

PROOF: Assume that there are but a finite number of primes, say $S = \{p_1, p_2, \ldots, p_n\}$, and consider the number:

$$m = p_1 p_2 \cdots p_n + 1$$

Since $m \notin S$, it is not prime. By Theorem 1.11, some prime must divide m. Let us assume, without loss of generality, that $p_1 | m$. Since p_1 divides both m and $p_1 p_2 \cdots p_n$: $p_1 | [m - (p_1 p_2 \cdots p_n)]$ [Theorem 1.5(b), page 28]. A contradiction, since $m - (p_1 p_2 \cdots p_n) = 1$.

CHECK YOUR UNDERSTANDING 1.30

Let a and b be relatively prime. Prove that if $a|n$ and $b|n$, then $ab|n$.

> Answer: See page A-9.

	EXERCISES	

Exercises 1-3. For given a and d, determine integers q and r, with $0 \leq r < d$, such that $a = dq + r$.

1. $a = 0, d = 1$
2. $a = -5, d = 133$
3. $a = -133, d = 5$

Exercises 4-6. Find the greatest common divisor of a and b, and determine integers s and t such that $\gcd(a, b) = s \cdot a + t \cdot b$

4. $a = 120, b = 880$
5. $a = -5, d = 133$
6. $a = -133, d = 5$

7. Prove that if 3 does not divide n, then $n = 3k + 1$ or $n = 3k + 2$ for some $k \in \mathbb{Z}$.

8. Let n be such that $3 \nmid (n^2 - 1)$. Show that $3 | n$.

9. Show that if n is not divisible by 3, then $n^2 = 3m + 1$ for some integer m.

10. Show that an odd prime p divides $2n$ if and only if p divides n.

11. Prove that if $a = 6n + 5$ for some n, then $a = 3m + 2$ for some m.

12. Show that $2 | (n^4 - 3)$ if and only if $4 | (n^2 + 3)$.

13. Prove that any two consecutive odd positive integers are relatively prime.

14. Let a and b not both be zero. Prove that there exist integers s and t such that $n = sa + tb$ if and only if n is a multiple of $\gcd(a, b)$.

15. Prove that the only three consecutive odd numbers that are prime are 3, 5, and 7.

16. Show that a prime p divides n^2 if and only if p divides n.

17. Prove that every odd prime p is of the form $4n + 1$ or of the form $4n + 3$ for some n.

18. Prove that every prime $p > 3$ is of the form $6n + 1$ or of the form $6n + 5$ for some n.

19. Prove that every prime $p > 5$ is of the form $10n + 1$, $10n + 3$, $10n + 7$, or $10n + 9$ for some n.

20. Prove that a prime p divides $n^2 - 1$ if and only if $p | (n - 1)$ or $p | (n + 1)$.

21. Prove that every prime of the form $3n + 1$ is also of the form $6k + 1$.

22. Prove that if n is a positive integer of the form $3k + 2$, then n has a prime factor of this form as well.

23. Prove that a and b are relatively prime if and only if the prime decompositions of a and b do not share a common prime.

24. Prove that $n > 1$ is prime if and only if n is not divisible by any prime p with $p \leq \sqrt{n}$.

| | **PROVE OR GIVE A COUNTEREXAMPLE** | |

25. There exists an integer n such that $n^2 = 3m - 1$ for some m.
26. If $a = 3m + 2$ for some m, then $a = 6n + 5$ for some n.
27. If m and n are odd integers, then either $m + n$ or $m - n$ is divisible by 4.
28. For any a, and b not both 0, there exist a **unique** pair of integers s and t such that $gcd(a, b) = s \cdot a + t \cdot b$.
29. For every n, $3 | (4^n - 1)$. GIO move to induction or state not to use induction
30. For every $n \in Z^+$, $3 | (4^n + 1)$.
31. There exists $n \in Z^+$ such that $3 | (4^n + 1)$.

CHAPTER 2
A Touch of Set Theory

Modern mathematics rests on set theory. Some of that foundation is introduced in the first four sections of the chapter, and a glimpse into the axiomatic construction of the theory is offered in Section 5.

> A bit of set notation has already been introduced in the previous chapter. (See page 10)

§1. BASIC DEFINITIONS

As you might expect, one defines two sets to be equal if each element of either set is also an element of the other:

> We remind you that the symbol $\in$ is read "*is an element of.*"

DEFINITION 2.1
SET EQUALITY

Two sets A and B are **equal**, written $A = B$ if:
$$x \in A \Rightarrow x \in B \quad \text{and} \quad x \in B \Rightarrow x \in A$$
$$(\text{or: } x \in A \Leftrightarrow x \in B)$$

Note that when it comes to set notation, **order is of no consequence**. For example: $\{1, 2, 3\} = \{2, 1, 3\}$. Moreover, to avoid an unpleasantness such as $\{1, 1, 2, 3\} = \{1, 2, 3\}$ we stipulate that elements cannot be listed more than once within the set notation.

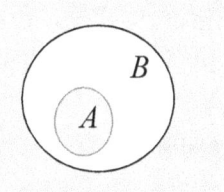

DEFINITION 2.2
SUBSET AND PROPER SUBSET

Let A and B be sets. A is said to be a **subset** of B, written $A \subseteq B$ if every element in A is also an element in B, i.e: $x \in A \Rightarrow x \in B$.

A is said to be a **proper subset** of B, written $A \subset B$, if A is a subset of B and $A \neq B$.

For example:
$$\{1, 2, 3\} \subseteq \{1, 2, 3\}, \{1, 2\} \subseteq \{1, 2, 3\}, \text{ and } \{1, 2\} \subset \{1, 2, 3\}$$

Just as you can add or multiply numbers to obtain other numbers, so then can sets be combined to obtain other sets:

DEFINITION 2.3
INTERSECTION AND UNION OF SETS

Let A and B be sets.

The **intersection** of A and B, written $A \cap B$, is the set consisting of the elements common to both A and B. That is:
$$A \cap B = \{x | x \in A \text{ and } x \in B\}$$
↑— read *such that*

> If someone asks you if you want tea or coffee, you are being offered one or the other, but not both: the **exclusive-or** is being used.
> In mathematics and science, however, the **inclusive-or** is generally used. In particular, to say that x is in A or B, allows for x to be both in A and in B.

The **union** of A and B, written $A \cup B$, is the set consisting of the elements that are in A or in B (see margin). That is:
$$A \cup B = \{x | x \in A \text{ or } x \in B\}$$

The adjacent visual representations of sets are called **Venn diagrams**.
[John Venn (1834-1923)].

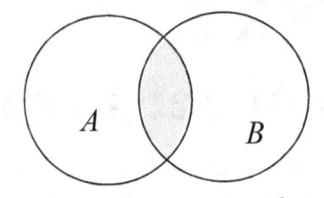
$A \cap B = \{x | x \in A \text{ and } x \in B\}$
(a)

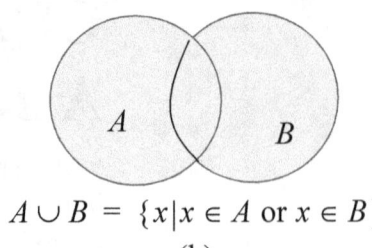
$A \cup B = \{x | x \in A \text{ or } x \in B\}$
(b)

Figure 2.1

For example, if $A = \{3, 5, 9, 11\}$ and $B = \{1, 2, 5, 6, 11\}$, then:
$$A \cap B = \{3, 5, 9, 11\} \cap \{1, 2, 5, 6, 11\} = \{5, 11\}$$
and: $A \cup B = \{3, 5, 9, 11\} \cup \{1, 2, 5, 6, 11\} = \{3, 5, 9, 11, 1, 2, 6\}$.

We note that the set that contains no elements is called the **empty set** (or the **null set**), and is denoted by $\emptyset$.

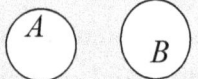

DEFINITION 2.4
DISJOINT SETS
Two sets A and B are **disjoint** if $A \cap B = \emptyset$.

In particular, since $\{1, 2, 3\} \cap \{4, 5\} = \emptyset$:
$$\{1, 2, 3\} \text{ and } \{4, 5\} \text{ are disjoint.}$$

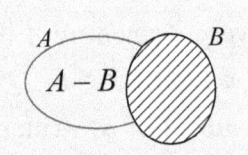

DEFINITION 2.5
A MINUS B
Let A and B be sets. **A minus B**, denoted by $A - B$, is the set of elements in A that are not in B:
$$A - B = \{x | x \in A \text{ and } x \notin B\}$$

As it is with numbers, set subtraction is not a commutative operation: $A - B$ need not equal $B - A$.

For example, if $A = \{1, 2, 5, 6, 10, 11\}$ and $B = \{3, 5, 9, 11\}$, then:
$$A - B = \{1, 2, 5, 6, 10, 11\} - \{3, 5, 9, 11\} = \{1, 2, 6, 10\}$$
and:
$$B - A = \{3, 5, 9, 11\} - \{1, 2, 5, 6, 10, 11\} = \{3, 9\}$$

When dealing with sets, one usually has a **universal set** in mind — a set that encompasses all elements under current consideration. The letter U is typically used to denote the universal set, and it generally takes on a rectangular form in Venn diagrams (see margin).

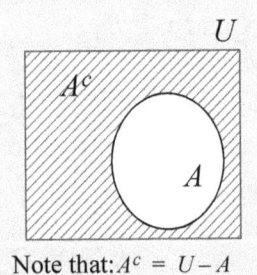
Note that: $A^c = U - A$

DEFINITION 2.6
COMPLEMENT OF A SET
Let A be a subset of the universal set U. The **complement** of A in U, written A^c, is the set of elements in U that are not contained in A:
$$A^c = \{x | x \in U \text{ and } x \notin A\}$$
(More simply: $\{x | x \notin A\}$, if U is understood)

For example, if $U = \{1, 2, 3, 4, 5, 6, 7, 8, 9\}$, then:
$$\{1, 3, 5, 7, 8\}^c = \{2, 4, 6, 9\}$$

CHECK YOUR UNDERSTANDING 2.1

(a) Let $U = \{1, 2, 3, 4, 5, 6, 7\}$, $A = \{1, 3, 5\}$, $B = \{2, 3, 4, 7\}$, and $C = \{3, 4, 5\}$. Determine:

(i) $(A \cup B)^c$ (ii) $(A \cap B^c) \cup (A \cup B)^c$ (iii) $(A - B)^c \cap C$

(iv) $[A - (B \cap C)]^c$ (v) $\{x \in U | x = y + 2, y \in B\}$

(b) Indicate True or False:

(i) $\{2\} \subseteq \{1, 2\}$ (ii) $2 \in \{1, 2\}$ (iii) $\{2\} \in \{1, \{2\}\}$

(iv) $\varnothing \subseteq \{1, 2\}$ (v) $\varnothing \in \{1, 2\}$ (vi) $\varnothing \in \{\{\varnothing\}\}$ (vii) $\varnothing \in \{\varnothing\}$

(a-1) $\{6\}$ (a-ii) $\{1, 5, 6\}$
(a-iii) $\{3, 4\}$
(a-iv) $\{2, 3, 4, 6, 7\}$
(a-v) $\{4, 5, 6\}$
True: (b-i), (b-ii), (b-iii), (b-iv), (b-vii)   False: (b-v), (b-vi)

Here is a link between the difference and complement concepts:

THEOREM 2.1 $A - B = A \cap B^c$

PROOF: $x \in A - B \Leftrightarrow x \in A$ and $x \notin B$

$\Leftrightarrow x \in A$ and $x \in B^c \Leftrightarrow x \in A \cap B^c$

We can also use a Membership Table (an approach reminiscent of Truth Tables) to establish the above result. As you can see, we labeled the first two columns of Row 1 in Figure 2.2: A and B. Other involved sets, built from those two sets, are acknowledged in the remaining three columns of Row 1.

Now, for any given x in the universal set, there are four possibilities:

$x \in A$ and $x \in B$	And so we positioned a **1** in the first two columns of Row 1 to indicate **containment**.
$x \in A$ and $x \notin B$	And so we positioned a **0** in the second column of Row 3 to indicate **non-containment**.
$x \notin A$ and $x \in B$	0 and then 1 appear in the first two columns of Row 4.
$x \notin A$ and $x \notin B$	0 and 0 appear in the first two columns of Row 5.

Since 1 appears under both A and B in Row 2 of Figure 2.2, a 0 appears in the third column of that row. Why? Because:

$$x \in A \text{ and } x \in B \Rightarrow x \notin A - B$$
$$\quad 1 \quad\quad\quad\quad 1 \quad\quad\quad\quad 0$$

Why does a 1 appear in the third column of Row 3? Because:

$$x \in A \text{ and } x \notin B \Rightarrow x \in A - B$$
$$\quad 1 \quad\quad\quad\quad 0 \quad\quad\quad\quad 1$$

You get the developing pattern, no? But just in case, let's finish up with Row 3:

The B^c column: $x \notin B \Rightarrow x \in B^c$
$$\quad\quad\quad 0 \quad\quad\quad 1$$

The $A \cap B^c$ column: $x \in A$ and $x \in B^c \Rightarrow x \in A \cap B^c$
$$\quad\quad\quad\quad\quad\quad 1 \quad\quad\quad\quad 1 \quad\quad\quad\quad 1$$

56 Chapter 2 A Touch of Set Theory

> You can turn this membership table into a truth table by replacing the sets A and B with the propositions
> $x \in A$ and $x \in B$ respectively.

Set Row 1:

A	B	$A-B$	B^c	$A \cap B^c$
1	1	0	0	0
1	0	1	1	1
0	1	0	1	0
0	0	0	0	0

Row 2:
Row 3:
Row 4:
Row 5:

$x \in A - B \Leftrightarrow x \in (A \cap B^c)$

Figure 2.2

EXAMPLE 2.1 Use a Membership Table to establish:
(a) $(A \cap B) \cup (A \cap B^c) = A$
(b) $(A-B) \cap (A-C) \subseteq A - (B \cap C)$

SOLUTION: (a)

A	B	$A \cap B$	B^c	$A \cap B^c$	$(A \cap B) \cup (A \cap B^c)$
1	1	1	0	0	1
1	0	0	1	1	1
0	1	0	1	0	0
0	0	0	0	0	0

$x \in A \Leftrightarrow x \in (A \cap B) \cup (A \cap B^c)$

(b)

A	B	C	$A-B$	$A-C$	$(A-B) \cap (A-C)$	$B \cap C$	$A-(B \cap C)$
1	1	1	0	0	0	1	0
1	1	0	0	1	0	0	1
1	0	1	1	0	0	0	1
0	1	1	0	0	0	1	0
1	0	0	1	1	1	0	1
0	1	0	0	0	0	0	0
0	0	1	0	0	0	0	0
0	0	0	0	0	0	0	0

if $x \in (A-B) \cap (A-C)$ then $x \in A - (B \cap C)$

EXAMPLE 2.2 Prove or give a counterexample.
(a) $(A \cup B)^c \cap C = C - (A \cup B)$
(b) $(A \cap B)^c \subseteq A \cup B$

SOLUTION: (a) True: $x \in (A \cup B)^c \cap C \Leftrightarrow x \in C$ and $x \notin (A \cup B)$
$\Leftrightarrow x \in C - (A \cup B)$

(b) A glance at the adjacent figure suggests that, in general, $(A \cap B)^c$ is not a subset of $A \cup B$. Suggestion aside, a specific counterexample is needed to nail down the claim. Here is one:

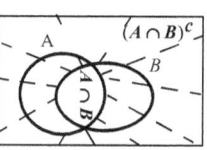

For $U = \{1, 2\}$, $A = \{1\}$, $B = \{1\}$:

$$(A \cap B)^c = \{2\} \text{ while } A \cup B = \{1\}$$

CHECK YOUR UNDERSTANDING 2.2

Prove or give a counterexample:
(a) $A - (B \cap C) = (A - B) \cap (A - C)$
(b) $(A \cap B^c)^c \cup B = A^c \cup B$

Answer: See page A-9.

Let p and q be propositions. Then:
(a) $\sim(p \wedge q) \Leftrightarrow \sim p \vee \sim q$
(b) $\sim(p \vee q) \Leftrightarrow \sim p \wedge \sim q$

The two DeMorgan's laws in the margin were established on page 6 (see margin). Here are the "set-versions" of those laws:

THEOREM 2.2 For any sets A and B:

DEMORGAN'S LAWS
(a) $(A \cap B)^c = A^c \cup B^c$
(b) $(A \cup B)^c = A^c \cap B^c$

PROOF: (a) In accordance with Definition 2.1, we are to show that:
$x \in (A \cap B)^c \Rightarrow x \in (A^c \cup B^c)$ and that $x \in (A^c \cup B^c) \Rightarrow (A \cap B)^c$.
Let's do them both at the same time:

$$x \in (A \cap B)^c \underset{\text{Definition 2.5}}{\Leftrightarrow} x \notin (A \cap B) \underset{\text{Definition 2.3}}{\Leftrightarrow} x \notin A \text{ or } x \notin B \underset{\text{Definition 2.5}}{\Leftrightarrow} x \in A^c \text{ or } x \in B^c$$

$$\text{Definition 2.3:} \Leftrightarrow x \in (A^c \cup B^c)$$

(b) We choose to use a Membership Table to establish $(A \cup B)^c = A^c \cap B^c$:

A	B	A^c	B^c	$A^c \cap B^c$	$A \cup B$	$(A \cup B)^c$
1	1	0	0	0	1	0
1	0	0	1	0	1	0
0	1	1	0	0	1	0
0	0	1	1	1	0	1

$$x \in A^c \cap B^c \Leftrightarrow x \in (A \cup B)^c$$

> **CHECK YOUR UNDERSTANDING 2.3**
>
> (a) Use a Membership Table to prove Theorem 2.1(a)
> (b) Prove Theorem 2.1(b) by a method similar to that used in the proof of Theorem 2.1(a).

Answer: See page A-10.

The union of n sets $\{S_i\}_{i=1}^{n} = \{S_1, S_2, \ldots, S_n\}$ can be denoted by $S_1 \cup S_2 \cup \ldots \cup S_n$ or by $\bigcup_{i=1}^{n} S_i$, and the union of the sets $\{S_i\}_{i=1}^{\infty}$ by $\bigcup_{i=1}^{\infty} S_i$ or $\bigcup_{i \in Z^+} S_i$. For any set A, $\{S_\alpha\}_{\alpha \in A}$ denotes a collection of sets indexed by A, and just as $\bigcup_{i \in Z^+} S_i$ can be used to denote the union of the sets in the collection $\{S_i\}_{i \in Z^+}$, so then does $\bigcup_{\alpha \in A} S_\alpha$ denote the union of the sets in the collection $\{S_\alpha\}_{\alpha \in A}$. For example, $\bigcup_{r \in \Re} S_r$ denotes the union of a collection of sets indexed by the set of real numbers $\Re$.

Similarly $\bigcap_{i=1}^{n} S_i$, $\bigcap_{i \in Z^+} S_i$, and $\bigcap_{\alpha \in A} S_\alpha$ represent the intersection of n sets, sets indexed by the positive integers, and sets indexed by the set A, respectively.

We are now in a position to address the most general form of DeMorgan's Laws:

THEOREM 2.3
DEMORGAN'S LAWS

For any collection $\{S_\alpha\}_{\alpha \in A}$ of sets:

(a) $\left(\bigcup_{\alpha \in A} S_\alpha\right)^c = \bigcap_{\alpha \in A} S_\alpha^c$ (b) $\left(\bigcap_{\alpha \in A} S_\alpha\right)^c = \bigcup_{\alpha \in A} S_\alpha^c$

PROOF: (a) $x \in \left(\bigcup_{\alpha \in A} S_\alpha\right)^c \Leftrightarrow x \notin \bigcup_{\alpha \in A} S_\alpha$

The containment table approach cannot be used to establish this result. Why not?

$\Leftrightarrow x \notin S_\alpha$ for every $\alpha \in A$
$\Leftrightarrow x \in S_\alpha^c$ for every $\alpha \in A$
$\Leftrightarrow x \in \bigcap_{\alpha \in A} S_\alpha^c$

As for (b):

> **CHECK YOUR UNDERSTANDING 2.4**
>
> Establish Theorem 2.2(b)

Answer: See page A-10.

INTERVAL NOTATION

Certain subsets of $\Re$, called intervals, warrant special recognition. As is indicated below, brackets are used to indicate endpoint-inclusions, while parenthesis denote endpoint-exclusions.

	Interval Notation	Geometrical Representation
All real numbers strictly between 1 and 5 (not including 1 or 5)	$(1, 5) = \{x \mid 1 < x < 5\}$ — excluding 1 and 5	
All real numbers between 1 and 5, including both 1 and 5.	$[1, 5] = \{x \mid 1 \le x \le 5\}$ — including 1 and 5	
All real numbers between 1 and 5, including 1 but not 5.	$[1, 5) = \{x \mid 1 \le x < 5\}$ — including 1 and excluding 5	
All real numbers between 1 and 5, including 5 but not 1.	$(1, 5] = \{x \mid 1 < x \le 5\}$ — excluding 1 and including 5	
All real numbers greater than 1.	$(1, \infty) = \{x \mid x > 1\}$ — the infinity symbol	
All real numbers greater than or equal to 1.	$[1, \infty) = \{x \mid x \ge 1\}$	
All real numbers strictly less than 5.	$(-\infty, 5) = \{x \mid x < 5\}$	
All real numbers less than or equal to 5.	$(-\infty, 5] = \{x \mid x \le 5\}$	
The set of all real numbers.	$(-\infty, \infty) = \{x \mid -\infty < x < \infty\}$	

Figure 2.3

In general, for $a < b$:

$[a, b] = \{r \in R \mid a \le r \le b\}$ is said to be a **closed interval**.

$(a, b) = \{r \in R \mid a < r < b\}$ is said to be an **open interval**.

$[a, b) = \{r \in R \mid a \le r < b\}$ and $(a, b] = \{r \in R \mid a < r \le b\}$ are said to be **half-open** (or **half-closed**) **intervals**.

In addition:

(a, ∞) and $(-\infty, b)$ are said to be open.

$[a, \infty)$ and $(-\infty, b]$ are said to be half-open (or half-closed).

$\Re = (-\infty, \infty)$ is said to be both closed and open (or clopen).

CHECK YOUR UNDERSTANDING 2.5

Simplify: (a) $(-2, 2) \cap [0, 5]$ (b) $[-1, 3]^c \cap (5, \infty)$

(c) $(-2, 0) \cup [-1, 2] \cup [3, 5]$

(a) $[0, 2)$ (b) $(5, \infty)$
(c) $(-2, 2] \cup [3, 5]$

	EXERCISES	

Exercises 1-19. For $U = \{1, 2, 3, \ldots\}$, $O = \{1, 3, 5, \ldots\}$, $E = \{2, 4, 6, \ldots\}$,
$A = \{5n | n \in U\}$, $B = \{3n | n \in U\}$, $C = \{1, 2, 3, \ldots, 15\}$,
$D = \{2, 4, 6, \ldots 10\}$, and $F = \{11, 12, 13, 14\}$, determine:

1. $O \cup E$
2. $O \cap E$
3. $A \cap B$
4. $A \cup B$
5. $B \cup C$
6. $B \cap C$
7. $C \cup D$
8. $C \cap D$
9. $O^c \cup E^c$
10. $O^c \cap A$
11. $C \cap O$
12. $(O \cap A)^c$
13. $(C \cap D) \cup F$
14. $C \cap (D \cup F)$
15. $(C \cup F) \cap D$
16. $(C \cap F^c) \cup F$
17. $(B^c \cap C) \cup (D \cap O)$
18. $[(O \cup E)^c \cup (A \cap B)]^c$
19. $(O \cap E)^c \cap C^c$

Exercises 20-23. Determine the set of all subsets of the given set A.

20. $A = \{1\}$
21. $A = \{1, 2\}$
22. $A = \{1, 2, 3\}$
23. $A = \{\emptyset\}$

24. Establish the following set identities (all capital letters represent subsets of a universal set U):

	$A \cup \emptyset = A$	**Domination Laws**	$A \cup U = U$
	$A \cap U = A$		$A \cap \emptyset = \emptyset$
	$A \cup A = A$	**Complementation Law**	$(A^c)^c = A$
Commutative Laws	$A \cup B = B \cup A$	**Associative Laws**	$A \cup (B \cup C) = (A \cup B) \cup C$
	$A \cap B = B \cap A$		$A \cap (B \cap C) = (A \cap B) \cap C$
		Distributive Laws	$A \cap (B \cup C) = (A \cap B) \cup (A \cap C)$
			$A \cup (B \cap C) = (A \cup B) \cap (A \cup C)$

Exercises 25-51. Prove that:

25. $(A - B) \subseteq B^c$
26. $[A^c \cup B]^c = A \cap B^c$
27. $A \cap (B - A) = \emptyset$
28. $(A - C) \cap (C \cap B) = \emptyset$
29. $[(A - B) - C] \subseteq (A - C)$
30. $(A \cap B) \cup (A \cap B^c) = A$
31. $A - (A \cap B) = A - B$
32. $(A - B)^c = B \cup A^c$
33. $U = (A \cap B) \cup (A \cup B)^c$
34. $[A^c \cup (A^c - B)]^c = A$
35. $[(A^c \cup B^c) - A]^c = A$
36. $A \cup (B - A) = A \cup B$
37. $(A - C) - (B - C) = (A - B) - C$
38. $(A - C) \cup (B - C) = (A \cup B) - C$
39. $(A - C) \cap (B - C) = (A \cap B) - C$
40. $(A - B) - (A - C) = A \cap (C - B)$
41. $(A - B) \cup (A - C) = A - (B \cap C)$
42. $(A - B) \cap (A - C) = A - (B \cup C)$
43. $(B - A) \cup (C - A) = (B \cup C) - A$
44. $(A \cup B) - C = (A - C) \cup (B - C)$
45. $(A - B)^c \cap (B - A)^c = (A \cup B)^c$
46. $(A \cap B) \cap C^c = (A \cap B) - C$

47. $(A - B) - (B - C) = A - B$ 48. $(A \cup B) - (C - A) = A \cup (B - C)$
49. $[(A \cup B) \cap (A \cup C)] \subseteq [A \cup (B \cap C)]$ 50. $(A \cap B^c)^c \cup B = A^c \cup B$
51. $(A \cap B \cap C) \cup (A^c \cap B \cap C) \cup B^c \cup C^c = U$

Exercises 52-59. Give a counterexample to show that each of the following statements is **False**.
52. $A - B = B - A$ 53. $(A - B)^c = B^c - A^c$ 54. $(A \cap B)^c = A^c \cap B^c$
55. $(A \cup B)^c = A^c \cup B^c$ 56. $(A \cup B)^c = A^c \cup B^c$ 57. $(A \cap B)^c = A^c \cap B^c$
58. $(A \cup B) - C = A \cup (B - C)$ 59. $(A \cap B)^c \cap C^c = A^c \cap (B \cap C)^c$

60. For n an integer distinct from 0, let $A_n = \{a \in Z | a \text{ is divisible by } n\}$. Determine the set:

   (a) $A_5 \cup A_{10}$ (b) $A_5 \cap A_{10}$ (c) $A_9 \cap (A_3)^c$ (d) $A_9 \cup (A_3)^c$

61. Prove that for any given set A, $\emptyset \subseteq A$ and $A \subseteq A$.

62. Prove that if $A \subseteq B$ and $B \subseteq C$, then $A \subseteq C$.

63. Prove that if $A \subseteq B$, $B \subseteq C$, and $C \subseteq A$ then $A = B = C$.

64. Prove that for any sets A and B of a universal set U: $A - B = A \cap B^c$.

65. Prove that $A \subseteq B$ if and only if $B^c \subseteq A^c$.

66. Prove that $A = (A \cap B) \cup (A - B)$ and $(A \cap B) \cap (A - B) = \emptyset$.
   [So, $(A \cap B) \cup (A - B)$ is a representation of A as a disjoint union]

67. Prove that $A \cup B = A \cup (B - A)$ and that $A \cap (B - A) = \emptyset$.
   (So, $A \cup (B - A)$ is a representation of $A \cup B$ as a disjoint union.)

68. Prove that $A \subseteq B$ if and only if $(A \cup C) \subseteq (B \cup C)$ for every $C \subseteq U$.

69. Prove that $A \subseteq B$ if and only if $(A \cap C) \subseteq (B \cap C)$ for every $C \subseteq U$.

70. Prove that the three statements

   (i) $A \subseteq B$ (ii) $A \cap B = A$ (iii) $A \cup B = B$

   are equivalent by sowing that $(i) \Rightarrow (ii) \Rightarrow (iii) \Rightarrow (i)$.

\# 71. Let $\{S_\alpha\}_{\alpha \in A}$ be any collection of sets. Prove that for any set X:

   (a) $\bigcup_{\alpha \in A} (S_\alpha \cap X) = \left(\bigcup_{\alpha \in A} S_\alpha\right) \cap X$ (b) $\bigcap_{\alpha \in A} (S_\alpha \cap X) = \left(\bigcap_{\alpha \in A} S_\alpha\right) \cap X$

PROVE OR GIVE A COUNTEREXAMPLE

72. If $A \cap B \neq \emptyset$ and $B \cap C \neq \emptyset$, then $A \cap C \neq \emptyset$.

73. If $A \cap B = \emptyset$ or $B \cap C = \emptyset$, then $A \cap C = \emptyset$.

74. If $A \subseteq (B \cap C)$ and $C \subseteq (B \cap A)$, then $A = C$.

75. If $A \cup B = A \cup C$, then $A = C$.

76. If $A \cap B = A \cap C$, then $A = C$.

77. If $A \cap B = A \cup B$, then either $A = \emptyset$ or $B = \emptyset$.

78. $A - (B - C) = (A - B) - C$.

79. $(A - B) - C = (A - C) - (B - C)$.

80. $A - B = B - A$ if and only if $A = B$.

81. If $A \subseteq (B \cap C)$ and $B \subseteq (C \cap D)$, then $(A \cap C) \subseteq (B \cap D)$.

82. $(A \cup B) \cap (A \cup C) = A \cup (B \cap C)$.

83. If no element of a set A is contained in a set B, then A cannot be a subset of B.

84. Two sets A and B are equal if and only if the set of all subsets of A is equal to the set of all subsets of B.

85. $\emptyset = \{\emptyset\}$.

§2. FUNCTIONS

You've dealt with functions in one form or another before, but have you ever been exposed to a definition? If so, it probably started off with something like:

A **function** is a rule...........

Or, if you prefer, a **rule** is a function.......

You are now too sophisticated to accept this sort of "circular definition." Alright then, have it your way:

All "objects" in mathematics are sets, and functions are no exceptions. The function f given by $f(x) = x^2$, is the subset $f = \{(x, x^2) | x \in \Re\}$ of the plane. Pictorially:

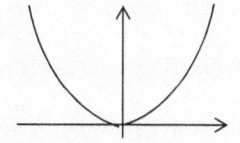

A function such as
$f = \{(x, x^2) | x \in \Re\}$
is often simply denoted by $f(x) = x^2$. Still, in spite of their dominance throughout mathematics and the sciences, functions that can be described in terms of algebraic expressions are truly exceptional. Scribble a curve in the plane for which no vertical line cuts the curve in more than one point and you have yourself a function. But what is the "rule" for the set g below?

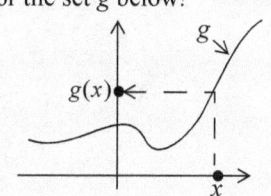

Note that the set S below, is not a function:

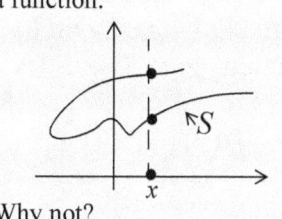

Why not?

DEFINITION 2.7
CARTESIAN PRODUCt

For given sets X and Y, we define the **Cartesian Product** of X with Y, denoted by $X \times Y$, to be the set of **ordered pairs**:

$$X \times Y = \{(x, y) | x \in X \text{ and } y \in Y\}$$

DEFINITION 2.8
FUNCTION

A **function** f from a set X to a set Y is a subset $f \subseteq X \times Y$ such that for every $x \in X$ there exists a <u>unique</u> $y \in Y$ with $(x, y) \in f$, and we write: $y = f(x)$.

DOMAIN
RANGE

The set X is said to be the **domain** of f, and

$$\{y \in Y | (x, y) \in f \text{ for some } x \in X\}$$

is said to be the **range** of f.

While the domain of f is all of X, the range of f need not be all of Y.

Moreover, for $A \subseteq X$ and $B \subseteq Y$:

IMAGE OF $A \subseteq X$

$f[A] = \{f(a) | a \in A\}$ is called the **image of A under f**, and $f^{-1}[B] = \{x \in X | f(x) \in B\}$ is called the **inverse image of B**.

INVERSE IMAGE OF $B \subseteq Y$

The notation $f: X \to Y$ is used to denote a function from a set X to a set Y. Moreover, as is customary, we write $y = f(x)$ to indicate that $(x, y) \in f$.

The function $f: X \to Y$ is also called a **mapping** from X to Y. In the event that $y = f(x)$, we will say that f **maps** x to y. Finally, we note that the symbols D_f and R_f are used to denote the domain and range of f, respectively.

Since functions are sets, the definition of equality is already at hand. Two functions f and g are equal if the set f is equal to the set g. In other words, if:

$$D_f = D_g \text{ and } f(x) = g(x) \text{ for every } x \in D_f = D_g.$$

> When not specified, the domain of a function expressed in terms of a variable x is **understood** to consist of all values of x for which the expression can be evaluated.

CHECK YOUR UNDERSTANDING 2.6

For $X = \{0, 1, 2, 3\}$ and $Y = \{2, 3, 4, 5, 6\}$ determine if the subset f of $X \times Y$ is a function $f: X \to Y$. If so, determine its range.

(a) $f = \{(0, 2), (1, 2), (2, 3), (2, 6)\}$

(b) $f = \{(0, 2), (1, 2), (2, 2), (3, 2)\}$

(c) $f = \{(0, 2), (1, 2), (2, 3)\}$

(d) $f = \{(2, 2), (3, 0), (4, 1), (5, 3), (6, 0)\}$

(a) No (b) Range: $\{2\}$
(c) No (d) No

COMPOSITION OF FUNCTIONS

Consider the schematic representation of the functions $f: X \to Y$ and $g: Y \to T$ in Figure 2.4, along with a third function $g \circ f: X \to T$.

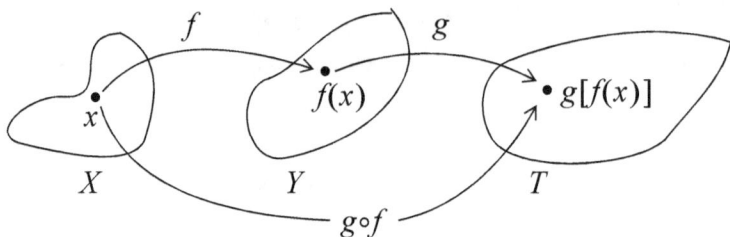

Figure 2.4

As is suggested in the above figure, the function $g \circ f: X \to T$ is given by:

$$(g \circ f)(x) = g[f(x)]$$
$$\uparrow \quad \uparrow$$
first apply f
and then apply g

Formally:

DEFINITION 2.9
COMPOSITION

Let $f: X \to Y$ and $g: Y \to T$ be such that the range of f is contained in the domain of g. The composite function $g \circ f: X \to T$ is given by:

$$(g \circ f)(x) = g[f(x)]$$

In set notation:
$$g \circ f = \{(x, t) | \exists y \in Y \ni (x, y) \in f \text{ and } (y, t) \in g\}$$

EXAMPLE 2.3 For $f(x) = x^2 + 1$ and $g(x) = 2x - 5$ find:

(a) $(g \circ f)(3)$ (b) $(f \circ g)(3)$

(c) $(g \circ f)(x)$ (d) $(f \circ g)(x)$

SOLUTION:

(a) $(g \circ f)(3) = g[f(3)] = g(3^2 + 1) = g(10) = 2 \cdot 10 - 5 = 15$

(b) $(f \circ g)(3) = f[g(3)] = f(2 \cdot 3 - 5) = f(1) = 1^2 + 1 = 2$

(c) $(g \circ f)(x) = g[f(x)] = g(x^2 + 1) = 2(x^2 + 1) - 5 = 2x^2 - 3$

(d) $(f \circ g)(x) = f[g(x)] = f(2x - 5) = (2x - 5)^2 + 1 = 4x^2 - 20x + 26$

CHECK YOUR UNDERSTANDING 2.7

(a) For $f(x) = x + 2$ and $g(x) = \dfrac{3x}{x + 1}$, determine:

  (i) $(g \circ f)(3)$ (ii) $(f \circ g)(3)$ (iii) $(g \circ f)(x)$ (iv) $(f \circ g)(x)$

(b) For $f = \{(1, a), (c, 5), (x, y)\}$ and
$$g = \{(a, a), (c, 3), (5, t), (y, 4)\}, \text{ determine: } g \circ f.$$
Is $f \circ g$ defined? Justify your answer.

(a-i) $\dfrac{5}{2}$ (a-ii) $\dfrac{17}{4}$

(a-iii) $\dfrac{3x+6}{x+3}$ (a-iv) $\dfrac{5x+2}{x+1}$

(b) $g \circ f = \{(1, a), (c, t), (x, 4)\}$

The time has come to break out of our comfort zone and to consider functions of the form $f: X \to Y$, where X or Y are no longer restricted to sets of real numbers. Here are some "exotic sets:"

DEFINITION 2.10
n-tuples

Let $n \geq 1$. The set of **n-tuples**, denoted by $\Re^n$, is given by:

$$\Re^n = \{(a_1, a_2, \ldots, a_n) \mid a_i \in \Re, 1 \leq i \leq n\}$$

Matrices

Let $n \geq 1$ and $m \geq 1$. The set of **n × m-matrices** (pronounced *n-by-m matrices*), denoted by $M_{n \times m}$, is given by:

$$M_{n \times m} = \begin{bmatrix} a_{11} & a_{12} & \cdots & a_{1m} \\ a_{21} & a_{22} & \cdots & a_{2m} \\ \vdots & \vdots & & \vdots \\ a_{n1} & a_{n2} & \cdots & a_{nm} \end{bmatrix}$$

To illustrate: $(2, \tfrac{1}{2}) \in \Re^2$ and $(3, \sqrt{2}, 5, 0) \in \Re^4$.

$$\begin{bmatrix} 2 & 5 \\ -1 & 9 \end{bmatrix} \in M_{2 \times 2}, \quad \begin{bmatrix} 4 \\ 0 \\ \pi \end{bmatrix} \in M_{3 \times 1}, \text{ and } \begin{bmatrix} 0 & 1 & -5 \\ 3 & 1 & 4 \end{bmatrix} \in M_{2 \times 3}$$

As might be expected: Two n tuples $(a_1, a_2, \ldots, a_n)$, $(b_1, b_2, \ldots, b_n)$ are said to be **equal** if $a_i = b_i$ for $1 \leq i \leq n$; in other words: if the are one and the same. The same goes for elements of $M_{n \times m}$.

EXAMPLE 2.4 Let $f: \Re^4 \to M_{2\times 2}$ and $g: M_{2\times 2} \to R^2$ be given by:

$$f(a,b,c,d) = \begin{bmatrix} -b & 2a \\ 3d & c \end{bmatrix} \text{ and } g\left(\begin{bmatrix} a & b \\ c & d \end{bmatrix}\right) = (a+d, bc)$$

(a) Determine:

(i) $f(5, 2, -9, 4)$ (ii) $g\left(\begin{bmatrix} 3 & 5 \\ 2 & 4 \end{bmatrix}\right)$ (iii) $(g \circ f)(1, 2, 5, 3)$

(b) Find $(g \circ f)(a, b, c, d)$

SOLUTION: (a-i) Just follow the "f-pattern:"

$$f(\overset{a}{5}, \overset{b}{2}, \overset{c}{-9}, \overset{d}{4}) = \begin{bmatrix} -2 & 2 \cdot 5 \\ 3 \cdot 4 & -9 \end{bmatrix} = \begin{bmatrix} -2 & 10 \\ 12 & -9 \end{bmatrix}$$

(a-ii) Follow the g-pattern: $g\left(\begin{bmatrix} 3 & 5 \\ 2 & 4 \end{bmatrix}\right) = (3+4, 5 \cdot 2) = (7, 10)$.

(a-iii) Both patterns come into play:

$$(g \circ f)(1, 2, 5, 3) = g[f(1, 2, 5, 3)] = g\left(\begin{bmatrix} -2 & 2 \\ 9 & 5 \end{bmatrix}\right) = (3, -18)$$

(b) $(g \circ f)(a, b, c, d) = g\left(\begin{bmatrix} -b & 2a \\ 3d & c \end{bmatrix}\right) = (-b+c, 6ad)$

ONE-TO-ONE AND ONTO FUNCTIONS

Our next goal is to single out some functions that count more than others; beginning with one-to-one functions:

> Equivalently, f is one-to-one if
> $x_1 \neq x_2 \Rightarrow f(x_1) \neq f(x_2)$
> In words:
> different x's are mapped to different y's.

DEFINITION 2.11
ONE-TO-ONE

A function $f: X \to Y$ is said to be **one-to-one** (or an injection) if

$$f(x_1) = f(x_2) \Rightarrow x_1 = x_2$$

In set notation:
$$(x_1, y), (x_2, y) \in f \Rightarrow x_1 = x_2$$

Figure 2.5 represents the action of two functions, f and g, from $X = \{0, 1, 2\}$ to $Y = \{a, b, c, d\}$. The function f on the left is one-to-one while the function g is not.

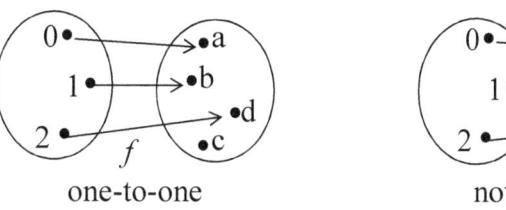

one-to-one   not one-to-one

Figure 2.5

EXAMPLE 2.5 (a) Show that the function $f(x) = \dfrac{x}{5x+2}$ is one-to-one.

(b) Show that the function
$$g(x) = x^3 + 3x^2 - x + 75 \text{ is not one-to-one.}$$

SOLUTION: (a) Appealing to Definition 2.10, we begin with $f(a) = f(b)$, and then go on to show that this can only hold if $a = b$:

$$f(a) = f(b)$$
$$\frac{a}{5a+2} = \frac{b}{5b+2}$$
$$a(5b+2) = b(5a+2)$$
$$5ab + 2a = 5ab + 2b$$
$$2a = 2b$$
$$a = b$$

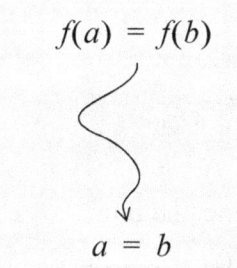

(b) Rather than $g(x) = x^3 + 3x^2 - x + 75$, consider the function $h(x) = x^3 + 3x^2 - x$ and corresponding equation $x^3 + 3x^2 - x = 0$:

$$x^3 + 3x^2 - x = 0$$
$$x(x^2 + 3x - 1) = 0$$
$$x = 0, \, x = \frac{-3 \pm \sqrt{13}}{2}$$
↑
using the quadratic formula

Recall that the solutions of the quadratic equation:
$$ax^2 + bx + c = 0$$
are given by the quadratic formula:
$$x = \frac{-b \pm \sqrt{b^2 - 4ac}}{2a}$$

We see that the function $h(x) = x^3 + 3x^2 - x$ is not one-to-one, for it maps three (need only two) different x's to zero: 0, $\frac{-3+\sqrt{13}}{2}$, and $\frac{-3-\sqrt{13}}{2}$. It follows that the function $g(x) = x^3 + 3x^2 - x + 75$ will map those same numbers to 75, and is therefore not one-to-one.

CHECK YOUR UNDERSTANDING 2.8

(a) Show that the function $f(x) = \dfrac{x}{x+1}$ is one-to-one.

Answer: See page A-11.   (b) Show that the function $f(x) = x^5 - x + 777$ is not one-to-one.

68 Chapter 2 A Touch of Set Theory

In words:
$f: X \to Y$ is onto if every element in Y is "hit" by some $f(x)$.

DEFINITION 2.12
ONTO

A function $f: X \to Y$ is said to be **onto** (or a surjection) if for every $y \in Y$ there exists $x \in X$ such that $f(x) = y$.

In set notation:
$$\forall y \in Y \, \exists x \in X \ni (x, y) \in f$$

Figure 2.6 represents the action of two functions, f and g, from $X = \{0, 1, 2, 3\}$ to $Y = \{a, b, c\}$. The function f on the left is onto, while g is not.

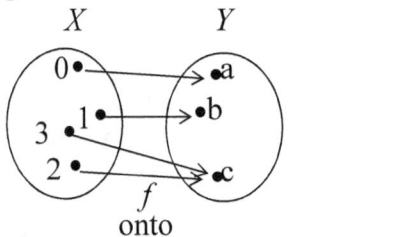

Figure 2.6

EXAMPLE 2.6 (a) Let $f: \Re^4 \to M_{2 \times 2}$ be given by:
$$f(x, y, z, w) = \begin{bmatrix} -y & 2x \\ 3w & z \end{bmatrix}$$

Show that f is both one-to-one and onto.

(a) Let $g: M_{2 \times 2} \to M_{2 \times 2}$ be given by:
$$g\left(\begin{bmatrix} a & b \\ c & d \end{bmatrix}\right) = \begin{bmatrix} b & a \\ c + d & 2b \end{bmatrix}$$

Show that f is neither one-to-one nor onto.

SOLUTION: (a) To show that f is one to one, we start with
$$f(x, y, z, w) = f(\bar{x}, \bar{y}, \bar{z}, \bar{w})$$
and go on to show that $(x, y, z, w) = (\bar{x}, \bar{y}, \bar{z}, \bar{w})$:

We need to consider two elements in $\Re^4$. They have to look different; and so we called one of the elements (x, y, z, w) and the other $(\bar{x}, \bar{y}, \bar{z}, \bar{w})$. (We could have labeled the other (A, B, C, D), or whatever. The two 4-tuples just have to look different, that's all.

$$f(x, y, z, w) = f(\bar{x}, \bar{y}, \bar{z}, \bar{w}) \Rightarrow \begin{bmatrix} -y & 2x \\ 3w & z \end{bmatrix} = \begin{bmatrix} -\bar{y} & 2\bar{x} \\ 3\bar{w} & z \end{bmatrix}$$

$$\Rightarrow \left.\begin{array}{r} -y = -\bar{y} \\ 2x = 2\bar{x} \\ 3w = 3\bar{w} \\ z = \bar{z} \end{array}\right\} \Rightarrow \left.\begin{array}{r} y = \bar{y} \\ x = \bar{x} \\ w = \bar{w} \\ z = \bar{z} \end{array}\right\}$$

$$\Rightarrow (x, y, z, w) = (\bar{x}, \bar{y}, \bar{z}, \bar{w})$$

To show that f is onto, we take an arbitrary element $\begin{bmatrix} a & b \\ c & d \end{bmatrix} \in M_{2 \times 2}$ and set our sights on finding $(x, y, z, w) \in R^4$ such that $f(x, y, z, w) = \begin{bmatrix} a & b \\ c & d \end{bmatrix}$:

$$f(x, y, z, w) = \begin{bmatrix} a & b \\ c & d \end{bmatrix} \Rightarrow \begin{bmatrix} -y & 2x \\ 3w & z \end{bmatrix} = \begin{bmatrix} a & b \\ c & d \end{bmatrix} \Rightarrow \left. \begin{array}{r} -y = a \\ 2x = b \\ 3w = c \\ z = d \end{array} \right\} \Rightarrow \left. \begin{array}{r} y = -a \\ x = b/2 \\ w = c/3 \\ z = d \end{array} \right\}$$

The above argument shows that f will map the element $\left(\dfrac{b}{2}, -a, d, \dfrac{c}{3} \right) \in R^4$ to $\begin{bmatrix} a & b \\ c & d \end{bmatrix} \in M_{2 \times 2}$. Let's check it out:

$$f\left(\dfrac{b}{2}, -a, d, \dfrac{c}{3} \right) = \begin{bmatrix} -(-a) & 2\left(\dfrac{b}{2}\right) \\ 3\left(\dfrac{c}{3}\right) & d \end{bmatrix} = \begin{bmatrix} a & b \\ c & d \end{bmatrix}$$

(b) $g\left(\begin{bmatrix} a & b \\ c & d \end{bmatrix} \right) = \begin{bmatrix} b & a \\ c+d & 2b \end{bmatrix}$ is not one-to one, since:

$$g\left(\begin{bmatrix} 0 & 0 \\ 1 & 0 \end{bmatrix} \right) = g\left(\begin{bmatrix} 0 & 0 \\ 0 & 1 \end{bmatrix} \right) = \begin{bmatrix} 0 & 0 \\ 1 & 0 \end{bmatrix}, \text{ and } \begin{bmatrix} 0 & 0 \\ 1 & 0 \end{bmatrix} \neq \begin{bmatrix} 0 & 0 \\ 0 & 1 \end{bmatrix}.$$

The function g is not onto, since no element $\begin{bmatrix} a & b \\ c & d \end{bmatrix}$ is mapped to $\begin{bmatrix} 1 & 0 \\ 0 & 3 \end{bmatrix}$: $f\left(\begin{bmatrix} a & b \\ c & d \end{bmatrix} \right) = \begin{bmatrix} b & a \\ c+d & \mathbf{2b} \end{bmatrix} \neq \begin{bmatrix} 1 & 0 \\ 0 & \mathbf{3} \end{bmatrix}$ since 3 is not 2 times 1.

CHECK YOUR UNDERSTANDING 2.9

Show that the function $f: M_{2 \times 2} \to R^4$ given by:

$$f\left(\begin{bmatrix} a & b \\ c & d \end{bmatrix} \right) = (d, -c, 3a, b)$$

is one-to-one and onto.

Answer: See page A-11.

Bijections and Their Inverses

> So:
> A bijection $f: X \to Y$ pairs off each element of X with an element of Y.

DEFINITION 2.13
BIJECTION
A function $f: X \to Y$ that is both one-to-one and onto is said to be a **bijection.**

Consider the bijection $f: \{0, 1, 2, 3\} \to \{a, b, c, d\}$ depicted in Figure 2.7(a) and the function $f^{-1}: \{a, b, c, d\} \to \{0, 1, 2, 3\}$ in Figure 2.7(b). Figuratively speaking, f^{-1}, read "f inverse," was obtained from f by "reversing" the direction of the arrows in Figure 2.7(a). On a more formal level:

$$f = \{(0, a), (1, b), (2, c), (3, d)\}$$

and: $f^{-1} = \{(a, 0), (b, 1), (c, 2), (d, 3)\}$

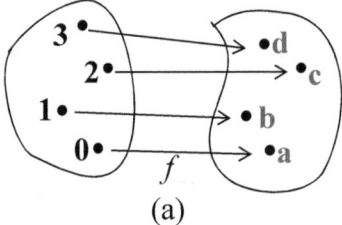 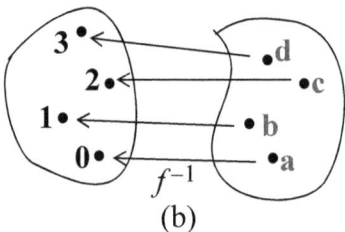

(a) (b)

Figure 2.7

In general:

DEFINITION 2.14
INVERSE FUNCTION
The **inverse** of a bijection $f: X \to Y$, is the function $f^{-1}: Y \to X$ given by:

$$f^{-1}(y) = x \text{ where } f(x) = y$$

More formally:

$$f^{-1} = \{(y, x) | (x, y) \in f\}$$

Figure 2.7 certainly suggests that if f is a bijection then so is f^{-1}, and that if you apply f and then f^{-1} you are back to where you started from (see margin). This is indeed the case:

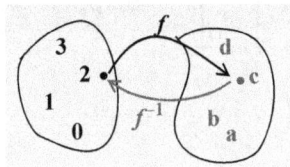

THEOREM 2.4 Let $f: X \to Y$ be a bijection. Then:

(a) $f^{-1}: Y \to X$ is also a bijection.

(b) $f^{-1}[f(x)] = x \ \forall x \in X$ and
$f[f^{-1}(y)] = y \ \forall y \in Y$

> Recall that to say that $f(x) = y$ is to say that $(x, y) \in f$ (see Definition 2.8).

PROOF: (a) f^{-1} **is one-to-one:** If $f^{-1}(y_1) = f^{-1}(y_2) = x$, then:

$$(y_1, x) \in f^{-1} \text{ and } (y_2, x) \in f^{-1}$$
$$\Rightarrow (x, y_1) \in f \text{ and } (x, y_2) \in f$$
$$\Rightarrow y_1 = y_2 \text{ (since } f \text{ is a function)}$$

f^{-1} **is onto:** Let $x \in X$. Since f is onto, there exists $y \in Y$ such that $(x, y) \in f$. Then: $(y, x) \in f^{-1} \Rightarrow f^{-1}(y) = x$.

(b) Let $x \in X$. Since $[x, f(x)] \in f$, $[f(x), x] \in f^{-1}$, which is to say: $x = f^{-1}[f(x)]$. As for the other direction:

CHECK YOUR UNDERSTANDING 2.10

Verify that for any bijection $f: X \to Y$:
$$f[f^{-1}(y)] = y \text{ for every } y \in Y$$

Answer: See page A-12.

EXAMPLE 2.7 (a) Find the inverse of the binary function of $f: \mathfrak{R}^4 \to M_{2 \times 2}$ given by:

$$f(x, y, z, w) = \begin{bmatrix} -y & 2x \\ 3w & z \end{bmatrix}$$

(see Example 2.6)

(b) Show, directly, that

$$f\left[f^{-1}\left(\begin{bmatrix} a & b \\ c & d \end{bmatrix}\right)\right] = \begin{bmatrix} a & b \\ c & d \end{bmatrix}$$

SOLUTION: (a) For given $\begin{bmatrix} a & b \\ c & d \end{bmatrix}$ we determine (x, y, z, w) such that $f(x, y, z, w) = \begin{bmatrix} a & b \\ c & d \end{bmatrix}$:

$$f(x, y, z, w) = \begin{bmatrix} a & b \\ c & d \end{bmatrix} \Rightarrow \begin{bmatrix} -y & 2x \\ 3w & z \end{bmatrix} = \begin{bmatrix} a & b \\ c & d \end{bmatrix} \Rightarrow \begin{matrix} -y = a \\ 2x = b \\ 3w = c \\ z = d \end{matrix} \Rightarrow \begin{matrix} y = -a \\ x = b/2 \\ w = c/3 \\ z = d \end{matrix}$$

Conclusion: $f^{-1}\left(\begin{bmatrix} a & b \\ c & d \end{bmatrix}\right) = \left(\frac{b}{2}, -a, d, \frac{c}{3}\right)$

We just observed that for $f(x, y, z, w) = \begin{bmatrix} -y & 2x \\ 3w & z \end{bmatrix}$:

$$f\left(\frac{b}{2}, -a, d, \frac{c}{3}\right) = \begin{bmatrix} a & b \\ c & d \end{bmatrix}$$

That being the case:

$$f^{-1}\left(\begin{bmatrix} a & b \\ c & d \end{bmatrix}\right) = \left(\frac{b}{2}, -a, d, \frac{c}{3}\right).$$

(b) $f\left[f^{-1}\left(\begin{bmatrix} a & b \\ c & d \end{bmatrix}\right)\right] = f\left(\dfrac{b}{2}, -a, d, \dfrac{c}{3}\right) = \begin{bmatrix} -(-a) & 2\left(\dfrac{b}{2}\right) \\ 3\left(\dfrac{c}{3}\right) & d \end{bmatrix} = \begin{bmatrix} a & b \\ c & d \end{bmatrix}$

since $f(x, y, z, w) = \begin{bmatrix} -y & 2x \\ 3w & z \end{bmatrix}$

CHECK YOUR UNDERSTANDING 2.11

Find the inverse of the bijection $f: M_{2 \times 2} \to R^4$ given by $f\left(\begin{bmatrix} a & b \\ c & d \end{bmatrix}\right) = (d, -c, 3a, b)$ and verify, directly, that:

$f[f^{-1}(x, y, z, w)] = (x, y, z, w)$ and that $f^{-1}\left[f\begin{bmatrix} a & b \\ c & d \end{bmatrix}\right] = \begin{bmatrix} a & b \\ c & d \end{bmatrix}.$

Answer:
$f^{-1}(x, y, z, w) = \begin{bmatrix} z/3 & w \\ -y & x \end{bmatrix}$
For the rest: See page A-11.

As it turns out, one-to-one and onto properties are preserved under composition:

THEOREM 2.5 Let $f: X \to Y$ and $g: Y \to W$ be functions with the range of f contained in the domain of g. Then:

(a) If f and g are one-to-one, so is $g \circ f$.

(b) If f and g are onto, so is $g \circ f$.

(c) If f and g are bijections, so is $g \circ f$.

PROOF: (a) Assume that both f and g are one-to-one, and that:

$$(g \circ f)(x_1) = (g \circ f)(x_2)$$

Which is to say: $g[f(x_1)] = g[f(x_2)]$

Since g is one-to-one: $f(x_1) = f(x_2)$

Since f is one-to-one: $x_1 = x_2$

(b) Assume that both f and g are onto, and let $w \in W$. We are to find $x \in X$ such that $(g \circ f)(x) = w$. Let's do it:

Since g is onto, there exists $y \in Y$ such that $g(y) = w$.
Since f is onto, there exists $x \in X$ such that $f(x) = y$.
It follows that $(g \circ f)(x) = g[f(x)] = g(y) = w$.

(c) If f and g are both bijections then, by (a) and (b), so is $g \circ f$.

Theorem 2.5(c) asserts that the composition $g \circ f$ of two bijections is again a bijection. As such, it has an inverse, and here is how it is related to the inverses of its components:

THEOREM 2.6 If $f: X \to Y$ and $g: Y \to W$ are bijections, then:
$$(g \circ f)^{-1} = f^{-1} \circ g^{-1}$$

> This is an example of a so-called "shoe-sock theorem." Why the funny name?
>
> One puts on socks then shoes: In the reverse process, the shoes come off and then the socks

PROOF: For given $w \in W$, let $x \in X$ be such that $(g \circ f)(x) = w$; which is to say, that $(g \circ f)^{-1}(w) = x$. We complete the proof by showing that $(f^{-1} \circ g^{-1})(w)$ is also equal to x:

$$(g \circ f)^{-1}(w) = x$$
$$w = (g \circ f)(x)$$
$$w = g[f(x)]$$
$$g^{-1}(w) = f(x)$$
$$f^{-1}[g^{-1}(w)] = x$$
$$(f^{-1} \circ g^{-1})(w) = x$$

EXERCISES

Exercises 1-6. For $X = \{a, b, c, d\}$ and $Y = \{A, B, C, D\}$ determine if the subset f of $X \times Y$ is a function $f: X \to Y$. If so, find its range.

1. $f = \{(a, A)\}$
2. $f = \{(a, A), (b, A), (c, A), (d, A)\}$
3. $f = \{(a, A), (a, B)(b, B), (c, D), (d, A)\}$
4. $f = \{(a, A), (b, B), (c, C), (d, D)\}$
5. $f = \{(a, B), (b, C), (c, D), (d, A)\}$
6. $f = \{(a, B), (b, A), (c, D), (d, C)\}$

Exercises 7-16. For $X = \{a, b, c, d\}$, $Y = \{A, B, C\}$, and $Z = \{1, 2, 3\}$, let

$f: X \to Y$, $g: Y \to Z$, $h: Z \to X$, and $k: Z \to Z$ be given by:

$f = \{(a, A), (b, C), (c, A), (d, B)\}$ $\quad g = \{(A, 1), (B, 3), (C, 2)\}$
$h = \{(1, b), (2, a), (3, c)\}$ $\quad k = \{(1, 2), (2, 3), (3, 1)\}$

Indicate whether or not the function is one-to-one, and whether or not it is onto.

7. $g \circ f$
8. $g \circ k$
9. $k \circ k$
10. $h \circ k$
11. $f \circ h$
12. $k \circ g$
13. $g \circ f$
14. $f \circ (h \circ k)$
15. $(k \circ k) \circ k$
16. $f \circ [h \circ (k \circ k)]$

Exercises 17-20. Is $f: \Re \to \Re$ (a) One-to-one? (b) Onto?

17. $f(x) = \dfrac{3x - 7}{x + 2}$
18. $f(x) = x^2 - 3$
19. $f(x) = \dfrac{x}{x^2 + 1}$
20. $f(x) = x^3 - x + 2$

Exercises 21-23. Is $f: \Re \to \Re^2$ (a) One-to-one? (b) Onto?

21. $f(x) = (x, x)$
22. $f(x) = (x, 1)$
23. $f(x) = (2x, -7x)$

Exercises 24-26. Is $f: \Re^2 \to \Re^2$ (a) One-to-one? (b) Onto?

24. $f(x, y) = (y, -x)$
25. $f(x, y) = (x, x + y)$
26. $f(x, y) = (2x, x + y)$

Exercises 27-28. Is $f: M_{2 \times 2} \to \Re^4$ (a) One-to-one? (b) Onto?

27. $f\left(\begin{bmatrix} a & b \\ c & d \end{bmatrix}\right) = (a, -2b, c, c - d)$
28. $f\left(\begin{bmatrix} a & b \\ c & d \end{bmatrix}\right) = (a - b, c, d, b - a)$

Exercises 29-30. Is $f: \Re^4 \to M_{2 \times 2}$ (a) One-to-one? (b) Onto?

29. $f(a, b, c, d) = \begin{bmatrix} a & b + a \\ c + b & d + a \end{bmatrix}$
30. $f(a, b, c, d) = \begin{bmatrix} ab & b + a \\ c + b & a^2 b^2 \end{bmatrix}$

Exercises 31-32. Is $f: M_{3 \times 1} \to M_{2 \times 2}$ (a) One-to-one? (b) Onto?

31. $f\left(\begin{bmatrix} a \\ b \\ c \end{bmatrix}\right) = \begin{bmatrix} a & b \\ c & 5 \end{bmatrix}$
32. $f\left(\begin{bmatrix} a \\ b \\ c \end{bmatrix}\right) = \begin{bmatrix} a & b \\ c & a + b + c \end{bmatrix}$

Exercises 33-41. Show that the given function $f: X \to Y$ is a bijection. Determine $f^{-1}: Y \to X$ and show, directly, that $(f^{-1} \circ f)(x) = x \ \forall x \in X$ and that $(f \circ f^{-1})(y) = y \ \forall y \in Y$.

33. $X = \Re, Y = \Re$, and $f(x) = 3x - 2$.

34. $X = (-\infty, 0) \cup (0, \infty)$, $Y = (-\infty, 1) \cup (1, \infty)$, and $f(x) = \dfrac{x+1}{x}$.

35. $X = (-\infty, -1) \cup (-1, \infty)$, $Y = (-\infty, 1) \cup (1, \infty)$, and $f(x) = \dfrac{2x}{x+1}$.

36. $X = Y = \Re^2$, and $f(a, b) = (-b, a)$.

37. $X = Y = \Re^2$, and $f(a, b) = (5a, b+3)$.

38. $X = Y = M_{2 \times 2}$, and $f\left(\begin{bmatrix} a & b \\ c & d \end{bmatrix}\right) = \begin{bmatrix} b & c \\ d & a \end{bmatrix}$.

39. $X = Y = M_{2 \times 2}$, and $f\left(\begin{bmatrix} a & b \\ c & d \end{bmatrix}\right) = \begin{bmatrix} c & 2d \\ a & -b \end{bmatrix}$.

40. $X = \Re^4$, $Y = M_{2 \times 2}$, and $f(a, b, c, d) = \begin{bmatrix} 2b & c+1 \\ d & -a \end{bmatrix}$.

41. $X = M_{3 \times 1}, Y = \Re^3$, and $f\left(\begin{bmatrix} a \\ b \\ c \end{bmatrix}\right) = (2a, a - b, b + c)$.

42. Prove that a function $f: \Re \to \Re$ is one-to-one if and only if the function $g: \Re \to \Re$ given by $g(x) = -f(x)$ is one-to-one.

43. Prove that for any given $f: X \to Y, g: Y \to S$, and $h: S \to T$: $h \circ (g \circ f) = (h \circ g) \circ f$.

44. Let $f: X \to Y, g: X \to Y$, and $h: Y \to W$ be given, with h a bijection.
    (a) Prove that if $h \circ f = h \circ g$, then $f = g$.
    (b) Show, by means of an example, that (a) need not hold when h is not a bijection.

45. Let $S \subseteq X, Y \neq \emptyset$, and $f: S \to Y$ be given. Prove that there exists a function $g: X \to Y$ such that $f(x) = g(x)$ for every $x \in S$. (That is, a function g which "extends" f to all of X.)

46. Let $S \subseteq X, Y \neq \emptyset$, and $f: X \to Y$ be given. Prove that there exists a function $g: S \to Y$ such that $f(x) = g(x)$ for every $x \in S$. (That is, a function g which is the "restriction" of f to the subset S.)

Exercise. 47-52. (Algebra of Functions) For any set X, and functions $f: X \to \Re$ and $g: X \to \Re$, we define $f+g: X \to \Re, f-g: X \to \Re, f \cdot g: X \to \Re$, and $\frac{f}{g}: X \to \Re$ as follows:

$$(f+g)(x) = f(x) + g(x) \qquad (f-g)(x) = f(x) - g(x)$$
$$(f \cdot g)(x) = f(x) \cdot f(x) \qquad \left(\frac{f}{g}\right)(x) = \frac{f(x)}{g(x)} \text{ if } g(x) \neq 0$$

47. Prove that for any $f: X \to \Re$ and $g: X \to \Re : f + g = g + f$ and $f \cdot g = g \cdot f$.

48. Exhibit $f: \Re \to \Re$, $g: \Re \to \Re$, such that $f - g \neq g - f$.

49. Exhibit one-to-one functions $f: \Re \to \Re$, $g: \Re \to \Re$, such that $f + g$ is not one-to-one.

50. Exhibit onto functions $f: \Re \to \Re$, $g: \Re \to \Re$, such that $f + g$ is not onto.

51. Exhibit one-to-one functions $f: \Re \to \Re$, $g: \Re \to \Re$, such that $f \cdot g$ is not one-to-one.

52. Exhibit onto functions $f: \Re \to \Re$, $g: \Re \to \Re$, such that $f \cdot g$ is not onto.

PROVE OR GIVE A COUNTEREXAMPLE

53. For $f: \Re \to \Re$ and $g: \Re \to \Re$, if $f + g$ is one-to-one, then both f and g must be one-to-one.

54. For $f: \Re \to \Re$ and $g: \Re \to \Re$, if $f + g$ is one-to-one, then f or g must be one-to-one.

55. For $f: \Re \to \Re$ and $g: \Re \to \Re$, if $f \cdot g$ is one-to-one, then both f and g must be one-to-one.

56. For $f: \Re \to \Re$ and $g: \Re \to \Re$, if $g \cdot f$ is one-to-one, then f or g must be one-to-one.

57. (a) If $f: X \to Y$ is an onto function, then so is the function $g \circ f: X \to W$ onto for any function $g: Y \to W$.

    (b) If $g: Y \to W$ is an onto function, then so is the function $g \circ f: X \to W$ onto for any function $f: X \to Y$.

58. (a) Let $f: X \to Y$ and $g: Y \to W$. If $g \circ f: X \to W$ is onto, then f must also be onto.

    (b) Let $f: X \to Y$ and $g: Y \to W$. If $g \circ f: X \to W$ is onto then g must also be onto.

59. (a) If $f: X \to Y$ is one-to-one, then so is the function $g \circ f: X \to W$ for any $g: Y \to W$.

    (b) If $g: Y \to W$ is one-to-one, then so is the function $g \circ f: X \to W$ for any $f: X \to Y$.

60. For any X and $Y \neq \emptyset$, there exists at least one function $f: X \to Y$.

61. For any $X \neq \emptyset$, there exists at least two functions $f: X \to X$.

§3 INFINITE COUNTING

Assume that you do not know how to count, and that you have two bags of marbles, one containing red marbles and the other blue. Without counting, you can still determine whether or not the two bags contain the same number of marbles:

> *Take a red marble from the one bag and a blue marble from the other, then put them aside. Continue this pairing process till one of the bags becomes empty. If, as the one bag becomes empty so does the other, then you can conclude that the two bags had the same number of marbles (without knowing the number).*

If you let the bag of red marbles be a set R, and the bag of blue marbles be a set B, and if you think of the paring off process as representing a function f which assigns to each red marble a unique blue marble, then you can say that the two bags will have the same number of marbles if the function f is a bijection.

Marbles aside:

Note that if there exists a bijection $f: A \to B$ then there also exists a bijection going the other way; namely: $f^{-1}: B \to A$ [see Theorem 2.4(a), page 70]

DEFINITION 2.15
CARDINALITY
Two sets A and B are of the **same cardinality**, written $\text{Card}(A) = \text{Card}(B)$, if there exists a bijection $f: A \to B$.

In a sense, the term "*same cardinality*" can be interpreted to mean "*same number of elements.*" The classier terminology is used since the expression "same number of elements" suggests that we have associated a number to each set, even those that are infinite.

The following theorem should not surprise you. Intuitively, it says that if a set A has the same number of elements as a set B, and if B has the same number of elements as a set C, then A and C have the same number of elements. Intuition aside:

THEOREM 2.7 If $\text{Card}(A) = \text{Card}(B)$ and $\text{Card}(B) = \text{Card}(C)$, then $\text{Card}(A) = \text{Card}(C)$.

PROOF: Applying Definition 2.15, we choose bijections $f: A \to B$ and $g: B \to C$. By Theorem 2.5(c), page 72, $g \circ f: A \to C$ is also a bijection. Returning to Definition 2.15, we conclude that A and B are of the same cardinality.

In our physical universe, if you take something away from a collection of objects you will certainly end up with a smaller number of objects in the collection. When it comes to infinite sets, however, this need no longer be the case:

EXAMPLE 2.8 Let $2Z^+$ denote the set of even positive integers. Show that $\text{Card}(2Z^+) = \text{Card}(Z^+)$.

SOLUTION: The diagram below illustrates how the elements of Z^+ can be paired with those of $2Z^+$:

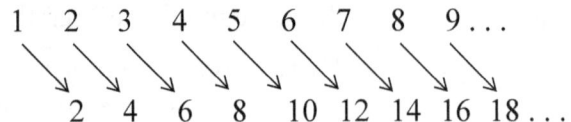

Let's formalize the above observation by showing that the function $f: Z^+ \to 2Z^+$ given by $f(n) = 2n$ is a bijection.

***f* is one-to-one:**
$$f(n_1) = f(n_2) \Rightarrow 2n_1 = 2n_2 \Rightarrow n_1 = n_2$$

***f* is onto:** Let $m \in 2Z^+$. Since m is even, $m = 2n$ for some $n \in Z^+$. Then: $f(n) = 2n = m$.

CHECK YOUR UNDERSTANDING 2.12

Answer: See page A-12.

Prove that:
$$\text{Card}(Z^+) = \text{Card}\{0, 1, 2, 3, \ldots\}$$

Roughly speaking, by a countable set we simply mean a set whose elements can actually be "counted:" one element, two elements, three, and so on until you have counted them all, even though it may require all of Z^+ to accomplish that feat. On a more formal footing:

DEFINITION 2.16 A set is **countable** if it is finite or has the **COUNTABLE SET** same cardinality as Z^+.

In particular, the set $\{1, 2, 3\}$ is countable, as is the set $2Z^+$ of Example 2.9. Below, we show that the set Z of all integers is also countable. This may not be too surprising, since the feeling is that there should certainly be enough positive integers to count anything. (That feeling is however wrong, as you will soon see.)

EXAMPLE 2.9 Prove that the set Z of all integers is countable.

SOLUTION: The diagram below illustrates how the elements of Z^+ can be paired off with those of Z:

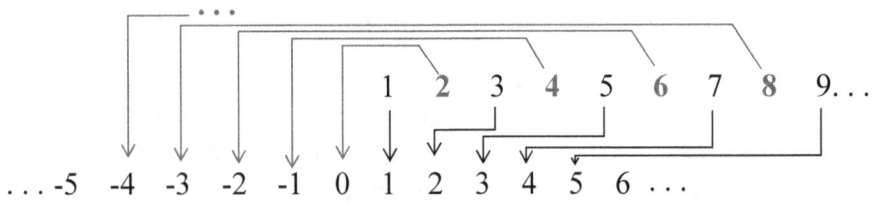

Can we explicitly pin down the function represented in the above diagram? Yes: $f(n) = \begin{cases} \dfrac{n+1}{2} & \text{if } n \text{ is odd} \\ -\left(\dfrac{n}{2} - 1\right) & \text{if } n \text{ is even} \end{cases}$

> To evaluate f, follow the stated instructions:
> If n is odd, say 7, then use the upper "rule:"
> $$f(7) = \frac{7+1}{2} = 4$$
> If n is even, say 8, then go with the bottom rule:
> $$f(8) = -\left(\frac{8}{2} - 1\right) = -3$$

f is one-to-one: Suppose that $f(n_1) = f(n_2)$. We first observe that since f takes even integers to non-positive integers, and odd integers to positive integers, n_1 and n_2 must both be even or both be odd. We consider both cases:

Case 1. n_1 and n_2 even. Taking the "low road" in the description of f, we have:
$$f(n_1) = f(n_2) \Rightarrow -\left(\frac{n_1}{2} - 1\right) = -\left(\frac{n_2}{2} - 1\right) \Rightarrow n_1 = n_2$$
$\uparrow$ details omitted

Case 2. n_1 and n_2 odd. Taking the "high road" in the description of f, we now have:
$$f(n_1) = f(n_2) \Rightarrow \frac{n_1 + 1}{2} = \frac{n_2 + 1}{2} \Rightarrow n_1 = n_2$$

f is onto: For given z we are to find a positive integer n such that $f(n) = z$, and again break the argument into a couple of cases:

Case 1. For $z \in Z$ and positive, consider the odd positive integer $n = 2z - 1$ (see margin). Applying f we have:
$$f(n) = f(2z - 1) = \frac{(2z-1)+1}{2} = \frac{2z}{2} = z$$

> $z = \dfrac{n+1}{2}$
> $2z = n+1$
> $n = 2z - 1$

Case 2. For $z \in Z$ and non-positive, consider the even positive integer $n = -2z + 2$ (see margin). Applying f we have:
$$f(n) = f(-2z + 2) = -\left(\frac{-2z+2}{2} - 1\right) = -(-z + 1 - 1) = z$$

> $z = -\left(\dfrac{n}{2} - 1\right)$
> $2z = -n + 2$
> $n = -2z + 2$

The set Q^+ of positive rational numbers (fractions) appears to have a lot more elements that does Z^+. Not really:

EXAMPLE 2.10 Show that Q^+ is countable.

SOLUTION: We call your attention to Figure 2.8. The positive integers occupy the first row of that figure. The positive rational numbers with denominator 2, which did not already appear in the first row, are listed in the second row. The positive rational numbers with a 3 in the denominator, which did not appear in either the first or second row, comprise the third row; and so on and so forth. Following the arrows in the figure, one is able to count each and every element of Q^+:

The <u>first</u> number counted is the number 1. Moving along the second arrow we arrive at the <u>second</u> number counted: 2, and then the third number: $\frac{1}{2}$. Following the third arrow we count the <u>fourth</u>, <u>fifth</u>, and <u>sixth</u> numbers: 3, $\frac{3}{2}$, and $\frac{1}{3}$, respectively. Can you follow our procedure and determine the next four numbers counted? Sure, they are pierced by the fourth arrow; namely: 4, $\frac{5}{2}$, $\frac{2}{3}$, and $\frac{1}{4}$.

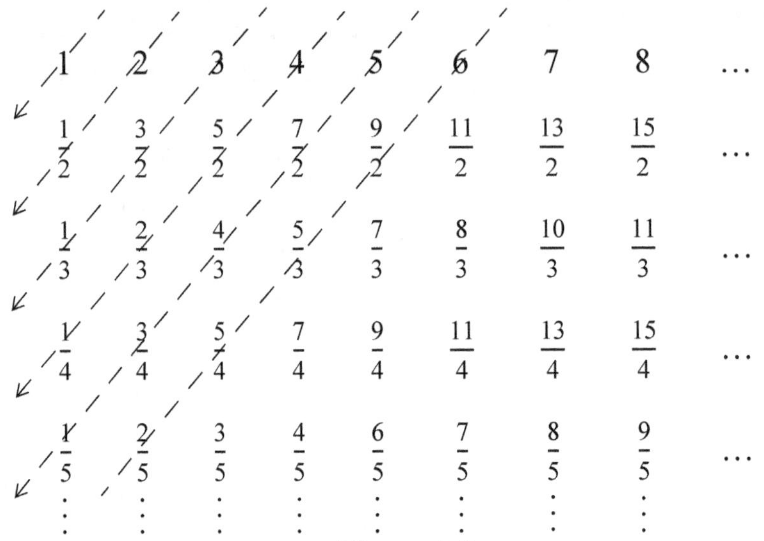

Figure 2.8

Though not explicitly associated with a nice compact rule, the above counting method does establish a function $f: N \to Q^+: f(1) = 1$, $f(2) = 2$, $f(3) = \frac{1}{2}$, $f(4) = 3$, $f(5) = \frac{3}{2}$; and, given enough time we could, by brute force, grind out the value of $f(336)$ and beyond. Moreover, since each element of Q^+ is ultimately pierced by one and only one arrow, f is a bijection, and Q^+ is seen to be countable.

In a more general setting:

THEOREM 2.8 The countable union of countable sets is again countable.

PROOF: Your turn:

CHECK YOUR UNDERSTANDING 2.13

Let $A_n = \{a_{n1}, a_{n2}, a_{n3}, ...\}$ for $n = 1, 2, 3, ...$. Show that

$$A = \bigcup_{n=1}^{\infty} A_n \text{ is countable.}$$

Note: You can assume, without loss of generality that $A_i \cap A_j = \emptyset$ if $i \neq j$ (see margin).

If: $\overline{A_n} = A_n - \bigcup_{i<n} A_i$

Then: $\bigcup_{n=1}^{\infty} A_n = \bigcup_{n=1}^{\infty} \overline{A_n}$

Answer: See page A-12.

2.3 Infinite Counting 81

All infinite sets thus far considered are countable. Are there **uncountable** sets? In other words, sets so large that there are not enough integers with which to count them? Yes:

THEOREM 2.9 The interval $J = [0, 1] = \{x | (0 \leq x \leq 1)\}$ is uncountable.

PROOF: Assume that J is countable (we will arrive at a contradiction). Being countable, we can list its elements in sequential form (a first, a second, a third, etc.):

$$J = \{r_1, r_2, r_3, r_4, \ldots\}$$

Below, we again list the elements of J, but now in decimal form (the first digit following the decimal point of the number r_n is a_{n1}, the second digit is a_{n2}, the third is a_{n3}, and so on).

$$r_1 = 0.a_{11}a_{12}a_{13}a_{14}\ldots$$
$$r_2 = 0.a_{21}a_{22}a_{23}a_{24}\ldots$$
$$r_3 = 0.a_{31}a_{32}a_{33}a_{34}\ldots$$
$$\vdots$$
$$r_n = 0.a_{n1}a_{n2}a_{n3}a_{n4}\ldots$$
$$\vdots$$

We now set to work on constructing a perfectly good number $r \in J$ which is not any of the numbers listed above — contradicting the assertion that the above list contains each and every element of J.

To be a bit more specific, we construct a number $r \in J$ which differs from the number r_i of the above list in its ith decimal place. To be entirely specific:

$$r = 0.a_1 a_2 a_3 a_4 \ldots \quad \text{where } a_i = \begin{cases} 0 \text{ if } a_{ii} \neq 0 \\ 1 \text{ if } a_{ii} = 0 \end{cases}$$

Note that r cannot be r_1, for if the first digit of r_1, namely a_{11} is anything but 0, then it cannot be r, as r's first digit is 0; and if the first digit of r_1 is zero, then it again cannot be r, as r's first digit would now be 1. By the same token, r cannot be r_2, since the second digit of r will surely differ from a_{22}, the second digit of r_2. The same argument can be used to establish the fact that r cannot be any r_i, since r's i^{th} digit differs from the ith digit of r_i.

CHECK YOUR UNDERSTANDING 2.14

For $a < b$ and $c < d$, let $L = [a, b]$ and $M = [c, d]$. Show that:

$$\text{Card}(L) = \text{Card}(M)$$

Suggestion: Consider line passing through L and M.

Answer: See page A-12.

82 Chapter 2 A Touch of Set Theory

It is reasonable to say that a set A has cardinality less than or equal to that of a set B if the elements of A can be paired off with those of a subset of B:

> We also say that the cardinality of B is **greater than or equal** to that of A, and write:
> $$\text{Card}(B) \geq \text{Card}(A)$$

DEFINITION 2.17 The cardinality of a set A is **less than or equal** to the cardinality of a set B, written $\text{Card}(A) \leq \text{Card}(B)$, if there exists a one-to-one function $f: A \to B$. (The function need no longer be onto.)

All of the following results might be anticipated, but only the first two are easily established. Alas, a proof of (c) lies outside the scope of this text.

THEOREM 2.10 Let A, B, and C denote sets.

(a) If $A \subseteq B$, then $\text{Card}(A) \leq \text{Card}(B)$.

(b) If $\text{Card}(A) \leq \text{Card}(B)$ and $\text{Card}(B) \leq \text{Card}(C)$ then $\text{Card}(A) \leq \text{Card}(C)$.

> Georg Cantor (1845-1916)
> Felix Bernstein (1878-1956)
> Ernst Schroeder (1841-1902)

CANTOR-BERNSTEIN-SCHROEDER THEOREM

(c) If $\text{Card}(A) \leq \text{Card}(B)$ and $\text{Card}(B) \leq \text{Card}(A)$ then $\text{Card}(A) = \text{Card}(B)$

PROOF: (a) The inclusion map $I_A: A \to B$ given by $I_A(a) = a$ is clearly one-to-one.

(b) From the given conditions, we know that there exist one-to-one functions $f: A \to B$ and $g: B \to C$. All we need do, to complete the proof, is to note that the composite function $g \circ f: A \to C$ is also one-to-one (Theorem 2.5(a), page 72).

CHECK YOUR UNDERSTANDING 2.15

Show that all finite intervals are of the same cardinality.
Suggestions: Use the result of CYU 2.15 and Theorem 2.10.

> Answer: See page A-13.

The above CYU assures us that all finite intervals are of the same cardinality. What about the set of all real numbers $\Re$? It is certainly bigger than every finite interval. Does its cardinality exceed that of the finite intervals? No:

THEOREM 2.11 The cardinality of $\Re$ is the same as that of a finite interval.

PROOF: An analytical proof: The tangent function maps the open interval $\left(-\frac{\pi}{2}, \frac{\pi}{2}\right)$ in a one-to-one fashion onto $\Re$ (see margin).

A geometrical argument: Figure 2.9(a) displays a bijection from the open interval $I = (-1, 1)$ to the semicircle S of radius 1 [excluding the end points of its diameter: $(-1, 0)$ and $(1, 0)$]. It follows that $\textbf{Card}(I) = \textbf{Card}(S)$.

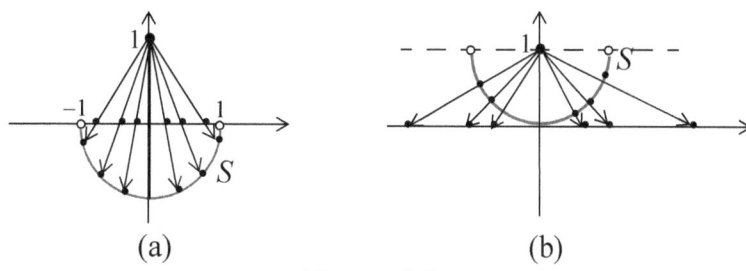

Figure 2.9

Figure 2.9(b) depicts a bijection from S (moved up one unit) to $\Re$ (the x-axis). It follows that **Card**(S) = **Card**$(\Re)$. Applying Theorem 2.7 we conclude that Card(I) = Card$(\Re)$.

We point out that $\Re$ is said to have **cardinality** c (for the $c$ontinuum). The assertion that there does not exists a set X such that Card(Z) < Card(X) < Card$(\Re)$ is called the **continuum hypothesis**.

CHECK YOUR UNDERSTANDING 2.16

(a) Show that the set $(-\infty, 3) \cup (5, 6] \cup [9, \infty)$ has cardinality c.

(b) Prove that there cannot exist a bijection from the set of real numbers to the set of rational numbers.

Answer: See page A-13.

At this point we only have two types of infinity: the countable type associated with Z^+, and the uncountable type associated with $\Re$. Is that all there is? Where should we look if we hope to find bigger infinities? Towards bigger sets, of course, and one that might come to mind is the familiar Cartesian plane ($\Re^2 = \{(x, y)|x, y \in \Re\}$. No luck, for Card$(\Re^2)$ = Card$(\Re)$ [Exercise 25]. And if you are tempted to try the three dimensional space: $\Re^3 = \{(x, y, z)|x, y, z \in \Re\}$ don't bother, for it too has cardinality c. Indeed, for any n, the set $\Re^n = \{(x_1, x_2, ..., x_n)|x_i \in \Re, 1 \le i \le n\}$ also has cardinality c [Exercise 26.] But there is another avenue we might consider, one that involves the following concept:

DEFINITION 2.18 The **power set** of a set A, denoted by $P(A)$, is the set of all subsets of A.

For example, if $A = \{a, b, c\}$, then:

$P(A) = \{\varnothing, \{a\}, \{b\}, \{c\}, \{a, b\}, \{a, c\}, \{b, c\}, \{a, b, c\}\}$

Observe that A has 3 element, and that $P(A)$ has $2^3 = 8$ elements. This is no fluke:

Note that A has 3 element, and that $P(A)$ has $2^3 = 8$ elements. This is no fluke:

Due to this theorem, the power set $P(A)$ of a given set A is often denoted by the symbol 2^A.

THEOREM 2.12 If the set A has n elements, then its power set $P(A)$ has 2^n elements.

PROOF: (By induction). Since we are only interested in counting, we will assume, without loss of generality, that our set A of n elements is the specific set $A = \{1, 2, 3, \ldots, n\}$.

I. The claim does hold at $n = 1$, for the set $P(\{1\}) = \{\emptyset, \{1\}\}$, contains $2 = 2^1$ elements.

II. Assume that $P(A_k)$ contains 2^k elements for $A_k = \{1, 2, 3 \ldots k\}$.

III. We complete the proof by showing that the set $P(A_{k+1})$ contains 2^{k+1} elements, where $A_{k+1} = \{1, 2, 3 \ldots k, k+1\}$.

Clearly, every subset of A_{k+1} either contains the number $k+1$ or it does not. Let X be the set of those subsets that do not contain $k+1$, and Y the set of those subsets that do contain $k+1$:

$$X = \{S \in P(A_{k+1}) | k+1 \notin S\} \text{ and } Y = \{S \in P(A_{k+1}) | k+1 \in S\}$$

Noting that X is the set $P(A_k)$, we conclude, by the induction hypothesis, that it contains 2^k elements. But every element in Y can be obtained by simply adjoining $k+1$ to some element in X:

$$Y = \{S \cup \{k+1\} | S \in X\}$$

As such, Y has the same number of elements as X, namely 2^k.

Noting that $P(A_{k+1}) = X \cup Y$ and that $X \cap Y = \emptyset$, we conclude that the number of elements in $P(A_{k+1})$ is $2^k + 2^k = 2^{k+1}$, and III is established.

The following wonderful theorem tells us that if you take the power set of any set A, even an infinite set, you will end up with a set whose cardinality is strictly greater than that of A. Wonderful, in that it tells us that there is no end to the different levels of infinity.

THEOREM 2.13 For any set A:

CANTOR $\qquad\qquad\qquad\text{Card}(A) < \text{Card}[P(A)]$

> The function $f: A \to P(A)$ given by $f(a) = \{a\}$ is certainly one-to-one.

PROOF: Clearly $\text{Card}(A) \leq \text{Card}[P(A)]$ (see margin). We complete the proof by showing that the assumption $\text{Card}(A) = \text{Card}[P(A)]$ leads to a contradiction:

Suppose there exists a bijection $g: A \to P(A)$ (we will arrive at a contradiction).

For any $a \in A$, $g(a)$ is a subset of A. As such, either $a \in g(a)$ or $a \notin g(a)$. Let us now focus on the following subset of A:

$$S_0 = \{a \in A | a \notin g(a)\}$$

Since g is onto, there must exist some element of A, say a_0, such that:

$$g(a_0) = S_0$$

Now, either $a_0 \in g(a_0) = S_0$ or $a_0 \notin g(a_0) = S_0$. Let's see which:

If $a_0 \in S_0$, then since $S_0 = \{a \in A | a \notin g(a)\}$: $a_0 \notin S_0$.

If $a_0 \notin S_0$, then since $S_0 = \{a \in A | a \notin f(a)\}$: $a_0 \in S_0$.

Since neither option can occur, $\text{Card}(A) = \text{Card}[P(A)]$ must be false.

> The assertion that for any infinite set X there does not exist a set Y for which $\text{Card}(X) < \text{Card}(Y) < \text{Card}[P(X)]$ is called the generalized continuum hypothesis.

CHECK YOUR UNDERSTANDING 2.17

Prove that there cannot exist a set of all sets.

Suggestion: Assume such a set S exists and consider the set $T = \bigcup_{A \in S} A$ along with Theorem 2.10(a) and Theorem 2.13 to arrive at a contradiction.

Answer: See page A-13.

EXERCISES

Exercises 1-12. Establish a specific bijection from the set A to the set B.

1. $A = Z^+$, $B = \{5n \mid n \in Z^+\}$
2. $A = \{5n \mid n \in Z^+\}$, $B = Z^+$
3. $A = Z^+$, $B = \{5n \mid n \in Z\}$
4. $A = \{5n \mid n \in Z\}$, $B = Z^+$
5. $A = Z^+$, $B = \{n + \sqrt{2} \mid n \in Z\}$
6. $B = \{n + \sqrt{2} \mid n \in Z\}$, $A = Z^+$
7. $A = \{5n \mid n \in Z^+\}$, $B = \{10n \mid n \in Z^+\}$
8. $A = Z^+$, $B = \{2^n \mid n \in Z\}$
9. For given $a, b \in Z^+$: $A = \{an \mid n \in Z^+\}$, $B = \{bn \mid n \in Z^+\}$.
10. For given $a, b \in Z^+$: $A = \{an \mid n \in Z\}$, $B = \{bn \mid n \in Z^+\}$.
11. For given $s, t \in \Re$, with $s \neq 0$ and $t \neq 0$: $A = \{ns \mid n \in Z^+\}$, $B = \{nt \mid n \in Z^+\}$.
12. For given $s, t \in \Re$, with $s \neq 0$ and $t \neq 0$: $A = \{sr \mid r \in \Re\}$, $B = \{tr \mid r \in Z^+\}$.

Exercises 13-20. For $a, b, c, d \in \Re$, with $a < b$ and $c < d$, exhibit a specific bijection, from:

13. $[a, b]$ to $[c, d]$
14. (a, b) to (c, d)
15. $[a, b)$ to $[c, d]$
16. $(a, b]$ to $(c, d]$
17. $[a, \infty)$ to $[b, \infty)$
18. (a, b) to (a, ∞)
19. (a, ∞) to (b, ∞)
20. (a, ∞) to $(-\infty, b)$

21. Prove that the set of intervals $\{(a, b) \mid a \text{ and } b \text{ are rational}\}$ is countable.
22. Prove that there are only countably many polynomials with rational coefficients.
23. Prove that there are only countably many solutions to the set of all polynomials with rational coefficients.
24. Prove that the set of irrational numbers is not countable.
25. Prove that there are uncountably many lines in the plane.
26. Prove that there are only countably many lines of the form $y = mx + b$, where $m, b \in Q$.
27. Prove that there are uncountably many circles in the plane.

Exercises 28-33. Prove that the given set is countable (see Definition 2.7, page 64).

28. $Z^+ \times Z^+$
29. $Z^+ \times Z^+ \times Z^+$
30. $Q \times Q$
31. $Q \times Q \times Q$

32. Prove that: $\text{Card}(\Re^2) = \text{Card}(\Re)$.

33. Use the Principle of Mathematical Induction to show that $\text{Card}(\Re^n) = \text{Card}(\Re)$ for any positive integer n.

34. Let $F = \{f: \{1, 2, 3\} \to \{0, 1\}\}$ (functions that assigns to each integer in the set $\{1, 2, 3\}$ the value of 0 or the value of 1. Prove that F contains $2^3 = 8$ elements, and that therefore $Card(F) = Card(P(\{1, 2, 3\}))$

35. Let $F = \{f: \{1, 2, ..., n\} \to \{0, 1\}\}$ (functions that assigns to each integer in the set $\{1, 2, ..., n\}$ the value of 0 or the value of 1). Use the Principle of Mathematical Induction to show that F contains 2^n elements.

36. Let $F = \{f: Z^+ \to \{0, 1\}\}$ (functions that assign to each positive integer either 0 or 1). Prove that $Card(F) = Card(\Re)$.

37. Prove that for any given set X, $Card(\{f: X \to \{0, 1\}\}) = Card(P(X))$.

	PROVE OR GIVE A COUNTEREXAMPLE	

38. If $Card(A) \ne Card(B)$ and $Card(B) \ne Card(C)$, then $Card(A) \ne Card(C)$.

39. If $Card(A) \ne Card(B)$ and $Card(B) = Card(C)$, then $Card(A) \ne Card(C)$.

40. If $Card(A) = Card(B)$ and $Card(C) = Card(D)$, then $Card(A \cup C) = Card(B \cup D)$.

41. If $Card(A) \ne Card(B)$ and $Card(C) \ne Card(D)$, then $Card(A \cup C) \ne Card(B \cup D)$.

42. If $Card(A) \ne Card(B)$ and $Card(C) = Card(D)$, then $Card(A \cup C) \ne Card(B \cup D)$.

43. If $Card(A) = Card(B)$ and $Card(C) = Card(D)$, then $Card(A \times C) = Card(B \times D)$.

44. If $Card(A) \ne Card(B)$ and $Card(C) \ne Card(D)$, then $Card(A \times C) \ne Card(B \times D)$.

45. If $Card(A) \ne Card(B)$ and $Card(C) = Card(D)$, then $Card(A \times C) \ne Card(B \times D)$.

46. The set of intervals $\{(a, b] | a \text{ and } b \text{ are rational}\}$ is countable.

47. The set of intervals $\{(a, b] | a \text{ is rational}\}$ is countable.

48. There can be at most countably many mutually disjoint circles (with positive radius) in the plane.

49. There can be at most countably many mutually disjoint lines in the plane.

§4. EQUIVALENCE RELATIONS

> Recall that $X \times Y$, called the **Cartesian Product** of X with Y, is the set of all ordered pairs (x, y), with $x \in X$ and $y \in Y$.

In Section 2 we defined a function from a set X to a set Y to be a subset $f \subseteq X \times Y$ such that:

For every $x \in X$ there exists a unique $y \in Y$ with $(x, y) \in f$.

Removing all restrictions, we arrive at a far more general concept than that of a function:

DEFINITION 2.19
RELATION

A **relation E from a set X to a set Y** is any subset $E \subseteq X \times Y$.
A relation from a set X to X is said to be a **relation on X.**

Each and every subset of $\Re \times \Re$, including the chaotic one in the margin, is a relation on $\Re$, suggesting that Definition 2.19 is a tad too general. Some restrictions are in order:

DEFINITION 2.20

A relation E on a set X is a subset $E \subseteq X \times X$ and is said to be:

REFLEXIVE

Reflexive: $(x, x) \in E$ for every $x \in X$.
(Every element of X is related to itself)

SYMMETRIC

Symmetric: If $(x, y) \in E$ then $(y, x) \in E$.
(If x is related to y, then y is related to x)

TRANSITIVE

Transitive: If $(x, y) \in E$ and $(y, z) \in E$ then $(x, z) \in E$.
(If x is related to y, and y is related to z, then x is related to z)

EQUIVALENCE RELATION

An **equivalence relation** on a set X is a relation that is reflexive, symmetric and transitive.

The notation $x \sim y$ is often used to indicate that x is related to y with respect to some understood relation E. Utilizing that option, we can rephrase Definition 2.20 as follows:

An **equivalence relation** $\sim$ on a set X is a relation which is

Reflexive: if $x \sim x$ for every $x \in X$.

Symmetric: if $x \sim y$, then $y \sim x$.

and Transitive: if $x \sim y$ and $y \sim z$, then $x \sim z$.

EXAMPLE 2.11 Show that the relation $\frac{a}{b} \sim \frac{c}{d}$ if $ad = bc$ is an equivalence relation on the set of rational numbers.

2.4 Equivalence Relations

SOLUTION:

Reflexive: $\frac{a}{b} \sim \frac{a}{b}$ since $ab = ba$.

Symmetric: $\frac{a}{b} \sim \frac{c}{d} \Rightarrow ad = bc \Rightarrow cb = da \Rightarrow \frac{c}{d} \sim \frac{a}{b}$.

Transitive: $\frac{a}{b} \sim \frac{c}{d}$ and $\frac{c}{d} \sim \frac{e}{f} \Rightarrow \underset{(*)}{ad = bc}$ and $\underset{(**)}{cf = de}$

We establish the fact that $\frac{a}{b} \sim \frac{e}{f}$ by showing that **af = be**:

$$af = \underset{\text{see }(*)}{\frac{bc}{d}} \cdot f = \frac{bc}{d} \cdot \underset{\text{see }(**)}{\frac{de}{c}} = be$$

As you know, when it comes to rational numbers, one simply writes $\frac{2}{3} = \frac{4}{6}$ rather than $\frac{2}{3} \sim \frac{4}{6}$.

> In general, equivalence relations enable one to establish a somewhat "fuzzy" sense of equality — a "fuzzyness" which is all but ignored in the above example; for, as you know, when it comes to the set of rational numbers, one simply writes $\frac{3}{7} = \frac{15}{35}$, even though the scribbles $\frac{3}{7}$ and $\frac{15}{35}$ certainly look different.

Recall that $a|b$ means that a divides b (see Definition 1.7, page 15).

EXAMPLE 2.12 Show that the relation $a \sim b$ if $2|(3a - b)$ is an equivalence relation on Z.

SOLUTION: The relation $a \sim b$ if $2|(3a - b)$ is:

Reflexive. $a \sim a$, since: $3a - \underset{\text{here, } a \text{ is playing the role of } b}{a} = 2a$

Symmetric. Assume that $a \sim b$, which is to say, that:
$$3a - b = 2h \text{ for some } h \in Z \quad (*)$$
We are to show that $b \sim a$, which is to say, that:
$$3b - a = 2n \text{ for some } n \in Z$$
Lets do it. From (*) $b = 3a - 2h$.
Hence: $3b - a = 3(3a - 2h) - a = 2\underset{n}{(4a - 3h)} = 2n$

An expression of the form $a = \frac{2h+b}{3}$ is unacceptable in the solution process, since we are involved with the set Z of integers and not "fractions."

TRANSITIVE: Assume that $a \sim b$ and $b \sim c$; which is to say, that:
(1) $3a - b = 2h$ and (2) $3b - c = 2k$ for $h, k \in Z$
We are to show that $a \sim c$; which is to say, that: $3a - c = 2n$.
Let's do it. From (2): $c = 3b - 2k$.
Hence: $3a - c = 3a - (3b - 2k)$
From (1): $= 2h + b - (3b - 2k) = 2\underset{n}{(h + k - b)} = 2n$

CHECK YOUR UNDERSTANDING 2.18

Show that the relation $(x_0, y_0) \sim (x_1, y_1)$ if $x_0 = x_1$ is an equivalence relation on $\Re^2$.

Answer: See page A-13

DEFINITION 2.21

EQUIVALENCE CLASS

Let ~ be an equivalence relation on X. For each $x_0 \in X$, the **equivalence class** of x_0, denoted by $[x_0]$, is the set:

$$[x_0] = \{x \in X | x \sim x_0\}$$

In words: The equivalence class of x_0 consists of all elements of X that are related to x_0. We now show that any element in $[x_0]$ will generate the same equivalence class:

THEOREM 2.14 Let ~ be an equivalence relation on X. For any $x_1, x_2 \in X$:

$$x_1 \sim x_2 \Leftrightarrow [x_1] = [x_2]$$

PROOF: Assume that $x_1 \sim x_2$. We show that $[x_1] \subseteq [x_2]$ (a similar argument can be used to show that $[x_2] \subseteq [x_1]$ and that therefore $[x_1] = [x_2]$):

$$x \in [x_1] \Rightarrow x \sim x_1$$

By transitivity, since $x_1 \sim x_2$: $x \sim x_2 \Rightarrow x \in [x_2]$

Conversely, if $[x_1] = [x_2]$, then, since $x_1 \in [x_2]$: $x_1 \sim x_2$.

EXAMPLE 2.13 Determine the set $\{[n]\}_{n \in Z}$ of equivalence classes corresponding to the equivalence relation $a \sim b$ if $2|(3a-b)$ of Example 2.12.

SOLUTION: Let's start off with $a = 0$. By definition:

$[0] = \{b \in Z | 2|(-b)\} = \{2n | n \in Z\}$ (the even integers)

Since 1 is not in $[0]$, $[1]$ will differ from $[0]$ (Theorem 2.14). Specifically:

$[1] = \{b \in Z | 2|(1-b)\} = \{2n+1 | n \in Z\}$ (the odd integers)

In the above example the give equivalence relation decomposed Z into disjoint equivalence classes; namely:

$$Z = [0] \cup [1] = \{\text{even integers}\} \cup \{\text{odd integers}\}$$

To put it another way: the equivalence classes in Example 2.13 effected a partition of Z, where:

DEFINITION 2.22

PARTITION

A set of nonempty subsets $\{S_\alpha\}_{\alpha \in A}$ of a set X is said to be a **partition** of X if:

(i) $X = \bigcup_{\alpha \in A} S_\alpha$

(ii) If $S_\alpha \cap S_{\bar{\alpha}} \neq \emptyset$ then $S_\alpha = S_{\bar{\alpha}}$

> To put it roughly:
> A partition of a set S chops S up into disjoint pieces.

Figure 2.10(a) displays a 5-subset partition $\{S_1, S_2, S_3, S_4, S_5\}$ of the indicated set. An infinite partition of $[0, \infty)$ is represented in Figure 2.10(b): $\{[n, n+1)\}_{n=0}^{\infty}$

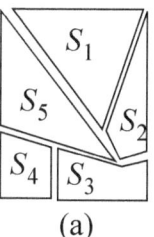
(a)

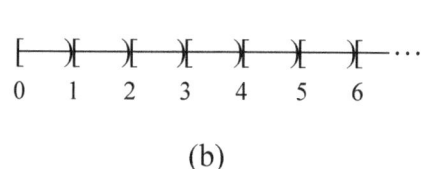
(b)

Figure 2.10

CHECK YOUR UNDERSTANDING 2.19

Determine if the given collection of subsets of $\Re$ is a partition of $\Re$?

(a) $\{[n, n+1]\}_{n \in Z}$

(b) $\{\{n\} | n \in Z\} \cup \{(i, i+1)\}_{i=0}^{\infty} \cup \{(-i-1, -i)\}_{i=0}^{\infty}$

(a): No   (b): Yes

There is an important connection between the equivalence relations on a set X and the partitions of X, and here it is:

THEOREM 2.15 (a) If $\sim$ is an equivalence relation on X, then the set of its equivalence classes, $\{[x]\}_{x \in X}$, is a partition of X.

(b) If $\{S_\alpha\}_{\alpha \in A}$ is a partition of X, then the relation $x_1 \sim x_2$ if $x_1, x_2 \in S_\alpha$ for some $\alpha \in A$ is an equivalence relation on X.

PROOF: (a) We Show that:

(i) $X = \bigcup_{x \in X} [x]$

and (ii) If $[x_1] \cap [x_2] \neq \varnothing$, then $[x_1] = [x_2]$.

(i): Since $\sim$ is an equivalence relation, $x \sim x$ for every $x \in X$. It follows that $x \in [x]$ for every $x \in X$, and that therefore $X = \bigcup_{x \in X} [x]$.

(ii): If $[x_1] \cap [x_2] \neq \varnothing$, then there exists $x_0 \in [x_1] \cap [x_2]$.

Since $x_0 \in [x_1]$ and $x_0 \in [x_2]$: $x_0 \sim x_1$ and $x_0 \sim x_2$.

By symmetry and transitivity: $x_1 \sim x_2$

By Theorem 2.14: $[x_1] = [x_2]$

(b) Let $\{S_\alpha\}_{\alpha \in A}$ be a partition of X. We show that the relation:

$x_1 \sim x_2$ if there exists $\alpha \in A$ such that $x_1, x_2 \in S_\alpha$ is an equivalence relation on X:

Reflexive: To say that $x \sim x$, is to say that x belongs to the same S_α as itself, and it certainly does.

Symmetric: $x \sim y \Rightarrow \exists \alpha \in A \ni x, y \in S_\alpha \Rightarrow y, x \in S_\alpha \Rightarrow y \sim x$

Transitive: Assume $x \sim y$ and $y \sim z$. We show that $x \sim z$:

Since $x \sim y: x, y \in S_\alpha$ for some $\alpha \in A$.

Since $z \sim y: y, z \in S_{\bar{\alpha}}$ for some $\bar{\alpha} \in A$.

Since $S_\alpha \cap S_{\bar{\alpha}} \neq \emptyset$ (y is contained in both sets): $S_\alpha = S_{\bar{\alpha}}$.

It follows that both x and z are in S_α (or in $S_{\bar{\alpha}}$ if you prefer), and that, consequently: $x \sim z$.

CONGRUENCE MODULO n

Here is a particularly important equivalence relation of the set of integers:

THEOREM 2.16 Let $n \in Z^+$. The relation $a \sim b$ if $n | (a - b)$ is an equivalence relation on Z.

PROOF: **Reflexive:** $a \sim a$ since $n | (a - a)$.

Symmetric: $a \sim b \Rightarrow n | (a - b) \Rightarrow n | (b - a) \Rightarrow b \sim a$.

Transitive:

$a \sim b$ and $b \sim c \Rightarrow n | (a - b)$ and $n | (b - c)$

Theorem 1.4(b), page 15: $\Rightarrow n | [(a - b) + (b - c)]$

$\Rightarrow n | (a - c) \Rightarrow a \sim c$

In the event that $n | (a - b)$, we say that:
a is congruent to b modulo n and write $a \equiv b$ mod n

THEOREM 2.17 Let $n \in Z^+$. If $a \equiv \bar{a}$ mod n and $b \equiv \bar{b}$ mod n, then:

(a) $a + b \equiv \bar{a} + \bar{b}$ mod n

(b) $ab \equiv \bar{a}\bar{b}$ mod n

PROOF: (a) If $n | (a - \bar{a})$ and $n | (b - \bar{b})$, then:

$$n|[(a-\bar{a}) + (b-\bar{b})] \Rightarrow n|[(a+b) - (\bar{a}+\bar{b})]$$

(a) If $n|(a-\bar{a})$ and $n|(b-\bar{b})$, then:

(1) $a - \bar{a} = hn$ and (2) $b - \bar{b} = kn$ for $h, k \in Z$

We are to show that $n|(ab - \bar{a}\bar{b})$; which is to say that $ab - \bar{a}\bar{b} = ns$

Lets do it: $ab - \bar{a}\bar{b} = (ab - \bar{a}b) + (\bar{a}b - \bar{a}\bar{b})$

$$= (a - \bar{a})b + \bar{a}(b - \bar{b})$$

$$= hnb + \bar{a}kn = n(\mathbf{hb} + \bar{\mathbf{a}}\mathbf{k}) = n\mathbf{s}$$

CHECK YOUR UNDERSTANDING 2.20

Let $n \in Z^+$. Let $a = d_a n + r_a$ and $b = d_b n + r_b$ with $0 \leq r_a < n$ and $0 \leq r_b < n$ (see Theorem 1.5, page 21). Prove that:

$$a \equiv b \bmod n \text{ if and only if } r_a = r_b$$
(same remainder when dividing by n)

Answer: See page A-13

Theorem 2.15 assures is that the equivalent classes associated with the equivalence relation of Theorem 2.17 partition the set of integers. Focusing on $n = 5$, we see that the equivalence class containing 0 consists of all multiples of 5, as the remainder of any multiple of 5 when divided by 5, is the same as that obtained by dividing 0 by 5 (see CYU 2.20). Specifically:

$$[0]_5 = \{\ldots, -20, -15, -10, -5, 0, 5, 10, 15, 20, \ldots\}$$

Note that the above equivalence class has many "names". It can also, be called the equivalent class containing 235, among infinitely many other choices:

$$[0]_5 = [125]_5 = [-15]_5 = \cdots$$

The same can be said about the four remaining equivalence classes:

$$[1]_5 = \{\ldots, -14, -9, -4, 1, 6, 11, 16, \ldots\}$$
$$[2]_5 = \{\ldots, -13, -8, -3, 2, 7, 12, 17, \ldots\}$$
$$[3]_5 = \{\ldots, -12, -7, -2, 3, 8, 12, 18, \ldots\}$$
$$[4]_5 = \{\ldots, -11, -6, -1, 4, 9, 13, 19, \ldots\}$$

Note that $[5] = [0]$.

Can we define a sum on the above five equivalence classes? Yes:

$$[a]_5 [+] [b]_5 = [a+b]_5$$

The above sum is well defined, in that it is independent of the chosen representatives in the two equivalence classes. Indeed

THEOREM 2.18 For given $n \in Z^+$, let $[Z]_n$ denote the set of equivalence classes associated with the equivalence relation $a \sim b$ if $n|(a-b)$; i.e:

$$[Z]_n = \{[0]_n, [1]_n, \ldots, [n-1]_n\}$$

Then:

(a) For any $[a]_n, [b]_n \in [Z]_n$, the operation
$$[a]_n [+] [b]_n = [a+b]_n$$
is well defined.

(b) For any $[a]_n, [b]_n, [c]_n \in [Z]_n$:
$$([a]_n [+] [b]_n)[+][c]_n = [a]_n [+]([b]_n [+] [c]_n)$$
(associative property)

PROOF: (a) We show that if $[a]_n = [\bar{a}]_n$ and $[b]_n = [\bar{b}]_n$, then $[a+b]_n = [\bar{a}+\bar{b}]_n$ (i.e the sum is independent of the chosen representatives for the equivalence classes $[a]_n$ and $[b]_n$):

$$[a]_n = [\bar{a}]_n \Rightarrow n|(a-\bar{a}) \Rightarrow a-\bar{a} = hn, \text{ for } h \in Z$$

and: $[b]_n = [\bar{b}]_n \Rightarrow n|(b-\bar{b}) \Rightarrow b-\bar{b} = kn$, for $k \in Z$.

Since $(a+b) - (\bar{a}+\bar{b}) = (a-\bar{a}) - (b-\bar{b}) = (h-k)n$:

$$[a+b]_n = [\bar{a}+\bar{b}]_n$$

(b) $([a]_n[+][b]_n)[+][c]_n = [a+b]_n[+][c]_n = [(a+b)+c]_n$
$$= [a+(b+c)]_n$$
$$= [a]_n[+]([b]_n[+][c]_n)$$

CHECK YOUR UNDERSTANDING 2.21

(a) Verify that the product $[a]_n[b]_n = [ab]_n$ in Z_n is well defined. That is: if $[a]_n = [\bar{a}]_n$ and $[b]_n = [\bar{b}]_n$, then: $[ab]_n = [\bar{a}\bar{b}]_n$.

(b) Prove that $[a]_n([b]_n[c]_n) = ([a]_n[b]_n)[c]_n$

(c) Prove that $[a]_n([b]_n[+][c]_n) = [a]_n[b_n][+][a]_n[c]_n$.

Answer: See page A-14.

EXERCISES

Exercises 1-3. Show that the given relation is an equivalence relation on Z.
1. $a \sim b$ if $|a| = |b|$.
2. $a \sim b$ if $2 \mid (a - 3b)$.
3. $a \sim b$ if $5 \mid (a - b)$.

Exercises 4-7. Show that the given relation is an equivalence relation on Q, the set of rational numbers.

4. $\frac{a}{b} \sim \frac{c}{d}$ if $\frac{a}{b} - \frac{c}{d} \in Z$.
5. $\frac{a}{b} \sim \frac{c}{d}$ if $2 \mid (b + d)$.
6. $\frac{a}{b} \sim \frac{c}{d}$ if $(ad - bc)(b^2 + d^2) = 0$.
7. $\frac{a}{b} \sim \frac{c}{d}$ if $(ad)^2 - (bc)^2 = 0$.

Exercises 8-13. Show that the given relation is an equivalence relation on $\Re$.

8. $x \sim y$ if $x^2 = y^2$.
9. $x \sim y$ if $|x| = |y|$.
10. $x \sim y$ if $|x + 1| = |y + 1|$.
11. $x \sim y$ if $x - y \in Z$.
12. $x \sim y$ if $\sin x = \sin(y + 2\pi)$.
13. $x \sim y$ if $x^2 - y^2 = 0$.

Exercises 14-17. Show that the given relation is an equivalence relation on $\Re^2$.

14. $(x_0, y_0) \sim (x_1, y_1)$ if $x_0 + y_0 = x_1 + y_1$.
15. $(x_0, y_0) \sim (x_1, y_1)$ if $x_0 y_0 = x_1 y_1$.
16. $(x_0, y_0) \sim (x_1, y_1)$ if $x_0^2 + y_0^2 = x_1^2 + y_1^2$.
17. $(x_0, y_0) \sim (x_1, y_1)$ if $x_0 = x_1$.

Exercises 18-21. Show that the given relation is not an equivalence relation on $\Re^2$.

18. $(x_0, y_0) \sim (x_1, y_1)$ if $x_0 = y_1$.
19. $(x_0, y_0) \sim (x_1, y_1)$ if $x_0 - y_1 = y_0 - x_1$.
20. $(x_0, y_0) \sim (x_1, y_1)$ if $x_0 x_1 = y_0 y_1$.
21. $(x_0, y_0) \sim (x_1, y_1)$ if $x_0 x_1 \leq 0$ and $y_0 y_1 \leq 0$.

Exercises 22-30. Determine whether or not the given relation is an equivalence relation on $\Re^3$.

22. $(x_0, y_0, z_0) \sim (x_1, y_1, z_1)$ if $y_0 = y_1$.
23. $(x_0, y_0, z_0) \sim (x_1, y_1, z_1)$ if $x_0 + y_0 + z_0 = x_1 + y_1 + z_1$.
24. $(x_0, y_0, z_0) \sim (x_1, y_1, z_1)$ if $x_0 = y_1 + z_1$.
25. $(x_0, y_0, z_0) \sim (x_1, y_1, z_1)$ if $x_0 z_0 + 2y_0 \leq x_1 z_1 + 2y_1$.
26. $(x_0, y_0, z_0) \sim (x_1, y_1, z_1)$ if $x_0 + 2y_0 - 3z_0 = x_1 + 2y_1 - 3z_1$.
27. $(x_0, y_0, z_0) \sim (x_1, y_1, z_1)$ if $(x_0 + y_0 + z_0)^2 = (x_1 + y_1 + z_1)^2$.
28. $(x_0, y_0, z_0) \sim (x_1, y_1, z_1)$ if $x_0^2 + y_0^2 + z_0^2 = x_1^2 + y_1^2 + z_1^2$.
29. $(x_0, y_0, z_0) \sim (x_1, y_1, z_1)$ if $x_0 + y_0 + z_0 + x_1 + y_1 + z_1 \geq 0$.
30. $(x_0, y_0, z_0) \sim (x_1, y_1, z_1)$ if $|y_0 z_0| = |y_1 z_1|$.

Exercises 31-34. Determine whether or not the given relation is an equivalence relation on $M_{2 \times 2}$.

31. $\begin{bmatrix} a & b \\ c & d \end{bmatrix} \sim \begin{bmatrix} \bar{a} & \bar{b} \\ \bar{c} & \bar{d} \end{bmatrix}$ if $a = \bar{d}$.

32. $\begin{bmatrix} a & b \\ c & d \end{bmatrix} \sim \begin{bmatrix} \bar{a} & \bar{b} \\ \bar{c} & \bar{d} \end{bmatrix}$ if $abc = \bar{a}\bar{b}\bar{c}$.

33. $\begin{bmatrix} a & b \\ c & d \end{bmatrix} \sim \begin{bmatrix} \bar{a} & \bar{b} \\ \bar{c} & \bar{d} \end{bmatrix}$ if $ad - bc = \bar{a}\bar{d} - \bar{b}\bar{c}$.

34. $\begin{bmatrix} a & b \\ c & d \end{bmatrix} \sim \begin{bmatrix} \bar{a} & \bar{b} \\ \bar{c} & \bar{d} \end{bmatrix}$ if $ad - \bar{b}\bar{c} = \bar{a}\bar{d} - bc$.

Exercises 35-41. Show that the given relation is an equivalence relation on $F(Z) = \{f : Z \to Z\}$ (the set of functions from Z to Z).

35. $f \sim g$ if $f(1) = g(1)$.

36. $f \sim g$ if $|f(n)| = |g(n)|$ for every $n \in Z$.

37. $f \sim g$ if $|f(n)| = |g(n)|$ for every $n \in Z$.

38. $f \sim g$ if $2|[f(n) + g(n)]$ for every $n \in Z$.

39. $f \sim g$ if $f(n+m) = g(n+m)$ for every $n, m \in Z$.

40. $f \sim g$ if $3|(2f(n) + g(n))$ for every $n \in Z$.

41. $f \sim g$ if $3|[2(g \circ f)(n) + f(n)]$ for every $n \in Z$.

Exercises 42-47. Describe the set of equivalence classes for the equivalence relation of:

42. Exercise 1
43. Exercise 3
44. Exercise 5
45. Exercise 9
46. Exercise 15
47. Exercise 17

Exercises 48-52. Show that the given collection S of subsets of the set X is a partition of X.

48. $X = \Re$, $S = \{(-\infty, 0) \cup \{0\} \cup (0, \infty)\}$.

49. $X = Z$, $S = \{\{3n | n \in Z\} \cup \{3n+1 | n \in Z\} \cup \{3n+2 | n \in Z\}\}$.

50. $X = Z^+ \times Z^+$, $S = \{(a, b) | \gcd(a, b) = n\}_{n \in Z^+}$.

51. $X = \Re \times \Re$, $S = \{(x, y) | y = x + b\}_{b \in R}$.

52. $X = \Re \times \Re$, $S = \{(x, y) | x^2 + y^2 = r^2\}_{r \in \Re}$.

Exercises 53-54. (Congruences) Let $n \in Z^+$. Use the Principle of Mathematical Induction to show that:

53. If $a_i \equiv \bar{a}_i \mod n$ for $1 \leq i \leq m$, then $a_1 + a_2 + \cdots + a_m \equiv \bar{a}_1 + \bar{a}_2 + \cdots + \bar{a}_m \mod n$.

54. If $a_i \equiv \bar{a}_i \mod n$ for $1 \leq i \leq m$, then $a_1 a_2 \cdots a_m \equiv \bar{a}_1 \bar{a}_2 \cdots \bar{a}_m \mod n$.

Exercises 55-60. Determine if the given relation on the set of all people is an equivalence relation. If not, specify the properties of an equivalence relation that are not satisfied.

55. $a \sim b$ if a and b are of the same sex.

56. $a \sim b$ if a is at least as old as b.

57. $a \sim b$ if a and b have the same biological parents.

58. $a \sim b$ if a and b have a common biological parent.

59. $a \sim b$ if a and b are of the same blood-type.

60. $a \sim b$ if a and b were born within three days of each other.

Exercises 61-64. Show that the given relation is an equivalence relation. Describe the set of its equivalence classes.

61. For $a, b \in \{1, 2, ..., 100\}$, $a \sim b$ if a and b end in the same digit.

62. For $a, b \in \{1, 2, ..., 101\}$, $a \sim b$ if a and b end in the same digit.

63. For $S, T \in P(\{1, 2, 3\})$, $S \sim T$ if the number of elements in S equals the number of elements in T.

64. For $S, T \in P(\{1, 2, 3\})$, $S \sim T$ if the sum of the elements in S equals the sum of the elements in T.

	PROVE OR GIVE A COUNTEREXAMPLE	

65. The union of any two equivalence relations on any given nonempty set X is again an equivalence relation on X.

66. The intersection of any two equivalence relations on any given nonempty set X is again an equivalence relation on X.

67. The union of any two reflexive relations on any given nonempty set X is again a reflexive relation on X.

68. The union of any two symmetric relations on any given set X is again a symmetric relation on X.

69. The union of any two transitive relations on any given set X is again a transitive relation on X.

70. For $a, b, n, m \in Z^+$, let S_n and S_m denote the set of equivalence classes associated with the equivalence relations $a \sim b$ if $n|(a-b)$ and $a \sim b$ if $m|(a-b)$, respectively. If $n < m$, then $S_n \subset S_m$.

71. If $C \subseteq X$, $A \sim B$ if $A \cap C = B \cap C$ is an equivalence relation on $P(X)$.

72. There exists an equivalence relation on the set $\{1, 2, 3, 4, 5\}$ for which each equivalence class contains an even number of elements.

73. For $a, b, n, m \in Z^+$, let S_n and S_m denote the set of equivalence classes associated with the equivalence relations $a \sim b$ if $n|(a-b)$ and $a \sim b$ if $m|(a-b)$, respectively. If $n < m$, then $S_n \subset S_m$.

74. If $n \geq 2$, then every integer is congruent modulo n to exactly one of the integers $0 \leq m < n$.

75. If $C \subseteq X$, $A \sim B$ if $A \cap C = B \cap C$ is an equivalence relation on $P(X)$ (see Definition 2.18, page 83).

76. There exists an equivalence relation on the set $\{1, 2, 3, 4, 5\}$ for which each equivalence class contains an even number of elements.

§5. WHAT IS A SET? (OPTIONAL)

The word "set" has been brandished about in previous sections. Here is an attempt to define that important concept:

DEFINITION A(?) A **set** is a specified collection of objects.
Attempt #1

Really? Who is to specify, and how? And anyway, what is a collection? Okay, the word "set" is not precise. Yet, there is an important idea lurking within the above would-be definition, a mathematically indispensable idea. Can we zero in on it? We can try.

Consider the variable proposition:

x is a person born of my mother

It appears that we can use the above to generate a well-defined collection of objects; namely, those objects which when substituted for *x* render the proposition True. To put it another way: those objects which satisfy the given proposition. This suggests another possible approach towards the definition of a set:

DEFINITION B(?) A **set** is a collection of objects satisfying a
Attempt #2 variable proposition.

Is this definition okay? Well, let's begin by comparing it with our previous attempt:

A set is a specified collection of objects.

First of all, note that the word collection is used in a different sense in the two would-be definitions. In our first attempt, we tried to get by with a circular definition: a set is a collection, or, if you prefer, a collection is a set. Now, however, the collection appears to be specifically determined by means of a variable proposition.

Another objection we had with "Definition A" is that it appeared a bit personal. Who is to specify the collection of objects? We no longer appear to have this problem. For example, the statement:

I can vote in a national election.

is True or False, depending on my age, nationality, and other specified factors. But all of these factors are simply used to determine whether or not *I* am an object satisfying the variable proportion:

x can vote in a national election.

There is, of course, another problem; namely, how does one phrase a precise variable proposition to begin with? Let us reconsider "*x can vote in a national election*". What does it mean to vote? What is a nation? The problem is that we are attempting to build something rigorous, a totally unambiguous statement, and we are trying to build it with non-rigorous material: fuzzy words from a common language. Still, the hope is that with extreme care one may be able to formulate a precise variable proposition, and by-pass that language barrier. But there is another fault with "Definition B" — a fatal fault, called the Russell Paradox:

Bertrand Russell 1872-1970). British Mathematician and Logician.

Assume that our second attempt does properly define the concept of a set. We could then consider the set S of objects satisfying the following variable proposition:

X is a set which does not contain itself as an element:

$$(*): \quad S = \{X | X \text{ is a set } \ni X \notin X\}$$

Now, one thing is clear, our set S either contains itself as an element ($S \in S$) or it does not ($S \notin S$). Let's see which:

$$\text{If } S \in S \text{ then, by } (*): S \notin S.$$

Okay, our assumption that $S \in S$ quickly leads us to the contradiction that $S \notin S$, and therefore cannot hold. The other option must hold. Let's check, just to make sure:

$$\text{If } S \notin S \text{ then, by } (*): S \in S.$$

What, again! This too cannot be.

Clearly, our would-be "Definition B" will not do, for it leads us to a set which neither contains itself as an element nor does it not contain itself as an element. It's back to the drawing board; but first, a parenthetical remark, for what its worth:

> We have spent considerable time and effort, and all for a couple of bad attempts in defining a concept. Not necessarily a waste of time. A conscientious attack on a problem may turn the investigator in wrong directions, some of which may lead to places more interesting than the intended destination. In the abstract universe of mathematics there is but one mode of transportation: thought. A lot of times it's fun to let it take us where it may.

Fine, but what is a set? Well, in the final analysis the concept of a set remains an undefined term. We can, however, tell you how certain sets have been created, and actually create a bunch of them right before your very eyes.

We want to call something a set, but then again do not want to offend anything that does not think itself as such by calling it a set. So, what we do is to consider a collection of nothing, absolutely nothing, and call it the **empty set**, and denote it by the symbol $\varnothing$. It is unquestionably the most important set of all mathematics.

We pass a formal law, lest there be any doubt as to the identity of $\varnothing$:

AXIOM 1 $\varnothing$ is a set.

We now have a legitimate set, by law, and there are going to be a lot more. Anticipating their existence, let us come to some agreement as to when two sets are to be considered the same, or equal.

(margin note: This argument is reminiscent of that found in Cantor's Theorem, page 84.)

Since sets are going to be collections of things, it seems reasonable to say that two sets are the same if they contain exactly the same things, in other words:

DEFINITION 2.23   Two sets A and B are **equal**, written $A = B$ if:
SET EQUALITY

$$x \in A \Rightarrow x \in B \quad \text{and} \quad x \in B \Rightarrow x \in A$$

$$(\text{or: } x \in A \Leftrightarrow x \in B)$$

This definition previously appeared on page 53.

The above definition is certainly reasonable, and we would not want to change it for the world. It is, however, going to impose a restriction on all sets, wherever they may be. To make this point, let us assume momentarily the existence of a set containing an element, say the set $\{a\}$. Suppose, in addition, that we allow $\{a, a\}$ also to be a set. Definition 2.23 would tell us that these two sets are equal, for every element in $\{a\}$ is also an element in $\{a, a\}$, and vice versa. This we do not want, right? And so we stipulate that no set can contain an element more than once. In particular $\{a, a\}$ is not an acceptable representation of a set.

At this point, there is a conspicuous lack of sets. Let us remedy the situation by passing another law:

AXIOM 2   For any set A, $\{A\}$ is also a set.

That's great! Prior to the above axiom we had but one set, and empty at that. We now have a lot of different sets. Witness a few:

$$\varnothing, \{\varnothing\}, \{\{\varnothing\}\}, \{\{\{\varnothing\}\}\}, \ldots$$

The above sets are all different. For example, $\varnothing$ and $\{\varnothing\}$ are not the same since there is an element in the latter set, namely the element $\varnothing$, while there is absolutely nothing in the set $\varnothing$. Also $\{\varnothing\} \neq \{\{\varnothing\}\}$, since the only element in the set $\{\varnothing\}$ is $\varnothing$, and the only element in the set $\{\{\varnothing\}\}$ is $\{\varnothing\}$, and we already noted that $\varnothing \neq \{\varnothing\}$; and so on down the line.

We've come a long way from our initial set $\varnothing$, but there is room for improvement. After all, each of our current sets is either empty or is a singleton set (contains but one element). Our set $\{\{\{\varnothing\}\}\}$, for example, contains only the element $\{\{\varnothing\}\}$. We are going to remedy this situation by passing other laws. But first, here is another parenthetical remark:

> What is our jurisdiction in passing these laws? In our abstract universe any one is entitled to create anything he or she wants. Some of those creations perish under an attack of reason, as was the case with our attempted creation labeled "Definition B". The laws that we are busy passing, or **axioms**, have survived in that they do not lead to any contradiction; or, at least, none has yet been unearthed.

We now introduce another law from which many more sets will evolve. First, however, let us agree to use the term **collection of sets** when referring to a set whose elements are themselves sets.

AXIOM 3 For any collection C of sets, there exists a set consisting precisely of those elements that belong to at least one of the sets in C.

The above set is called the **union of the sets in C**, and is denoted by the symbol:

$$\bigcup_{A \in C} A$$

When dealing with a collection of two sets, say $C = \{A_1, A_2\}$, we may choose to represent $\bigcup_{A \in C} A$ in the form $A_1 \cup A_2$. Similarly, the notation $A_1 \cup A_2 \cup A_3$ can be used to denote the union of the sets A_1, A_2, and A_3, and so on.

Back to our set safari, and we begin by recalling our collection stemming from the first two axioms:

$$\varnothing, \{\varnothing\}, \{\{\varnothing\}\}, \{\{\{\varnothing\}\}\}, \ldots$$

Axiom 3 now assures us that the following is also a set:

$$\{\varnothing\} \cup \{\{\varnothing\}\} = \{\varnothing, \{\varnothing\}\}$$

We can now employ Axiom 2 to arrive at yet another new set:

$$\{\{\varnothing, \{\varnothing\}\}\}$$

Back to Axiom 3:

$$\{\varnothing, \{\varnothing\}\} \cup \{\{\varnothing, \{\varnothing\}\}\} = \{\varnothing, \{\varnothing\}, \{\varnothing, \{\varnothing\}\}\}$$

And then Axiom 2 assures us that the set containing the above set; namely:

$$\{\{\varnothing, \{\varnothing\}, \{\varnothing, \{\varnothing\}\}\}\}$$

is also a set, and we could union it with our set $\{\varnothing, \{\varnothing\}, \{\varnothing, \{\varnothing\}\}\}$ to obtain yet another set.

There is a rather interesting development above which may be lost within the maze of commas and brackets. Let us try to clarify the situation with the introduction of some notation.

We start off by letting the symbol 0 denote our empty set:

$$0 = \varnothing$$

(the above is not a set equality but simply designates a new symbol, 0 for the empty set)

We then denote the set $\{\varnothing\}$ by the symbol 1:

$$1 = \{\varnothing\} = \{0\}$$

(once more, a matter of notation)

Our set $\{\varnothing, \{\varnothing\}\}$ will be denoted by 2:

$$2 = \{\varnothing, \{\varnothing\}\} = \{0, 1\} = \{0\} \cup \{1\} = 1 \cup \{1\}$$

The symbol 3 will be used to denote the set consisting of our sets 0,1, and 2:
$$3 = \{0, 1, 2\} = \{0, 1\} \cup \{2\} = 2 \cup \{2\}$$
And in a similar manner we arrive at:
$$4 = \{0, 1, 2, 3\} = 3 \cup \{3\}$$
$$5 = \{0, 1, 2, 3, 4\} = 4 \cup \{4\}$$
$$6 = \{0, 1, 2, 3, 4, 5\} = 5 \cup \{5\}$$
$$\vdots$$
$$135 = \{0, 1, 2, ..., 134\} = 134 \cup \{134\}$$
$$\vdots$$

Let us give names to the above sets. We will call the set 0, *zero*. The set 1 will be called *one*; 2 will be called *two*; 3 is *three*, and so on.

Wait a minute, what's going on? Are we talking about sets or are we talking about numbers? Both. What we have done is to define the number 3, among others. We can now pick up this number 3 and cuddle it, if we want. It is no longer a vague concept, but is something concrete, at least in our abstract mathematical universe. Three is the set $\{\varnothing, \{\varnothing\}, \{\varnothing, \{\varnothing\}\}\}$ or, equivalently, the set $\{0, 1, 2\}$. Either way, there it is, something.

> We hastily remark that a mathematician counts potatoes the same way everybody else does. He or she does not go around continuously aware of the fact that 3 is actually a set which is also an element of the set 9. Still, 3 is now something well-defined, and that is nice.

We can feel rather proud of ourselves. Just a few pages ago we had but one set, and empty at that. We now have infinitely many sets, and sets containing many elements. We now have, for example, sets containing 1,036,752 elements, among them is the set 1,036,752. But there is room for improvement. We do not, for instance, as yet have an infinite set. A particularly nice set we would like to construct is the one consisting of all of the sets we have already constructed: 0, 1, 2, 3, 4, and so on (which we will call the whole numbers). It is pretty clear, however, that our current axioms will not yield that collection since none of those axioms can boost us into the infinite. We need another law, and here is the most obvious one to pass:

There exists a set consisting of all the whole numbers.

Well, that's right to the point alright; but, unfortunately, it will not do. We know exactly what the number zero is: $\varnothing$. We also know that:
$$1 = 0 \cup \{0\}, \text{ and that } 943 = 942 \cup \{942\}$$

We can build like mad, for as long as we like, with as many gaps as we wish, but there will always be larger whole numbers yet to be constructed. This concept of <u>all</u> the whole numbers is quite firm at one end, at the number 0, but kind of hangs loose at the other end. Some new ammunition is called for, and we start off with a definition:

> In particular, if $A = \{0, 1\}$, then the successor of A is the set $\{0, 1, \{0, 1\}\}$.

DEFINITION 2.24 For any set A, the set $A \cup \{A\}$ is called the **successor of** A.

In particular, 1 is the successor of 0, 2 is the successor of 1, and 943 is the successor of 942 (recall that $943 = 942 \cup \{942\}$).

We are now in a position to pass our next law:

AXIOM 4 There exists a set which contains 0 and contains the successor of each of its elements.

Let us agree to call any set which satisfies the condition of the above axiom, a **successor set**. Clearly:

ANY SUCCESSOR SET MUST CONTAIN EVERY WHOLE NUMBER

Now we've gone ahead and done it, we created an infinite set, and maybe a lot of them. We also know that each of these successor sets contains all of the whole numbers. What we do not know, yet, is whether or not there exists a successor set which contains nothing other that the whole numbers. All successor sets might just be too big.

Glancing back at our previous axioms, we see that none of them provides us with the means of plucking smaller sets from a given set. And so we set to work on constructing an appropriate plucking mechanism, beginning with a definition:

> This definition previously appeared on page 53.

DEFINITION 2.25
SUBSET AND PROPER SUBSET

A set A is said to be a **subset** of a set B, written $A \subseteq B$, if each element in A is also contained in B.

If A is a subset of B, but is not equal to B, then A is said to be a **proper subset** of B; written: $A \subset B$.

And so we have:
$$\{2, 6, 3, 8\} \subset \{1, 9, 6, 3, 2, 8\}$$
$$\{2, 6, 3, 8\} \subseteq \{1, 9, 6, 3, 2, 8\}$$
$$\{2, 6, 3, 8\} = \{2, 6, 3, 8\}$$
$$\{2, 6, 3, 8\} \subseteq \{2, 6, 3, 8\}$$
$$\{2, 6, 3, 8\} \not\subset \{2, 6, 3, 8\}$$

Let us denote this thing $\{0, 1, 2, 3, \ldots\}$ which we are **trying** to create by the letter N and ask ourselves the following question:

If N were a set, how would it be distinguishable from all other sets?

Answer: it should be a smallest successor set. Can we make the phrase *smallest successor set* precise? Yes:

DEFINITION 2.26 A **minimal** (or **smallest**) **successor set** is a successor set which is a subset of every successor set.

Now, if we can agree that N should satisfy the property that it is a minimal successor set, and if we can show that there is exactly one minimal successor set, then we will have our set N.

Alright, let us first show that there is at most one minimal successor set, and then proceed to find it.

THEOREM 2.16 There exists at most one minimal successor set.

PROOF: Let A and B be two minimal successor sets.

Since B is a successor set, and since A is a minimal successor set:
$$A \subseteq B$$
Since A is a successor set, and since B is a minimal successor set:
$$B \subseteq A$$
Employing Definition 2.23, we conclude that $A = B$.

We now know that there can be at most one minimal successor set. Still to be demonstrated is that one exists. All in good time. First, another law is passed:

AXIOM 5 For any given set A and any variable proposition $p(x)$, there exists a set (possibly empty) consisting of those elements of A which satisfy $p(x)$.

What is the difference between Axiom 5 and our defamed "Definition B:" *A set is a collection of objects satisfying a variable proposition*

The main distinction is that while "Definition B" led to an absurdity, via Russell's paradox, Axiom 5 does not (or, at least, none has as yet been uncovered). The specific distinction between Axiom 5 and "Definition B" is the phrase "*those elements of A*" appearing in Axiom 5. We are no longer permitting everything to be a candidate for satisfying a variable proposition $p(x)$, but are now restricting candidacy to those elements which are themselves elements of some existing set.

Unlike the union concept which stemmed from Axiom 3, the intersection concept rests on Axiom 5:

DEFINITION 2.27 Let C be a non-empty collection of sets. The **intersection of the sets in** C, denoted by $\bigcap_{A \in C} A$, is the set consisting of those elements contained in every set in C:
$$\bigcap_{A \in C} A = \{x | x \in A \text{ for every } A \in C\}$$

Back to our search for a minimal successor set. Axiom 4 assures us of the existence of at least one successor set $\bar{S}$:
$$0 \in \bar{S} \text{ and } x \in \bar{S} \Rightarrow x \cup \{x\} \in \bar{S}$$

There is a distinction between a Theorem and an Axiom. Axioms are dictated. They are the initial building blocks from which mathematics is constructed. Theorems, on the other hand, are mathematical constructions, built from axiomatic bricks and logical cement. Once established, they too can be used as building blocks in the construction of other theorems. And so it goes, until there are so many blocks hanging around that one loses track of which axiom is needed for the proof of a given result. But it really does not matter, providing, of course, that we don't permit a faulty block to creep into our collection. If one does, then we will be in serious trouble, for any argument built in part with a faulty block may itself be faulty.

Now, there are some subsets of $\bar{S}$ which are themselves successor sets ($\bar{S}$ itself, for example). The collection:
$$C = \{S \subseteq \bar{S} \mid S \text{ is a successor set}\}$$
is therefore not empty, and we may consider the intersection of all the sets in C, which we optimistically call N:
$$N = \bigcap_{S \in C} S$$
So, N is the intersection of all the subsets of the successor set $\bar{S}$ which are themselves successor sets. We now show that N is itself a successor set:

Since each $S \in C$ is a successor set, $0 \in S$ and $x \cup \{x\} \in S$ for every $x \in S$. Since $N = \bigcap_{S \in C} S$, $0 \in N$; and for every $x \in N = \bigcap_{S \in C} S$, $x \cup \{x\} \in N$.

NOTE: The adjacent argument shows that any intersection of successor sets is again a successor set.

Alright, we have just established the fact that the intersection, N, of all subsets of the successor set $\bar{S}$ which are themselves successor sets is itself a successor set. We now define this particular set, N, to be the **set of whole numbers.**

Objections! The above set N, being the intersection of all subsets of $\bar{S}$ and being itself a successor set, is clearly the smallest successor set in the particular set $\bar{S}$. But how about successor sets other than $\bar{S}$ which may exist already, or others that may evolve if we continue creating additional sets? How do we know that no new successor set will evolve which is smaller than N?

Objection overruled:

THEOREM 2.17 If A is a successor set, then $N \subseteq A$.

PROOF: Consider once more the successor set $\bar{S}$ which led us to the definition of N above. Since, as we have noted, the intersection of successor sets is again a successor set, $A \cap \bar{S}$ is also a successor set, and it is contained in $\bar{S}$. But N sits inside every successor set contained in $\bar{S}$. Thus: $N \subseteq A \cap \bar{S} \subseteq A$

We now have the set $N = \{0, 1, 2, 3, \ldots\}$ and could proceed towards a rigorous set-theoretical construction of the real number system, and beyond. Instead, we chose to end the chapter with some general set theory remarks. First, however, a final result:

THEOREM 2.18 $\qquad Math_{god} = \emptyset$

PROOF: Clear.

SOME HISTORICAL REMARKS:

The formal birth of set theory occurred in 1874 with the publication of the first purely set theoretic work: *Uber ene Eigenschaft des Inbegriffes aller Reelen Algebraishen Zahlen* (On a Property of the Collection of all Real Algebraic Numbers). Within that paper, Georg Cantor (1845-1916) distinguishes two infinite subsets of the reals that are not of the same cardinality, thereby pointing out the existence of different levels of infinity. He was not, however, the first to discover this staggering fact. Indeed, Galileo Galilei (1564-1642) as early as 1632 recorded the following interesting observation: *There are as many squares as there are numbers because they are just as numerous as their roots*. Later, Bernhard Bolzano (1781-1848), familiar with Galileo's work, gave additional examples of bijections between infinite sets and some of their proper subsets. But these were isolated instances, and the creation of set theory as well as its fundamental development is justly accredited to Cantor.

Cantor's valuable contributions were far from universally accepted by the mathematical community of his time. Many disagreed vehemently with his work which, to them, appeared to rest on little more than intuition and empty fabrications based on nonconstructive reasoning. Some suggested that his work encroached on the domain of philosophers, while others even accused him of violating religious principles. Most noteworthy among his numerous critics of the day was one of Cantor's former professors, Leopold Kronecher (1823-1891). Kronecher objected loudly and strenuously to Cantor's uninhibited use of infinite sets. According to that noteworthy mathematician: Definitions *must contain the means of reaching a decision in a finite number of steps, and existence proofs must be conducted so that the quantity in question can be calculated with any required degree of accuracy.* Even the irrational numbers fail to satisfy his imposed criteria, and, along with the infinite, they too were disregarded by Kronecher whose mathematical philosophy may best be read within his often quoted statement: *God created the natural numbers, and all the rest is the work of man.*

And then came the paradoxes, which naturally served to further fan the flames of discontent. Cantor's reliance on precise statements to generate his sets simply would not do, and Bertrand Russell (1872-1970), in 1902 demonstrated that Cantor's own definition of sets leads to a contradiction. Other similar paradoxes emerged, and they generated such turmoils that in 1908 Henri Poincare (1845-1912), a leading mathematician at the time, made the following statement at the International Congress of Mathematics: *Later mathematicians will regard set theory as a disease from which one has recovered.*

The sensitive Cantor did not recover from the onslaught of criticisms from his peers. Particularly disturbed by what he felt to be Kronecher's malicious and unjust persecutions, he suffered a complete nervous breakdown in 1884, and, to some degree, mental illness plagued him for the rest of his life. He died in the psychiatric clinic at Halle on January 6, 1918, but not before witnessing the beginning of the tremendous role his theory would play in mathematics, and realizing a belated recognition which he so justly deserved.

In 1908, Ernst Zermelo (1871-1953) published his *Untersuchungen ueber die Grundlagen der Mengenlehre* (Investigations into the Foundations of Set Theory). In that work, an axiomatic system for set theory is presented. After postulating the existence of certain sets, and with an additional undefined notion, that of membership ($\in$), he required but seven axioms to set the formal foundation for set theory and, indeed, for most of mathematics. These original axioms were later amended by Abraham Fraenkel (1891-1790), John Von Neumann (1903-1957), and Kurt Gödel (1906-1978), among others, and has come to be called the **Zermelo-Fraenkel Axiomatic System**. It remains the most widely used axiomatic system of the day.

Yes, there is no universal axiomatic system. Though it is certainly true that the overwhelming majority of mathematics is globally accepted, there remain important results which depend on axioms accepted by some and rejected by others. In an attempt to clarify this statement, we turn to a few remarks concerning the axiomatic foundations of mathematics.

Two axiomatic systems are **equivalent** if each axiom in either system is a consequence of those in the other. It follows that though the axioms in one system might be quite distinct from those in the other, both in form and number, any proposition which can be established to be True in one system can also be established to be True in the other. Thus, equivalent systems lead to the same theory. One system, however, might possess certain properties which makes it more appealing than another.

One appealing property is that the system be efficient in the sense that each of its axioms is really needed — that none of its axioms is a consequence of the others. Such a system is said to be **independent**.

Another nice property, admittedly more vague than the previous one, is that the system should, as much as possible, consist of "intuitively valid" axioms. After all, axioms are the building blocks from which the theory is developed, and, as such, they should be "believable" and "fundamental in nature."

The five axioms of this chapter are independent, and fundamental in nature. We began by stipulating the existence of a set, the empty set. Then, with the introduction of other axioms, arrived at an axiomatic system sufficiently rich to allow for the construction of the set of natural numbers. We also established a few of the surface results founded on our meager collection of axioms. A richer theory might, however, evolve from an axiomatic system which properly contains ours. But how do we go about expanding our system?

We expanded all along. Very little theory could be based on our first axiom: $\varnothing$ **is a set**. And so we expanded by introducing the axiom on sets of sets, thus assuring the existence of a lot of singleton sets. There was then a need for a greater variety of sets, a need which was partially fulfilled by the introduction of the union axiom. We continued the process of satisfying needs, arriving finally at the five-axiom system of this section, which we now denote by the Greek letter omega: Ω.

If we continue to play with Ω, we may very well come across some statement S which, for some reason or other, we would like to be True. If it can be determined that neither S nor its negation is a consequence of the axioms in Ω, then we say that S is **independent of** Ω, and may choose to add S to Ω. In particular, the statement: *For any given set A there exists a set consisting of all the subsets of A* can be shown to be independent of the five axioms in Ω. We could therefore add it to the five axioms in Ω and in so doing arrive at a richer system Ω.

We now turn our attention to a couple of interesting statements that are independent of Ω. The first of these appeared as one of the axioms posed in Zermelo's 1908 paper on the Foundations of Sets:

THE AXIOM OF CHOICE: Given any non-empty collection C of non-empty sets, there exists a set consisting of exactly one element from each set in C.

(This amounts to being able to "choose" and element from each of the sets in C)

The motivation for the above axiom may, in part, be attributed to Cantor. In 1883 he asserted that every set can be **well-ordered**; which is to say, that an order relation can be imposed on any set, under which each of its non-empty subsets contains a smallest or first element. This, for example, is already the situation with any countable set. Cantor indicated that he would substantiate this fundamental and truly remarkable fact at a future date. He did not, and, as Zermelo subsequently showed, for good reasons.

The above **WELL-ORDERING PRINCIPLE** has such a wide range of applications that it was one of the famous twenty-three unsolved problems formally offered for consideration to the mathematical community by David Hilbert (1862-1943), at the 1900 International Congress of Mathematics.

It was for Zermelo, in 1904, to offer a proof in the affirmative, a proof that depended on the principle set forth within his Axiom of Choice. In other words, given the Axiom of Choice, the Well-Ordering Principle follows. Moreover, it is easy to see that if the Well-Ordering Principle holds, then so does the Axiom of Choice (well order each of the sets in question, and select the first element of each). Thus, in the Zermelo-Fraenkel Axiomatic System, the Axiom of Choice is **equivalent** to the Well-Ordering Principle, in that the validity of either implies that of the other; or, if you prefer, if you don't have the one then you don't have the other (nor any of its other numerous equivalent formulations).

In 1931, a twenty five year old student at the University of Vienna, Kurt Gödel (1906-1978) showed that if the Zermelo-Fraenkel axiom system excluding the Axiom of Choice is consistent, then adding the Axiom of Choice will not lead to a contradiction. It was not until 1966, however, that Paul Cohen (1934-2007) succeeded in proving that the addition of its negation would also not lead to a contradiction. Thus, the Axiom of Choice is independent of the other axioms within the Zermelo-Fraenkel system, and whether to accept it or not boils down to a matter of personal inclinations.

> *This conviction of the solvability of every mathematical problem is a powerful incentive to the worker. We hear within us the perpetual call: There is the problem. Seek its solution. You can find it by pure reason, for in mathematics there is no: **we will not know**.*

Oh yes there is! For in 1931 Gödel published his famous **Incompleteness Theorem**:

> *In any mathematical system rich enough to encompass the natural numbers, there is an assertion expressible within the system that is true, yet is not provable within that system.*

And just in case that was not bad enough, he then went on to prove that the consistency of such a system is itself an undecidable proposition.

What a double whammy! First, there will always be undecidable propositions. And, worse than that, we can never gain assurance that our mathematics is based on a firm foundation. It should be underlined, however, that just because we are not able to prove that our axiomatic system is consistent, that does not mean that it is not. Indeed, not many mathematicians lose much sleep over this issue, as most have the utmost faith that our Zermelo-Fraenkel axiomatic system, along with its variations, are indeed consistent systems.

Yes, there are potential flaws in modern mathematics. Perhaps some drastic fundamental changes, possibly in the field of logic, will remedy the situation. Or, perhaps, imperfection is within the very nature of things; not only in our physical universe, but in our expanding mathematical universe as well. At any rate, as things stand now, mathematics is not what we would call perfect, but it may very well be the closest thing to perfection around.

CHAPTER 3
A Touch of Analysis

The axiomatic structure of the real number system is introduced in Section 1 wherein the completion axiom taking center stage. Sequences, including Cauchy sequences, are featured in Section 2. The metric structure of the real number system is discussed in Section 3, and the important concept of continuity is featured in Section 4.

§1. THE REAL NUMBER SYSTEM

If it looks like a duck, walks like a duck, and quacks like a duck, then it probably is a duck.

Unlike ducks, which appear to require but three defining characteristics to distinguish them from all other worldly creatures, eleven characteristics (axioms) are needed to distinguish the set of real numbers from all other mathematical creatures.

DEFINITION 3.1
COMPLETE ORDERED FIELD

A **complete ordered field** is a set F, along with two operations, called **addition** ($+$) and **multiplication** ($\cdot$), along with an order relation ($\geq$), such that:

Commutative Axiom: (1) $x + y = y + x \quad \forall x, y \in F$.

Associative Axiom: (2) $x + (y + z) = (x + y) + z$ and $x \cdot (y \cdot z) = (x \cdot y) \cdot z \quad \forall x, y, z \in F$.

Distributive Axiom: (3) $x \cdot (y + z) = x \cdot y + x \cdot z \quad \forall x, y, z \in F$.

Additive Identity Axiom: (4) There exists an element which we will label $0 \in F$ such that $x + 0 = x$ for all $x \in F$.

Additive Inverse Axiom: (5) For each $x \in F$ there exists an element which we will label $-x \in F$ such that $x + (-x) = 0$.

Multiplicative Identity Axiom: (6) There exists an element which we will label $1 \in F$ with $1 \neq 0$ such that $1 \cdot x = x$ for all $x \in F$.

Multiplicative Inverse Axiom: (7) For each $x \neq 0$ there exists an element $x^{-1} \in F$ such that $x \cdot x^{-1} = 1$.

Trichotomy Axiom: (8) For any $x, y \in F$, either $x \geq y$ or $y \geq x$.

Additive Inequality Axiom: (9) If $x \geq y$, then $x + z \geq y + z$ for every $z \in F$.

Multiplicative Inequality Axiom: (10) If $x \geq y$ and $z \geq 0$, then $x \cdot z \geq y \cdot z$

Completeness Axiom: (See Definition 3.2 below) (11) Every nonempty subset of F that is bounded from above has a least upper bound.

It can be shown that the set of real numbers $\Re$, with standard addition and standard multiplication, is a complete ordered field. It can also be shown that any other complete ordered field can only differ from $\Re$ superficially (the number 5, for example, might be written as V).

FOCUSING ON THE COMPLETION AXIOM

You may be able to anticipate the meaning of the terminology in the completeness axiom of Definition 3.1 on your own; but just in case:

DEFINITION 3.2 Let S be a subset of $\Re$.

UPPER BOUND a is an **upper bound** of S if $a \geq s$ for every $s \in S$.

S is **bounded from above** if S has an upper bound.

LOWER BOUND b is a **lower bound** of S if $b \leq s$ for every $s \in S$.

S is **bounded from below** if S has a lower bound.

S is **bounded** if S has both a lower and an upper bound.

LEAST UPPER BOUND α is the **least upper bound** (or supremum) of $S \neq \emptyset$ if it is an upper bound of S and if it is less than or equal to every upper bound of S.

GREATEST LOWER BOUND β is the **greatest lower bound** (or infimum) of $S \neq \emptyset$ if it is a lower bound of S and if it is greater than or equal to every lower bound of S.

In Exercise 12 you are invited to verify that if a least upper bound (or greatest lower bound) exists, then it is unique.

The notation lub S and glb S is used to denote the least upper bound and greatest lower bound of S, respectively. For example:

$$\text{glb}\{1, 2, 5\} = 1 \text{ and } \text{lub}\{1, 2, 5\} = 5$$

$$\text{glb}\,(-2, 4] = -2 \text{ and } \text{lub}\,(-2, 4] = 4$$

$$\text{lub}(-\infty, 7) = \text{lub}(-\infty, 7] = 7 \text{ and } \text{glb}\,(3, \infty) = \text{glb}\,[3, \infty) = 3$$

As is illustrated above, neither the least upper bound nor the greatest lower bound of a set S need be an element of that set. If it is, then it is said to be the **maximum** or greatest member of S, and the **minimum** or smallest member of S, respectively.

CHECK YOUR UNDERSTANDING 3.1

Determine the least upper bound, the greatest lower bound, the maximum, and the minimum element of the given set, if they exist.

(a) $(3, 5) \cup [4, 7]$ (b) $(-\infty, 0) \cup [1, 3] \cup \{9\}$

(c) $\{x < 0 \mid x^2 < 2\} \cup \{x \geq 0 \mid x^2 \leq 2\}$

(a) lub: 7, glb: 3, Max: 7
(b) lub: 9, Max: 9
(c) lub: $\sqrt{2}$, glb: $-\sqrt{2}$, Max: $\sqrt{2}$

The following result provides a useful characterization for the least upper bound and greatest lower bound of a given set.

THEOREM 3.1 (a) A number α is the least upper bound of $S \neq \varnothing$ if and only if it satisfies the following two properties:

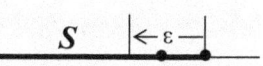

(i) α is an upper bound of S.

(ii) For any given $\varepsilon > 0$ there exists some $s \in S$ (which depends on ε) such that $s > \alpha - \varepsilon$.

(That is: $\alpha - \varepsilon$ is not itself an upper bound)

(b) A number β is the greatest lower bound of $S \neq \varnothing$ if and only if it satisfies the following two properties:

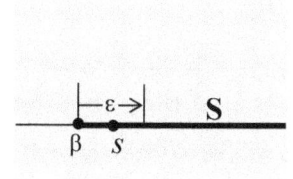

(i) β is a lower bound of S.

(ii) For every $\varepsilon > 0$ there exists some $s \in S$ (which depends on ε) such that $s < \beta + \varepsilon$.

(That is: $\beta + \varepsilon$ is not itself a lower bound)

The Greek letter epsilon ε *is generally used to denote a "small" unspecified number.*
Note that:
ε is **FIRST** given.
THEN: s is to be found to accommodate that particular ε.

PROOF: (a) If $\alpha = \text{lub } S$ then α is, in particular, an upper bound. To show that (ii) also holds, we consider a given $\varepsilon > 0$.

Since $\alpha - \varepsilon < \alpha$, and since nothing smaller than α can be an upper bound of S, there must exist some $s \in S$ to the right of $\alpha - \varepsilon$ (see margin).

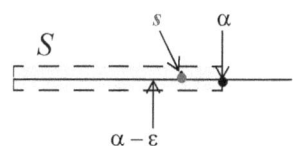

Conversely, suppose α satisfies (i) and (ii). To show that α is the least upper bound of S we consider an arbitrary upper bound a of S, and show that $\alpha \leq a$:

Assume, to the contrary, that $\alpha > a$. Let $\varepsilon = \alpha - a > 0$. By (ii):
$$\exists\, s \in S \ni s > \alpha - \varepsilon = \alpha - (\alpha - a) = a$$

Contradicting the assumption that a is an upper bound of S.

CHECK YOUR UNDERSTANDING 3.2

Answer: See page A-14.

Prove Theorem 3.1(b).

EXAMPLE 3.1

We remind you that:
$Z^+ = \{1, 2, 3, ...\}$

Let $S = \left\{ \dfrac{1}{8}, \dfrac{2}{9}, \dfrac{3}{10}, \dfrac{4}{11}, ... \right\} = \left\{ \dfrac{n}{n+7} \,\Big|\, n \in Z^+ \right\}$

Show that lub $S = 1$.

SOLUTION: (a) Since the denominator of $\frac{n}{n+7}$ is always greater than the numerator, 1 is seen to be an upper bound of S. To show that it also satisfies (ii) of Theorem 3.1(a), we observe that for any given $\varepsilon > 0$:

$$\frac{n}{n+7} > 1 - \varepsilon \Leftrightarrow n > (n+7)(1-\varepsilon)$$
$$\Leftrightarrow n > n - n\varepsilon + 7 - 7\varepsilon$$
$$\Leftrightarrow n\varepsilon > 7 - 7\varepsilon$$
$$\text{Since } \varepsilon > 0: \quad \Leftrightarrow n > \frac{7 - 7\varepsilon}{\varepsilon}$$

Cutting out the middle steps in the above development we conclude that for any integer $n > \frac{7-7\varepsilon}{\varepsilon}$, the element $\frac{n}{n+7}$ of S will lie to the right of $1 - \varepsilon$.

> Note that the smaller ε is, the larger $\frac{7-7\varepsilon}{\varepsilon}$ becomes, and that therefore no one element of S will lie to the right of $1 - \varepsilon$ for all $\varepsilon > 0$. But for any **given** $\varepsilon > 0$ an element of S does exist to "accommodate" that ε.

CHECK YOUR UNDERSTANDING 3.3

Find the smallest element of the set S of Example 3.1 which lies to the right of:

(a) $\frac{99}{100}$ (b) $\frac{99.9}{100}$

Suggestion: Write the given number in the form $1 - \varepsilon$.

(a) $\frac{694}{694+7}$ (b) $\frac{6994}{6949+7}$

THEOREM 3.2
ARCHIMEDEAN PRINCIPLE
For any given $a, b \in \Re$ with $a > 0$, there exists a positive integer n such that $na > b$.

PROOF: Let $S = \{na \mid n \in Z^+\}$. We first show that S is not bounded above:

Assume, to the contrary, that S is bounded above. By the completion axiom, S has a least upper bound α. Letting a play the role of ε in Theorem 3.1(a-ii), we conclude that there exists an element of S, say $n_0 a$, which lies to the right of $\alpha - a$. But if $\alpha - a < n_0 a$, then $\alpha < n_0 a + a = (n_0 + 1)a \in S$ — contradicting the fact that α is an upper bound of S.

To complete the proof we need but note that since S is not bounded above, the given b in the statement of the theorem is not an upper bound of S. It follows that some element na of S must lie to the right of b.

CHECK YOUR UNDERSTANDING 3.4

(a) Prove that for any given number $b \in \Re$ there exists a positive integer n such that $n > b$.

(b) Prove that for any given $\varepsilon > 0$ there exists a positive integer n such that $\frac{1}{n} < \varepsilon$.

Answer: See page A-14.

Here is particularly useful consequence of the completion axiom:

THEOREM 3.3
NESTED CLOSED INTERVAL PROPERTY

If $\{I_n = [a_n, b_n]\}_{n=1}^{\infty}$ is a collection of nonempty closed intervals such that

$[a_{n+1}, b_{n+1}] \subseteq [a_n, b_n]$, then $\bigcap_{n=1}^{\infty} I_n \neq \varnothing$.

PROOF: Consider the set $L = \{a_n | n \in Z^+\}$ of left endpoints of the given closed intervals $[a_n, b_n]$:

$$a_1 \quad a_2 \quad a_3 \cdots \quad \cdots b_3 \quad b_2 \quad b_1$$

Since L is bounded above (by b_1, for example), it has a least upper bound α. We complete the proof by showing that α is contained in each I_n:

Since each b_n is an upper bound of L, and since $\alpha = \text{lub } L$, $\alpha \leq b_n$ for every n. We also know that $\alpha \geq a_n$ for ever n (α is an upper bound of L). It follows that $a_n \leq \alpha \leq b_n$ for every n, and that therefore $\alpha \in \bigcap_{n=1}^{\infty} I_n$.

CHECK YOUR UNDERSTANDING 3.5

Note that these nested intervals are not closed.

Let $J_n = \left(0, \frac{1}{n}\right]$ for $n \in Z^+$. Show that $\bigcap_{n=1}^{\infty} J_n = \varnothing$.

Answer: See page A-15.

Suggestion: Use CYU 3.4(b).

So, in the real number system: $\sqrt{x}$ exists for every $x \geq 0$.

THEOREM 3.4 For any given $x \geq 0$ there exists $\alpha \geq 0$ such that $\alpha^2 = x$.

PROOF: The set $S = \{s \geq 0 | s^2 \leq x\}$ is nonempty as it contains 0. It is also bounded above (by 1 if $0 \leq x \leq 1$, and by x if $x > 1$). As such

$\alpha = \text{lub } S$ exists. We show that $\alpha^2 = x$ by eliminating the two other possibilities, $\alpha^2 < x$ and $\alpha^2 > x$:

Assume that $\alpha^2 < x$. We will exhibit an element of S that is greater than α, thereby contradicting the fact that α is an upper bound of S.

Our first step is to consider the square of numbers a bit larger than α. For any $n > 1$:

$$\left(\alpha + \frac{1}{n}\right)^2 = \alpha^2 + \frac{2\alpha}{n} + \frac{1}{n^2}$$

$$< \alpha^2 + \frac{2\alpha}{n} + \frac{1}{n} = \alpha^2 + \frac{1}{n}(2\alpha + 1) \quad (*)$$

> Since $\alpha \geq 0$ and $\alpha^2 < x$:
> $\frac{x - \alpha^2}{2\alpha + 1} > 0$
> It follows, from CYU 3.4(b), that $\frac{1}{N} < \frac{x - \alpha^2}{2\alpha + 1}$ for some N.

Choosing N sufficiently large so that $\frac{1}{N} < \frac{x - \alpha^2}{2\alpha + 1}$ (see margin), and appealing to (*), we have:

$$\left(\alpha + \frac{1}{N}\right)^2 < \alpha^2 + \frac{1}{N}(2\alpha + 1) < \alpha^2 + \left(\frac{x - \alpha^2}{2\alpha + 1}\right)(2\alpha + 1) = x$$

So, $\alpha + \frac{1}{N} \in S$, and is larger than α — a contradiction.

Having ruled out the possibility $\alpha^2 < x$, we now do the same for $\alpha^2 > x$, and do so by looking at the square of numbers a bit smaller than x:

$$\left(\alpha - \frac{1}{n}\right)^2 = \alpha^2 - \frac{2\alpha}{n} + \frac{1}{n^2} > \alpha^2 - \frac{2\alpha}{n} \quad (**)$$

Choosing N, sufficiently large, so that $\frac{1}{N} < \frac{\alpha^2 - x}{2\alpha}$ and appealing to (**), we have:

$$\left(\alpha - \frac{1}{N}\right)^2 > \alpha^2 - \frac{1}{N}(2\alpha) > \alpha^2 - \left(\frac{\alpha^2 - x}{2\alpha}\right)(2\alpha) = x$$

Since $\left(\alpha - \frac{1}{N}\right)^2 > x$, and since (by assumption) $s^2 \leq x$ for every element in S, $\alpha - \frac{1}{N}$ is an upper bound of S — contradicting the fact that α is the least upper bound of S.

We remind you that a **rational number** is a number of the form $\frac{a}{b}$, where a and b are integers, with $b \neq 0$. A number that is not rational is said to be **irrational**.

> You are not insulting a number by calling it irrational — you are just saying that it is not the ratio of two integers.

Not every number is rational:

THEOREM 3.5 There is no rational number $\frac{a}{b}$ such that $\left(\frac{a}{b}\right)^2 = 2$. (In other words: $\sqrt{2}$ is irrational).

PROOF: (By contradiction) Assume there exists a rational number $\frac{a}{b}$ in lowest terms (margin) such that $\left(\frac{a}{b}\right)^2 = 2$. Then:

> Any rational number $\frac{a}{b}$ can be expressed in **lowest terms** (a and b share no common factor).

$$\left(\frac{a}{b}\right)^2 = 2 \Rightarrow \frac{a^2}{b^2} = 2$$
$$\Rightarrow a^2 = 2b^2 \quad (*)$$
$$\Rightarrow a^2 \text{ is even}$$

Theorem 1.10, page 47: $\Rightarrow a$ is even

$\Rightarrow a = 2k$ for some integer k

Substituting $2k$ for a in (*) we have:

$$(2k)^2 = 2b^2 \Rightarrow 4k^2 = 2b^2 \Rightarrow 2b^2 = 4k^2$$
$$\Rightarrow b^2 = 2k^2$$
$$\Rightarrow b^2 \text{ is even}$$

Theorem 1.10: $\Rightarrow b$ is even

$\Rightarrow b = 2h$ for some integer h

Bringing us to: $\frac{a}{b} = \frac{2k}{2h}$ — contradicting our stated condition that $\frac{a}{b}$ is in lowest terms.

Since the assumption that $\sqrt{2}$ is rational led to a contradiction, we conclude that $\sqrt{2}$ must be irrational.

EXAMPLE 3.2 Show that the sum of any irrational number and any rational number is irrational.

SOLUTION: (By contradiction) Let x be irrational. Assume there exists a rational number $\frac{a}{b}$ such that $x + \frac{a}{b}$ is rational, say:

$$x + \frac{a}{b} = \frac{c}{d}$$

Then: $x = \frac{c}{d} - \frac{a}{b}$

$$x = \frac{cb - ad}{db} \quad (*)$$

Since $cb - ad$ is an integer and db is an integer distinct from 0 (why?), (*) tells us that x is rational — contradicting our stated condition that x is irrational.

118 Chapter 3 A Touch of Analysis

CHECK YOUR UNDERSTANDING 3.6

(a) Prove that the product of any irrational number with any nonzero rational number is irrational

(b) Can the sum of two irrational number be rational? Justify your answer.

Answer: See page A-15.

DENSE SUBSETS OF $\Re$

In other words: between any two distinct real numbers, one can always find an element of D.

DEFINITION 3.3
DENSE SUBSET

A set D of real numbers is **dense** in $\Re$ if for any given real numbers a and b with $a < b$ there exists $d \in D$ such that $a < d < b$.

Here are two important dense subsets of $\Re$:

THEOREM 3.6 The set of rational numbers and the set or irrational numbers are dense in $\Re$.

PROOF: To show that the rationals are dense in $\Re$ we exhibit, for any given $a, b \in \Re$ with $a < b$, a rational number lying between a and b:

Case 1. $a < 0 < b$: Not much to be done here, since $0 = \frac{0}{1}$ is a rational number.

Case 2. $0 \leq a < b$: Our goal is to find a rational number $\frac{m}{n}$ such that:

$$a < \frac{m}{n} < b$$

We begin by choosing a positive integer n such that:

$$\frac{1}{n} < b - a \quad \text{or:} \quad \boldsymbol{a < b - \frac{1}{n}} \quad (*)$$

See CYU 3.4(b)

We then consider the non-empty set:

$$S = \left\{ t \in Z^+ \mid \frac{t}{n} > a \right\}$$

The Well-Ordering Principle (page 39), assures us that S has a smallest (first) element, which we will call m.

Since $m \in S$: $\frac{m}{n} > a$. If we can show that $\frac{m}{n} < b$ then we will be done. We can:

since $m - 1$ is less than m: $m - 1 \notin S$ (*)

$$\frac{m-1}{n} \leq a < b - \frac{1}{n}$$

$$\frac{m}{n} - \frac{1}{n} < b - \frac{1}{n}$$

$$\frac{m}{n} < b$$

Case 3. $a < b \leq 0$: Noting that $0 \leq -b < -a$, we appeal to the previous case and choose a rational number $\frac{m}{n}$ such that $-b < \frac{m}{n} < -a$. Multiplying through by -1 yields the desired result: $a < -\frac{m}{n} < b$.

Using the established fact that the rationals are dense in $\Re$ we now show that the irrationals are also dense in $\Re$:

Let $a < b$ be given. Choose a rational number $\frac{m}{n}$ such that:

$$a - \sqrt{2} < \frac{m}{n} < b - \sqrt{2}$$

or: $\quad a < \frac{m}{n} + \sqrt{2} < b$

This completes the proof, since the sum of a rational number and an irrational number is itself irrational (Example 3.2).

CHECK YOUR UNDERSTANDING 3.7

(a) Prove that no finite subset of $\Re$ is dense in $\Re$.

(b) Let S be dense in $\Re$. Show that any open interval (a, b) must contain infinitely many elements of S.

Answer: See page A-15.

EXERCISES

Exercises 1-6. Determine the least upper bound and greatest lower bound of the given set, and its maximum and minimum element, if they exist.

1. $(0, 5) \cup \{7\}$
2. $[0, 5] \cup \{7\}$
3. $(-\infty, 3) \cup [4, 10)$
4. $(-\infty, 3) \cup [4, \infty)$
5. $Z^+ \cap \{x \in \Re | x \geq 7\}$
6. $Z^+ \cap \{x \in \Re | x \geq 7\}$

Exercises 7-11. Determine the least upper bound of the set S. Justify your claim.

7. $S = \left\{ 1 - \dfrac{1}{n} \middle| n \in Z^+ \right\}$
8. $S = \left\{ 1 - \dfrac{1}{n^2} \middle| n \in Z^+ \right\}$
9. $S = \left\{ \dfrac{2n}{n+1} \middle| n \in Z^+ \right\}$
10. $S = \left\{ \dfrac{n^2 - 1}{2n^2} \middle| n \in Z^+ \right\}$
11. $S = \left\{ \dfrac{20 \sin n}{5n} \middle| n > 10 \right\} \cup \left\{ 5 - \dfrac{1}{n^2} \middle| 1 \leq n \leq 5 \right\}$

12. (a) Prove that if a subset S of $\Re$ has a least upper bound, then it is unique.
    (b) Prove that if a subset S of $\Re$ has a greatest lower bound, then it is unique.

13. (a) Prove that if a subset S of $\Re$ has a maximum element, then it equals lub S.
    (b) Prove that if a subset S of $\Re$ has a minimum element, then it equals glb S.

14. Prove that any finite subset of $\Re$ contains a maximum and minimum element.

15. Prove that every nonempty subset of $\Re$ that is bounded from below has a greatest lower bound.

16. Prove that if A is a nonempty bounded subset of $\Re$, then glb $A \leq$ lub A.

17. (a) Let A and B be nonempty subsets of $\Re$, bounded above, with $A \subseteq B$. Show that lub $A \leq$ lub B.
    (b) Give an example of nonempty sets A and B, bounded above, with $A \subset B$ and lub $A =$ lub B.

18. For $A \subseteq \Re$ and $B \subseteq \Re$, let $A + B = \{a + b | a \in A, b \in B\}$. Prove that:
    (a) If A and B are bounded above, then so is $A + B$.
    (b) If A and B are bounded below, then so is $A + B$.

19. (a) Prove that if a subset S of $\Re$ has a least upper bound α, then the set of all upper bounds of S is $[\alpha, \infty)$.
    (b) Prove that if a subset S of $\Re$ has a greatest lower bound β, then the set of all lower bounds of S is $(-\infty, \beta]$.

20. Prove that every number is both an upper bound and a lower bound of $\varnothing$.

21. Prove that A is bounded if and only $|A| = \{|a|\}_{a \in A}$ is bounded.

22. (a) Prove that if $A \subseteq B$, with $A \neq \emptyset$, then lub $A \leq$ lub B.
    (b) Give an example where $A \subset B$ and lub $A =$ lub B.

23. (a) Prove that if $A \subseteq B$, with $A \neq \emptyset$, then glb $A \geq$ glb B.
    (b) Give an example where $A \subset B$ and glb $A =$ glb B.

24. (a) Let A and B be nonempty bounded sets of real numbers such that for every $a \in A$ there exists $b \in B$ such that $a < b$, and for every $b \in B$ there exists $a \in A$ such that $b < a$. Prove that lub $A =$ lub B and glb $A =$ glb B.
    (b) Give an example of sets A and B satisfying the conditions of part (a), with $A \neq B$.

25. (a) Let A and B be nonempty bounded sets of real numbers such that $a < b$ for every $a \in A$ and every $b \in B$. Prove that lub $A \leq$ glb A.
    (b) Give an example of sets A and B satisfying the conditions of part (a), with lub $A =$ glb A.

26. Let A be bounded above, and let $x \in \Re$. Prove that:
    (a) lub $\{x + a | a \in A\} = x +$ lub A
    (b) lub $\{xa | a \in A\} = x($lub $A)$ if $x \geq 0$, and that glb$\{xa | a \in A\} = x($glb$A)$ if $x < 0$.

27. Let A be a nonempty subset of $\Re$ which is bounded above but does not have a maximum element. Prove that A cannot be finite.

28. Let A be a nonempty subset of $\Re$ which is bounded below but does not have a minimum element. Prove that A cannot be finite.

29. Let A be a nonempty subset of $\Re$ which is bounded above but does not have a maximum element. Prove that for any $a \in A$, lub $A =$ lub $(A - \{a\})$.

30. Let A be a nonempty subset of $\Re$ which is bounded below but does not have a minimum element. Prove that for any $a \in A$, lglb $A =$ glb $(A - \{a\})$.

31. Let A be a nonempty subset of $\Re$ which is bounded above but does not have a maximum element, and let $\{a_1, a_2, ..., a_n\} \subseteq A$. Show that lub $A =$ lub $(A - \{a_1, a_2, ..., a_n\})$.

32. Let A be a nonempty subset of $\Re$ which is bounded below but does not have a minimum element, and let $\{a_1, a_2, ..., a_n\} \subseteq A$. Show that glb $A =$ glb $(A - \{a_1, a_2, ..., a_n\})$.

33. Show that any subset of $\Re$ that contains a dense subset of $\Re$ is itself dense.

34. Give an example of an infinite subset of $\Re$ that is not dense in $\Re$.

35. Prove that if x and y are rational and z is irrational then $xy + z$ is irrational.

36. Prove that for any positive integer n, $\sqrt{n}$ is rational if and only if $\sqrt{n}$ is an integer.

37. Show that there exists irrational numbers x and y such that x^y is rational.
    Suggestion: Consider the number $(\sqrt{2})^{\sqrt{2}}$. If it is rational, then the claim is seen to hold. If it is irrational, then consider raising it to the power $\sqrt{2}$.

PROVE OR GIVE A COUNTEREXAMPLE

38. If $A \subseteq B$, and if B is bounded, then A is bounded.

39. If $A \subset B$, and if B is bounded, then lub $A <$ lub B or glb $A <$ glb B.

40. If lub $A \leq$ lub B, then there exists an element $b \in B$ that is an upper bound of A.

41. If glb $A >$ glb B, then there exists an element $a \in B$ that is a lower bound of B.

42. For $A \subseteq \Re$ and $B \subseteq \Re$, let $A + B = \{a + b | a \in A, b \in B\}$. If $\alpha_A =$ lub A and $\alpha_B =$ lub B, then $\alpha_A + \alpha_B =$ lub $(A + B)$.

43. For $A \subseteq \Re$ and $B \subseteq \Re$, let $A + B = \{a + b | a \in A, b \in B\}$. If $\beta_A =$ glb A and $\beta_B =$ glb B, then $\beta_A + \beta_B =$ glb $(A + B)$.

44. For $A \subseteq \Re$ and $B \subseteq \Re$, let $A \cdot B = \{ab | a \in A, b \in B\}$. If A and B are bounded, then so is $A \cdot B$.

45. For $A \subseteq \Re$ and $B \subseteq \Re$, let $A \cdot B = \{ab | a \in A, b \in B\}$. If $\alpha_A =$ lub A and $\alpha_B =$ lub B, then $\alpha_A \cdot \alpha_B =$ lub $(A \cdot B)$.

46. For $A \subseteq \Re$ and $B \subseteq \Re$, let $A \cdot B = \{ab | a \in A, b \in B\}$. If $\beta_A =$ glb A and $\beta_B =$ glb B, then $\beta_A \cdot \beta_B =$ glb $(A \cdot B)$.

47. Every infinite subset of $\Re$ is dense in $\Re$.

48. Let A and B be subsets of $\Re$. If $A \cup B$ is dense in $\Re$ then A or B must be dense in $\Re$.

49. Let A and B be subsets of $\Re$. If $A \cap B$ is dense in $\Re$ then A and B must be dense in $\Re$.

50. Let A and B be subsets of $\Re$. If A and B are dense in $\Re$ then $A \cup B$ is also dense in $\Re$.

51. Let A and B be subsets of $\Re$. If A and B are dense in $\Re$ then $A \cap B$ is also dense in $\Re$.

52. The product of any irrational numbers and any rational number is again irrational.

53. If x and y are irrational and z is rational then $x + y + z$ is irrational.

54. If x and y are irrational and z is rational then $xy + z$ is irrational.

55. If x and y are irrational and z is rational then $xy + z$ is rational.

56. If $\dfrac{ax + b}{cx + d} = 1$ then x must be rational.

57. (a) If one solution of the quadratic equation $ax^2 + bx + c = 0$ is rational then the other solution is also rational.

    (b) If one solution of the quadratic equation $ax^2 + bx + c = 0$ is rational, and if the coefficients a, and c are integers, then the other solution is also rational.

§2. SEQUENCES

Formally:

DEFINITION 3.4
SEQUENCE
A sequence of real numbers is a real-valued function with domain the set of positive integers.

Formality aside, one seldom represents a sequence in the function-form $f: Z^+ \to \Re$ but, rather, as an infinite string of numbers, or **terms**

$$(a_1, a_2, a_3, \ldots) \quad \text{or} \quad (a_n)_{n=1}^{\infty}$$

with n^{th}-term a_n.

> Unlike the set $\{a_n\}_{n=1}^{\infty}$, elements in a sequence $(a_n)_{n=1}^{\infty}$ can appear more than once, as is the case with the sequence $(0, 1, 0, 1, 0, 1, \ldots)$.

Consider the sequences:

(a) $\left(1, \frac{1}{2}, \frac{1}{3}, \ldots\right)$ (b) $\left(\frac{n+1}{n}\right)_{n=1}^{\infty}$ and (c) $(1, 2, 1, 2, 1, 2, \ldots)$

While the sequence in (a) appears to be heading to 0, and that of (b) to 1, the sequence in (c) does not look to be going anywhere in particular, as its terms keep jumping back and forth between 1 and 2. Appearances are well and good, but mathematics demands precision:

DEFINITION 3.5
CONVERGENT SEQUENCE
A sequence $(a_n)_{n=1}^{\infty}$ **converges** to the number α if for any given $\varepsilon > 0$ there exists a positive integer N (which depends on ε) such that:

$$n > N \Rightarrow |a_n - \alpha| < \varepsilon$$

A sequence that does not converge is said to **diverge**.

> In words: By going far enough in the sequence, $(n > N)$, you can get the terms of the sequence to be as close as you want to α, $(|a_n - \alpha| < \varepsilon)$.

The next theorem asserts that a sequence cannot converge to two different numbers. That being the case, if $(a_n)_{n=1}^{\infty}$ converges to α, we are justified in saying that α is **the** limit of $(a_n)_{n=1}^{\infty}$, and write:

$$\lim_{n \to \infty} a_n = \alpha, \text{ or } \lim a_n = \alpha, \text{ or simply } a_n \to \alpha.$$

> We remind you that the **absolute value function**
> $$|a| = \begin{cases} a & \text{if } x \geq 0 \\ -a & \text{if } x < 0 \end{cases}$$
> denotes the distance between the number a and the origin on the number line, and that $|a - b|$ represents the distance between the numbers a and b. For example: $|2 - 7| = 5$ is the distance between 2 and 7, while $|3 + 4| = |3 - (-4)| = 7$ is the distance between 3 and -4.

Additional Notation: For given $a \in \Re$ and $\varepsilon > 0$, we will use the symbol $S_\varepsilon(a)$ to denote the set of numbers that lie within ε units of a:

$$S_\varepsilon(a) = \{x \in \Re \mid |x - a| < \varepsilon\}$$

In anticipation of higher dimensional spaces, we call $S_\varepsilon(a)$ the (open) **sphere of radius ε about a**.

THEOREM 3.7 A sequence can have at most one limit.

PROOF: Assume that a sequence $(a_n)_{n=1}^{\infty}$ converges to two different numbers α and β (we will arrive at a contradiction). Letting β denote the larger of the two numbers, we consider $\varepsilon = \frac{\beta - \alpha}{2}$:

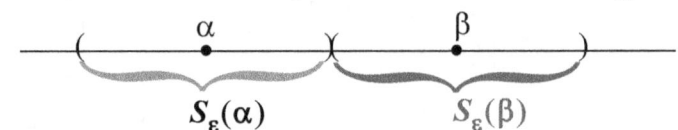

Since $(a_n)_{n=1}^{\infty}$ converges to α and to β, there exist integers N_α and N_β such that $n > N_\alpha \Rightarrow |a_n - \alpha| < \varepsilon$ and $n > N_\beta \Rightarrow |a_n - \beta| < \varepsilon$. Let N be the larger of N_α and N_β. As such, the term a_{N+1} of the sequence must lie in both $S_\varepsilon(\alpha)$ and $S_\varepsilon(\beta)$ — a contradiction, since $S_\varepsilon(\alpha) \cap S_\varepsilon(\beta) = \varnothing$.

EXAMPLE 3.3
(a) Prove that $\lim_{n \to \infty} \frac{n+1}{n} = 1$.

(b) Prove that for any constant c the sequence $(c, c, c, c, \ldots)$ converges to c.

(c) Show that the sequence $(1, 2, 1, 2, 1, 2, \ldots)$ diverges.

SOLUTION: (a) Let $\varepsilon > 0$ be given. We are to find N such that $n > N \Rightarrow \left|\frac{n+1}{n} - 1\right| < \varepsilon$. Let's do it:

We want: $n > N \Rightarrow \left|\frac{n+1}{n} - 1\right| < \varepsilon$

Let's rewrite the goal: $n > N \Rightarrow \left|\frac{n+1-n}{n}\right| < \varepsilon$

again: $n > N \Rightarrow \left|\frac{1}{n}\right| < \varepsilon$

and again: $n > N \Rightarrow \frac{1}{n} < \varepsilon$

and finally: $n > N \Rightarrow n > \frac{1}{\varepsilon}$

So, to find an N such that $n > N \Rightarrow \left|\frac{n+1}{n} - 1\right| < \varepsilon$ is to find an N such that $n > N \Rightarrow n > \frac{1}{\varepsilon}$. Piece of cake: let N be the first integer greater than $\frac{1}{\varepsilon}$.

> Note how N is dependent on ε — the smaller the given ε, the larger the N.

(b) For $(c_n) = (c, c, c, c, \ldots)$ and any given $\varepsilon > 0$ let $N = 1$. Then:
$$n > N \Rightarrow |c_n - c| = |c - c| = 0 < \varepsilon.$$

(c) We show that $(1, 2, 1, 2, 1, 2, \ldots)$ diverges by demonstrating that no fixed but **arbitrary** $r \in \Re$ can be the limit of the sequence:

Let N be **any** positive integer. Since any two numbers in $S_{\frac{1}{2}}(r) = \left\{ x \mid |x - r| < \frac{1}{2} \right\}$ are less than one unit apart (see margin), both a_{N+1} and a_{N+2} cannot be contained in $S_{\frac{1}{2}}(r)$, as one of the numbers is 1 while the other is 2. This shows that no N "works" for $\varepsilon = \frac{1}{2}$, and that, consequently, the **arbitrarily** chosen number r cannot be a limit of $(1, 2, 1, 2, 1, 2, \ldots)$.

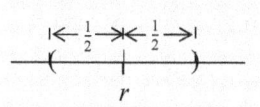

CHECK YOUR UNDERSTANDING 3.8

(a) Let $(a_n)_{n=1}^{\infty} = \left(7 - \dfrac{101}{n} \right)_{n=1}^{\infty}$.

   (i) Prove that $\lim\limits_{n \to \infty} a_n = 7$.

   (ii) Find the smallest positive integer N such that $n > N \Rightarrow |a_n - 7| < \dfrac{1}{100}$.

(b) Show that the sequence $(a_n)_{n=1}^{\infty} = \left(\dfrac{n - 5}{333} \right)_{n=1}^{\infty}$ diverges.

Answer: See page A-15.

DEFINITION 3.6 A sequence $(a_n)_{n=1}^{\infty}$ is:

INCREASING Increasing if $a_n \leq a_{n+1}$

DECREASING Decreasing if $a_n \geq a_{n+1}$

MONOTONE Monotone if it is either increasing or decreasing.

BOUNDED Bounded if there exists a real number M such that $|a_n| \leq M$ for every n.

Example 3.3(c) shows that not every bounded sequence converges. However:

THEOREM 3.8 Every increasing (decreasing) sequence that is bounded from above (below) converges.

PROOF: Let $(a_n)_{n=1}^{\infty}$ be increasing and bounded from above. The completion axiom assures us that the set $\{a_n\}$ has a least upper bound: α. We show that $(a_n)_{n=1}^{\infty}$ converges to α:

Let $\varepsilon > 0$ be given. Theorem 3.1 (a-ii), page 113, tells us that there exists a term a_N such that $a_N > \alpha - \varepsilon$. Since $(a_n)_{n=1}^{\infty}$ is an increasing sequence, $a_n > \alpha - \varepsilon$ for every $n > N$. Since $\alpha = \text{lub}\{a_n\}$, $a_n \leq \alpha < \alpha + \varepsilon$ for every n. It follows that $|a_n - \alpha| < \varepsilon$ for every $n > N$ (see margin).

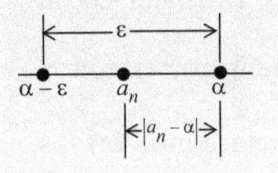

A similar argument can be used to show that every decreasing sequence bounded from below converges (Exercise 35).

CHECK YOUR UNDERSTANDING 3.9

Answer: See page A-16.

Prove that if a sequence converges, then it is bounded.

THE ALGEBRA OF SEQUENCES

When it comes to sums, differences, products, and quotients, sequences behave nicely:

THEOREM 3.9 If $\lim a_n = \alpha$ and $\lim b_n = \beta$, then:

(a) $\lim c a_n = c\alpha$, for any $c \in \mathfrak{R}$.

(b) $\lim (a_n + b_n) = \alpha + \beta$
(The limit of a sum equals the sum of the limits)

(c) $\lim (a_n b_n) = \alpha\beta$
(The limit of a product equals the product of the limits)

(d) $\lim \dfrac{a_n}{b_n} = \dfrac{\alpha}{\beta}$, providing no $b_n = 0$ and $\beta \neq 0$.
(The limit of a quotient equals the quotient of the limits...)

PROOF:
(a) Case 1. $c = 0$.
$$\lim c a_n = \lim (0 \cdot a_n) = 0 = 0 \cdot \alpha = c\alpha$$

Case 2. $c \neq 0$. For given $\varepsilon > 0$ we are to exhibit an N such that
$$n > N \Rightarrow |c a_n - c\alpha| < \varepsilon$$
i.e: $n > N \Rightarrow |c||a_n - \alpha| < \varepsilon$
i.e: $n > N \Rightarrow |a_n - \alpha| < \dfrac{\varepsilon}{|c|}$

Since $\lim a_n = \alpha$, we know that for **any** $\bar{\varepsilon} > 0$ there exists an N such that $n > N \Rightarrow |a_n - \alpha| < \bar{\varepsilon}$. In particular, for $\bar{\varepsilon} = \dfrac{\varepsilon}{|c|}$ we can choose N such that $n > N \Rightarrow |a_n - \alpha| < \dfrac{\varepsilon}{|c|}$, and we are done.

(b) Let $\varepsilon > 0$. We are to find N such that
$$n > N \Rightarrow |(a_n + b_n) - (\alpha + \beta)| < \varepsilon \quad (*)$$
Note that:
$$|(a_n + b_n) - (\alpha + \beta)| = |(a_n - \alpha) + (b_n - \beta)| \leq |a_n - \alpha| + |b_n - \beta|$$
$$\uparrow \text{ triangle inequality}$$

So, if we can arrange things so that both $|a_n - \alpha|$ and $|b_n - \beta|$ are less than $\dfrac{\varepsilon}{2}$, then (*) will hold. Let's arrange things:

Since $a_n \to \alpha$, there exists N_α such that: $n > N_\alpha \Rightarrow |a_n - \alpha| < \dfrac{\varepsilon}{2}$.

Since $b_n \to \beta$, there exists N_β such that: $n > N_\beta \Rightarrow |b_n - \beta| < \dfrac{\varepsilon}{2}$.

Letting $N = \max\{N_\alpha, N_\beta\}$ (the larger of N_α and N_β), we find that for $n > N$:
$$|(a_n + b_n) - (\alpha + \beta)| \leq |a_n - \alpha| + |b_n - \beta| < \frac{\varepsilon}{2} + \frac{\varepsilon}{2} = \varepsilon$$

(c) Let $\varepsilon > 0$. We are to find N such that:
$$n > N \Rightarrow |a_n b_n - \alpha\beta| < \varepsilon$$
In order to get $|a_n - \alpha|$ and $|b_n - \beta|$ into the picture (for we have control over those two expressions), we insert the clever zero $-a_n\beta + a_n\beta$ in the expression $|a_n b_n - \alpha\beta|$:
$$|a_n b_n - \alpha\beta| = |a_n b_n - a_n\beta + a_n\beta - \alpha\beta|$$
$$= |(a_n b_n - a_n\beta) + (a_n\beta - \alpha\beta)|$$
$$\leq |a_n b_n - a_n\beta| + |a_n\beta - \alpha\beta| = \underbrace{|a_n||b_n - \beta|}_{(i)} + \underbrace{|\beta||a_n - \alpha|}_{(ii)}$$

The next step is to find an N such that **both** (i) and (ii) are less than $\dfrac{\varepsilon}{2}$.

Focusing on (i): $|a_n||b_n - \beta|$:

The temptation is to let N_β be such that $n > N_\beta \Rightarrow |b_n - \beta| < \dfrac{\varepsilon}{2|a_n|}$ (yielding $|a_n||b_n - \beta| < |a_n|\dfrac{\varepsilon}{2|a_n|} = \dfrac{\varepsilon}{2}$). No can do. For one thing, if $a_n = 0$, then the expression $\dfrac{\varepsilon}{2|a_n|}$ is undefined. More importantly:

[Margin note: Let the positive number $\frac{\varepsilon}{2M}$ play the role of ε in Definition 3.5.]

$\frac{\varepsilon}{2|a_n|}$ is **NOT A CONSTANT!** We can, however, take advantage of the fact that there exists an $M > 0$ such that $|a_n| < M$ for every n (see CYU 3.9), and choose N_β such that $n > N_\beta \Rightarrow |b_n - \beta| < \frac{\varepsilon}{2M}$ (see margin). Then: $n \geq N_\beta \Rightarrow |a_n||b_n - \beta| < M\frac{\varepsilon}{2M} = \frac{\varepsilon}{2}$.

Focusing on (ii): $|\beta||a_n - \alpha|$.

Wanting $|\beta||a_n - \alpha|$ to be less than $\frac{\varepsilon}{2}$, one might be tempted to choose N_α such that $n > N_\alpha \Rightarrow |a_n - \alpha| < \frac{\varepsilon}{2|\beta|}$. But what if $\beta = 0$? To get around this potential problem we choose N_α such that $n > N_\alpha \Rightarrow |a_n - \alpha| < \frac{\varepsilon}{2|\beta| + 1}$. No problem now:

$$n \geq N_\alpha \Rightarrow |\beta||a_n - \alpha| < |\beta|\frac{\varepsilon}{2|\beta| + 1} < \frac{\varepsilon}{2}$$

since $\frac{|\beta|}{2|\beta| + 1} < \frac{|\beta|}{2|\beta|}$

Letting $N = \max\{N_\alpha, N_\beta\}$, we see that, for $n > N$:

$$|a_n b_n - \alpha\beta| \leq |a_n||b_n - \beta| + |\beta||a_n - \alpha| < \frac{\varepsilon}{2} + \frac{\varepsilon}{2} = \varepsilon$$

(d) Appealing to (c), we establish the fact that $\lim \frac{a_n}{b_n} = \frac{\alpha}{\beta}$, by showing that $\lim \frac{1}{b_n} = \frac{1}{\beta}$:

Let $\varepsilon > 0$ be given. We are to find N such that:

$$n > N \Rightarrow \left|\frac{1}{b_n} - \frac{1}{\beta}\right| = \frac{|\beta - b_n|}{|b_n||\beta|} < \varepsilon \quad (*)$$

Since $b_n \to \beta \neq 0$, we can choose N_1 such that:

$$n > N_1 \Rightarrow |b_n - \beta| < \frac{|\beta|}{2}.$$

[Margin: $|\beta| - |b_n| \leq |\beta - b_n|$
$= |b_n - \beta| < \frac{|\beta|}{2}$
$\Rightarrow |b_n| > \frac{|\beta|}{2}$]

For $n > N_1$ we also have: $|b_n| > \frac{|\beta|}{2}$ (see margin). Consequently, For $n > N_1$:

$$\frac{|\beta - b_n|}{|b_n||\beta|} = \frac{1}{|b_n|} \cdot \frac{|\beta - b_n|}{|\beta|} < \frac{1}{\frac{|\beta|}{2}} \cdot \frac{|\beta - b_n|}{|\beta|} = \frac{2}{|\beta|^2}|b_n - \beta| \quad (**)$$

Since $b_n \to \beta$, we can choose N_2 such that:

$$n \geq N_2 \Rightarrow |b_n - \beta| < \frac{|\beta|^2}{2}\varepsilon \quad (***)$$

Letting $N = \max\{N_1, N_2\}$, we find that, for $n \geq N$:

$$\underbrace{\frac{|\beta - b_n|}{|b_n||\beta|}}_{\text{By (**)}} < \frac{2}{|\beta|^2}\underbrace{|b_n - \beta|}_{\text{By (***)}} < \frac{2}{|\beta|^2} \cdot \frac{|\beta|^2}{2}\varepsilon = \varepsilon$$

thereby establishing (*).

CHECK YOUR UNDERSTANDING 3.10

(a) Let $(a_n)_{n=1}^{\infty}$ and $(b_n)_{n=1}^{\infty}$ be such that $0 \leq a_n \leq b_n$ for every n. Show that if $a_n \to \alpha$ and $b_n \to \beta$, then $\alpha \leq \beta$.

(b) Give an example of two sequences $(a_n)_{n=1}^{\infty}$ and $(b_n)_{n=1}^{\infty}$ with $0 \leq a_n < b_n$ and such that $\lim a_n = \lim b_n$.

Answer: See page A-16.

SUBSEQUENCES

Roughly speaking, to get a subsequence of $(a_n)_{n=1}^{\infty}$, simply discard some of its terms in an orderly fashion. Formally:

DEFINITION 3.7
SUBSEQUENCE
$(a_{n_k})_{k=1}^{\infty} = (a_{n_1}, a_{n_2}, a_{n_3}, \ldots)$ is a **subsequence** of $(a_n)_{n=1}^{\infty}$ if each a_{n_k} is a term of (a_n), and $n_1 < n_2 < n_3 < \cdots$.

For example, $(9, 11, 13, 15, \ldots)$ is a subsequences of $(1, 2, 3, \ldots)$, while $(4, 2, 8, 6, \ldots)$ is not.

CHECK YOUR UNDERSTANDING 3.11

Coin a function-form definition of a subsequence of a sequence (see Definition 3.4). We'll get you started:

The sequence $g: Z^+ \to \Re$ is a subsequence of the sequence $f: Z^+ \to \Re$ if ....

Answer: See page A-16.

There are sequences, like $(1, 2, 3, 4, \ldots)$, that do not contain any convergent subsequences. However:

Bernhard Bolzano (1781-1841), Karl Weierstrass (1815-1897).

THEOREM 3.10 Every bounded sequence contains a convergent subsequence.
BOLZANO-WEIERSTRASS

PROOF: Let $(a_n)_{n=1}^{\infty}$ be a bounded sequence. Being bounded, there exists $M > 0$ such that $a_n \in [-M, M]$ for every n.

Cut $[-M, M]$ into two equal pieces: $[-M, 0]$ and $[0, M]$. Select one of those two intervals (of length $\frac{M}{2}$) which contains an infinite number of elements of the sequence $(a_n)_{n=1}^{\infty}$, and call it I_1. Next cut I_1 in half and let I_2 be one of those halves (now of length $\frac{1}{2} \cdot \frac{M}{2} = \frac{M}{2^2}$) which still contains infinitely many elements of $(a_n)_{n=1}^{\infty}$. Continue in this fashion to generate a nested sequence of closed intervals I_n of length $\frac{M}{2^n}$, each containing infinitely many terms of $(a_n)_{n=1}^{\infty}$.

Theorem 3.3, page 115, assures us that $\bigcap_{n=1}^{\infty} I_n \neq \emptyset$. Let $x_0 \in \bigcap_{n=1}^{\infty} I_n$.

We now construct a subsequence of $(a_n)_{n=1}^{\infty}$ converging to x_0:

Let: a_{n_1} be any term of $(a_n)_{n=1}^{\infty}$ in I_1

a_{n_2} be any term in I_2 with $n_2 > n_1$

(this we can do since I_2 contains infinitely many entries of (a_n))

a_{n_3} be any term in I_3 with $n_3 > n_2$

Proceeding in the above fashion we arrive at a subsequence $(a_n)_{n=1}^{\infty}$ with $a_{n_k} \in I_k$. As for convergence:

Let $\varepsilon > 0$ be given. Choose N such that $\frac{M}{2^N} < \varepsilon$. Since x_0 and a_{n_k} are contained in I_k, for $n_k > N$: $|a_{n_k} - x_0| < \varepsilon$.

Augustin Louis Cauchy (1789-1857).

CAUCHY SEQUENCES

To say that $\lim_{n \to \infty} a_n = \alpha$ is to say that the a_n's eventually get arbitrarily close to α. We now consider sequences whose terms eventually get arbitrarily close to **each other** (with no mention made of any limit whatsoever):

DEFINITION 3.8
CAUCHY SEQUENCE

A sequence $(a_n)_{n=1}^{\infty}$ is a **Cauchy sequence** if, for any given $\varepsilon > 0$ there exists a positive integer N (which depends on ε) such that if $n, m > N$, then $|a_n - a_m| < \varepsilon$.

CHECK YOUR UNDERSTANDING 3.12

Answer: See page A-16.

Prove that every Cauchy sequence is bounded.

It may not be surprising to find that every convergent sequence is Cauchy. After all, if the terms of a sequence are "bunching up" around a number α, then they must certainly be getting close to each other. A bit more surprising is that the converse also holds:

THEOREM 3.11 A sequence converges **if and only if** it is Cauchy.

PROOF: Assume that $(a_n)_{n=1}^{\infty}$ converges to α. We show $(a_n)_{n=1}^{\infty}$ is Cauchy:

Let $\varepsilon > 0$ be given. Choose N such that $n > N$ implies $|a_n - \alpha| < \frac{\varepsilon}{2}$. Then, for $n, m > N$:

$$|a_n - a_m| = |a_n - \alpha + \alpha - a_m| \leq |a_n - \alpha| + |a_m - \alpha| < \frac{\varepsilon}{2} + \frac{\varepsilon}{2} = \varepsilon.$$

Conversely, assume that $(a_n)_{n=1}^{\infty}$ is Cauchy.

By CYU 3.12, $(a_n)_{n=1}^{\infty}$ is bounded and, as such it has a convergent subsequence $(a_{n_k})_{k=1}^{\infty}$ (Theorem 3.10). Let $\lim_{k \to \infty} a_{n_k} = \alpha$. We complete the proof by showing that $\lim_{n \to \infty} a_n = \alpha$:

Let $\varepsilon > 0$ be given. Since $(a_n)_{n=1}^{\infty}$ is Cauchy, there exists N such that $|a_n - a_m| < \frac{\varepsilon}{2}$ for all $n, m > N$ (in particular for all $n, n_k > N$). Since $a_{n_k} \to \alpha$, we can choose a term a_{n_K} of (a_{n_k}), with $n_K > N$, such that: $|a_{n_K} - \alpha| < \frac{\varepsilon}{2}$. For $n > N$ we then have:

$$|a_n - \alpha| = |a_n - a_{n_K} + a_{n_K} - \alpha|$$
$$\leq |a_n - a_{n_K}| + |a_{n_K} - \alpha| < \frac{\varepsilon}{2} + \frac{\varepsilon}{2} = \varepsilon$$

CHECK YOUR UNDERSTANDING 3.13

Let X be a nonempty subset of $\Re$. A sequence $(a_n)_{n=1}^{\infty}$ is said to be a **sequence in X** if each $a_n \in X$. A sequence in X is said to **converge in X** if $\lim a_n = \alpha \in X$.

Construct a Cauchy sequence in $X = (0, \infty)$ which does not converge in X.

Answer: See page A-16.

EXERCISES

Exercises 1-6. Find a formula for the nth term of the given sequence.

1. $\left(\frac{1}{2}, \frac{2}{3}, \frac{3}{4}, \frac{4}{5}, \ldots\right)$
2. $\left(\frac{2}{3}, -\frac{3}{9}, \frac{4}{27}, -\frac{5}{81}, \ldots\right)$
3. $(1, 2, 1, 2, 1, 2, \ldots)$
4. $(1, 2, 3, 2, 3, 4, 5, 6, 7, 8, 9, 10, \ldots)$
5. $(1, 1, 2, 4, 3, 9, 4, 16, \ldots)$
6. $(a, b, a+b, a^2, b^2, a^2+b^2, a^3, b^3, a^3+b^3, \ldots)$

Exercises 7-15. Find the limit of the given sequence. Use Definition 3.5 to justify your claim.

7. $\left(\frac{1}{n}\right)_{n=1}^{\infty}$
8. $\left(5 + \frac{1}{2n}\right)_{n=1}^{\infty}$
9. $\left(-1 - \frac{1}{\sqrt{n}}\right)_{n=1}^{\infty}$
10. $\left(\frac{n}{n+1}\right)_{n=1}^{\infty}$
11. $\left(\frac{n^2}{n^2+1}\right)_{n=1}^{\infty}$
12. $\left(\frac{3n^2+n}{n^2}\right)_{n=1}^{\infty}$
13. $\left(\frac{2n^2+10}{n^2+1}\right)_{n=1}^{\infty}$
14. $(r^n)_{n=1}^{\infty}$, for $r \leq 1$
15. $\left(\frac{n^2+10n}{n^2+1}\right)_{n=1}^{\infty}$

Exercises 16-21. Show that the given sequence diverges.

16. $\left(\frac{1}{10}, 1, \frac{1}{20}, 1, \frac{1}{30}, 1, \frac{1}{40}, 1, \ldots\right)$
17. $(r^n)_{n=1}^{\infty}$, for $r > 1$
18. $((-1)^n)_{n=1}^{\infty}$
19. $\left(\frac{(-1)^n n}{n+1}\right)_{n=1}^{\infty}$
20. $\left(\frac{n^2}{n+100}\right)_{n=1}^{\infty}$
21. $\left(\frac{n!}{2^n}\right)_{n=1}^{\infty}$

22. Prove that $\lim_{n \to \infty} a_n = 0$ if and only if $\lim_{n \to \infty} |a_n| = 0$.

Exercises 23-31. Find the limit of the given sequence if it exists. You may use Theorem 3.9 and the result of Exercise 22 to justify your claim.

23. $\left(\frac{(-1)^n}{n+5}\right)_{n=1}^{\infty}$
24. $\left(\frac{\sin n}{\sqrt{n}}\right)_{n=1}^{\infty}$
25. $\left(\frac{(-1)^n 2n}{n+1}\right)_{n=1}^{\infty}$
26. $\left(\frac{\sqrt{n^2+5}}{n^2+5}\right)_{n=1}^{\infty}$
27. $\left(\frac{n^3 + 10n^2 - n + 1}{n^3}\right)_{n=1}^{\infty}$
28. $\left(\frac{n^2 \cos n - n \sin 2n}{n^3 + 5}\right)_{n=1}^{\infty}$
29. $\left(\frac{\sin n \cos n}{n+1}\right)_{n=1}^{\infty}$
30. $\left(\frac{\cos n}{n+1} - \frac{(-1)^n}{n}\right)_{n=1}^{\infty}$
31. $\left(\left(\frac{n+1}{n-3}\right)^3 \left(\frac{n^3+n-1}{2n^3}\right)\right)_{n=1}^{\infty}$

32. Let $0 < a_n \leq b_n$ for $n > N$. Prove that if $b_n \to 0$, then $a_n \to 0$.

Chapter 3 A Touch of Analysis

Exercise 33-34. Find the limit of the given sequence. You may use Exercises 22 and 32 to justify your claim.

33. $\left(\dfrac{\sin n}{n!}\right)_{n=1}^{\infty}$

34. $\left(\dfrac{(-100)^n}{n!}\right)_{n=1}^{\infty}$

35. Prove that every decreasing sequence, bounded below, converges.

36. (a) Give an example of two converging sequences $(a_n)_{n=1}^{\infty}$ and $(b_n)_{n=1}^{\infty}$ such that
$\lim\limits_{n \to \infty} (a_n + b_n) = 5$.

   (b) Give an example of two divergent sequences $(a_n)_{n=1}^{\infty}$ and $(b_n)_{n=1}^{\infty}$ such that
$\lim\limits_{n \to \infty} (a_n + b_n) = 5$.

37. Prove that the sequence $(r^n)_{n=1}^{\infty}$ converges if and only for $-1 < r \le 1$.

38. Prove that $\lim\limits_{n \to \infty} a_n = \alpha$ if and only if, for any given $\varepsilon > 0$, $S_\varepsilon(\alpha)$ contains all but finitely many terms of $(a_n)_{n=1}^{\infty}$.

39. Prove that if $\lim\limits_{n \to \infty} a_n = \alpha$ and if $a_n \ge 0$ for all $n > N$, then $\alpha \ge 0$.

40. Prove that a sequence $(a_n)_{n=1}^{\infty}$ converges if and only if $(a_n)_{n=N}^{\infty}$ converges for any positive integer N.

41. Prove that if $\lim\limits_{n \to \infty} a_n = 0$ and if $|r - b_n| \le a_n$ then $\lim\limits_{n \to \infty} b_n = r$.

42. Write down the first four terms of the sequence $(a_n)_{n=1}^{\infty}$, if $a_1 = 2$ and $a_{n+1} = \dfrac{1}{5 - a_n}$. Show that the sequence converges. Suggestion: Consider Theorem 3.8.

43. (a) Prove that if $\lim\limits_{n \to \infty} a_n = 0$ and if $(b_n)_{n=1}^{\infty}$ is bounded, then $\lim\limits_{n \to \infty} a_n b_n = 0$.

   (b) Give an example of sequences $(a_n)_{n=1}^{\infty}$ and $(b_n)_{n=1}^{\infty}$, such that $\lim\limits_{n \to \infty} a_n = 0$ and $(a_n b_n)_{n=1}^{\infty}$ diverges.

   (c) Give an example of a convergent sequence $(a_n)_{n=1}^{\infty}$ and a bounded sequence $(b_n)_{n=1}^{\infty}$, such that $(a_n b_n)_{n=1}^{\infty}$ diverges.

44. Establish the following "Squeeze Theorem:"
If $a_n \le b_n \le c_n$ for $n > N$, and if $\lim\limits_{n \to \infty} a_n = \lim\limits_{n \to \infty} c_n = \alpha$, then $\lim\limits_{n \to \infty} b_n = \alpha$.

45. Prove that if $\lim\limits_{n \to \infty} a_n = \alpha$, then every subsequence of $(a_n)_{n=1}^{\infty}$ also converges to α.

46. Prove that every subsequence of a Cauchy sequence is itself a Cauchy sequence.

47. Prove that if a sequence contains two subsequences with different limits, then the sequence diverges.

48. Give an example of a divergent sequence that contains:

    (a) Two subsequences with different limits.

    (b) Three subsequences with different limit.

    (c) Infinitely subsequences with different limits.

PROVE OR GIVE A COUNTEREXAMPLE

49. Every bounded sequence converges.

50. If $(a_n)_{n=1}^{\infty}$ converges, then so does $(|a_n|)_{n=1}^{\infty}$.

51. If $(|a_n|)_{n=1}^{\infty}$ converges, then so does $(a_n)_{n=1}^{\infty}$.

52. If $(a_n)_{n=1}^{\infty}$ and $(a_n + b_n)_{n=1}^{\infty}$ converge, then $(b_n)_{n=1}^{\infty}$ must converge.

53. If $(a_n + b_n)_{n=1}^{\infty}$ converges, then $(a_n)_{n=1}^{\infty}$ and $(b_n)_{n=1}^{\infty}$ must both converge.

54. If $(a_n + b_n)_{n=1}^{\infty}$ diverges, then $(a_n)_{n=1}^{\infty}$ and $(b_n)_{n=1}^{\infty}$ must both diverge.

55. If $\lim_{n \to \infty} a_n = \alpha$, then $\lim_{n \to \infty} |a_n| = |\alpha|$.

56. If $\lim_{n \to \infty} |a_n| = |\alpha|$, then $\lim_{n \to \infty} a_n = \alpha$.

57. If $\lim_{n \to \infty} a_n = \alpha$ and if $a_n > 0$ for all $n > N$, then $\alpha > 0$.

58. If $\lim_{n \to \infty} a_n = \alpha$ and $\lim_{n \to \infty} b_n = \beta$, and if $a_n > b_n$ for all $n > N$, then $\alpha > \beta$.

59. If $(a_n)_{n=1}^{\infty}$ and $(b_n)_{n=1}^{\infty}$ are Cauchy sequences, then $(a_n + b_n)_{n=1}^{\infty}$ is a Cauchy sequence.

60. If $(a_n)_{n=1}^{\infty}$ and $(b_n)_{n=1}^{\infty}$ are Cauchy sequences, then $(a_n b_n)_{n=1}^{\infty}$ is a Cauchy sequence.

§3. METRIC SPACE STRUCTURE OF $\Re$

Roughly speaking, a metric space is a set with an imposed notion of distance. Our concern in this section is with the **standard or Euclidean metric** d on $\Re$; that function which assigns to any two numbers in $\Re$ the distance between them:

$$d(x, y) = |x - y|$$

The following distance-properties of the real number system will evolve into the defining axioms of an abstract metric space in the next chapter:

THEOREM 3.12 (i) $\forall x, y \in \Re : |x - y| \geq 0$, with $|x - y| = 0$ if and only if $x = y$.
> The distance between two numbers is never negative, and is 0 only if the two numbers are one and the same.

(ii) $\forall x, y \in \Re : |x - y| = |y - x|$
> The distance between x and y is the same as that from y to x.

(iii) $\forall x, y, z \in \Re : |x - y| \leq |x - z| + |z - y|$
> The triangle inequality.

> If $a \geq 0$ and $b \geq 0$ then:
> $a^2 \leq b^2 \Leftrightarrow a \leq b$
> Here: $a = |x - y|$ and $b = |x - z| + |z - y|$.

PROOF: The first two properties are direct consequences of the definition of the absolute value function. We establish (iii) by showing that $|x - y|^2 \leq (|x - z| + |z - y|)^2$ (see margin):

$$(|x-z| + |z-y|)^2 = |x-z|^2 + 2|x-z||z-y| + |z-y|^2$$

$|a|^2 = a^2$: $\quad = (x-z)^2 + 2|x-z||z-y| + (z-y)^2$

$a \leq |a|$: $\quad \geq (x-z)^2 + 2(x-z)(z-y) + (z-y)^2$

expanding: $\quad = x^2 - 2xz + z^2 + 2xz - \mathbf{2xy} - 2z^2 + 2zy + z^2 - 2zy + y^2$

$\quad = x^2 - \mathbf{2xy} + y^2 = (x-y)^2 = |x-y|^2$

CHECK YOUR UNDERSTANDING 3.14

> Answer: See page A-16.

Prove that for any $x, y \in \Re$, $|x - y| \geq |x| - |y|$

DEFINITION 3.9
OPEN SET

A set $O \subseteq \Re$ is **open** if for every $a \in O$ there exists an $\varepsilon > 0$ such that:

$$S_\varepsilon(a) \subseteq O$$

> We remind you that
> $S_\varepsilon(a) = \{x \mid |x - a| < \varepsilon\}$

EXAMPLE 3.4 (a) Show that the interval $(1, 5)$ is open.

(b) Show that the interval $(1, 5]$ is not open.

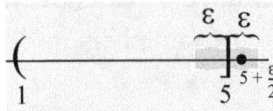

SOLUTION: (a) For any given $a \in (1, 5)$ we are to find $\varepsilon > 0$ such that $S_\varepsilon(a) \subseteq (1, 5)$. This is easy to do: take ε to be the smaller of the two numbers $a - 1$ and $5 - a$ (see margin).

Any smaller $\varepsilon > 0$ will do just as well; but no larger ε will work.

(b) For **any** given $\varepsilon > 0$, $5 + \frac{\varepsilon}{2} > 5$. It follows that there does **not** exist an $\varepsilon > 0$ such that $S_\varepsilon(5) \subseteq (1, 5]$.

CHECK YOUR UNDERSTANDING 3.15

(a) Show that $(a, b) = \{x | a < x < b\}$ is open for any $a < b$.

(b) Show that $(-\infty, a)$ and (a, ∞) are open for any $a \in \Re$.

(c) Show that no finite subset of $\Re$ is open.

Answer: See page A-16.

If you peek ahead to page 171 you will see that the following three properties of open sets are transformed into the three defining axioms of a topological space.

THEOREM 3.13 (i) $\Re$ and $\varnothing$ are open sets.

(ii) Arbitrary unions of open sets are open.

(iii) Finite intersections of open sets are open.

PROOF:

(i) $\Re$ is open: For any $x \in \Re$, $S_1(x) \subseteq \Re$.

$\varnothing$ is open (by default): For any $a \in \varnothing$ there exists an $\varepsilon > 0$ such that $S_\varepsilon(a) \subseteq \varnothing$ — for the simple reason that no such a exists.

(ii) Let $\{O_\alpha\}_{\alpha \in A}$ be a collection of open sets, and let $x \in \bigcup_{\alpha \in A} O_\alpha$.

See the indexing remarks that follow CYU 2.3, page 58.

Since x is in the union of the O_α's, there must exist some $\alpha_0 \in A$ such that $x \in O_{\alpha_0}$. Since O_{α_0} is open, we can choose $\varepsilon > 0$ such that $S_\varepsilon(x) \subseteq O_{\alpha_0}$. Then:

$$x \in S_\varepsilon(x) \subseteq O_{\alpha_0} \subseteq \bigcup_{\alpha \in A} O_\alpha$$

(iii) Let $\{O_i\}_{i=1}^n$ be a collection of open sets.

For $x \in \bigcap_{i=1}^n O_i$ and $1 \leq i \leq n$, choose ε_i such that $S_{\varepsilon_i}(x) \subseteq O_i$. and let $\varepsilon = \min\{\varepsilon_1, \varepsilon_2, \ldots, \varepsilon_n\}$. Since $S_\varepsilon(x) \subseteq O_i$ for each i,

$$x \in S_\varepsilon(x) \subseteq \bigcap_{i=1}^n O_i.$$

CHECK YOUR UNDERSTANDING 3.16

Answer: See page A-17.

Give an example illustrating the fact that an infinite intersection of open sets need not be open.

We remind you that for a given set A, A^c denotes the complement of A.

DEFINITION 3.10
CLOSED SET

A set $H \subseteq \Re$ is **closed** if its complement $H^c = \{x \in \Re | x \notin H\}$ is open.

EXAMPLE 3.5 (a) Show that $[1, 5] = \{x|(1 \leq x \leq 5)\}$ is closed.
(b) Show that $[1, \infty)$ is closed.
(c) Show that $(-\infty, 1] \cup (5, \infty)$ is not closed.

SOLUTION: (a) Since $[1, 5]^c = (-\infty, 1) \cup (5, \infty)$ is open [see CYU 3.15(b) and Theorem 3.13(ii)], $[1, 5]$ is closed.

(b) Since $[1, \infty)^c = (-\infty, 1)$ is open, $[1, \infty)$ is closed.

(c) Since $[(-\infty, 1] \cup (5, \infty)]^c = (1, 5]$ is not open [Example 3.4(b)], $(-\infty, 1] \cup (5, \infty)$ is not closed.

CHECK YOUR UNDERSTANDING 3.17

Answer: See page A-17.

(a) Show that $[a, b] = \{x | a < x < b\}$ is closed for any $a < b$.
(b) Is $(1, 3]$ closed? Justify your answer.

Here is the "closed-version" of Theorem 3.13:

THEOREM 3.14 (i) $\Re$ and $\varnothing$ are closed sets.

(ii) Any intersection of closed sets is again a closed set.

(iii) Any finite union of closed sets is again a closed set.

PROOF: (i) Since both $\Re^c = \varnothing$ and $\varnothing^c = \Re$ are open [Theorem 3.13(i)], both $\Re$ and $\varnothing$ are closed.

(ii) Let $\{H_\alpha\}_{\alpha \in A}$ be a collection of closed sets. Since:

$$\left(\bigcap_{\alpha \in A} H_\alpha\right)^c = \bigcup_{\alpha \in A} (H_\alpha)^c, \text{ and since each } (H_\alpha)^c \text{ is}$$

↑ Theorem 2.3(b), page 58

open (Definition 3.10), $\bigcap_{\alpha \in A} H_\alpha$ is closed [Theorem 3.13(ii) and Definition 3.10].

(iii) Let $\{H_i\}_{i=1}^n$ be a collection of closed sets. Since $\left(\bigcup_{i=1}^n H_i\right)^c = \bigcap_{i=1}^n (H_i)^c$ is open [Theorem 2.3(a), page 58], $\bigcup_{i=1}^n H_i$ is closed.

CHECK YOUR UNDERSTANDING 3.18

Answer: See page A-17.

Give an example illustrating the fact that the infinite union of closed sets need not be closed.

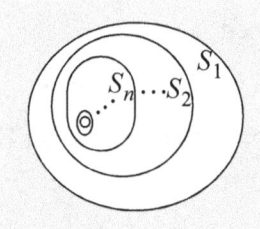

A sequence $(S_1, S_2, S_3, \ldots)$ of sets is said to be **nested** if $S_i \subseteq S_j$ for $i \geq j$ (see margin).

In the exercises you are asked to show that the intersection of a nested sequence of nonempty bounded sets may turn out to be empty, and that the intersection of a nested sequence of nonempty closed sets may also be empty. However:

This is a generalization of Theorem 3.3, page 115.

THEOREM 3.15 If $(H_i)_{i=1}^\infty$ is a nested sequence of **nonempty closed bounded** sets, then $\bigcap_{i=1}^\infty H_i \neq \emptyset$.

PROOF: For each i, select an element $a_i \in H_i$. The Bolzano-Weierstrass theorem of page 129 assures us that the bounded sequence $(a_i)_{i=1}^\infty$ contains a subsequence $(a_{i_k})_{k=1}^\infty$ that converges to some number α. We show that $\alpha \in H_i$ for every i:

Suppose, to the contrary, that $\alpha \notin H_{i_0}$ for some i_0. Since $\alpha \in (H_{i_0})^c$, and since $(H_{i_0})^c$ is open, we can find $\varepsilon > 0$ such that $S_\varepsilon(\alpha) \subseteq (H_{i_0})^c$. It follows that $S_\varepsilon(\alpha) \cap H_{i_0} = \emptyset$ (see margin). Since the H_i's are nested, $a_{i_k} \in H_{i_0}$ for every $i_k > i_0$, which implies that $S_\varepsilon(\alpha)$ contains only finitely many terms of the sequence $(a_{i_k})_{k=1}^\infty$, contradicting the assumption that $(a_{i_k})_{k=1}^\infty$ converges to α (see Exercise 38, page 133).

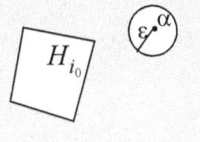

DEFINITION 3.11
OPEN COVER
$\{O_\alpha\}_{\alpha \in A}$ is an **open cover** of $S \subseteq \Re$ if each O_α is open and $S \subseteq \bigcup_{\alpha \in A} O_\alpha$.

FINITE SUBCOVER
An open cover $\{O_\alpha\}_{\alpha \in A}$ of S is said to contain a **finite subcover** if there exists $\{\alpha_1, \alpha_2, \ldots, \alpha_n\} \subseteq A$ such that $S \subseteq \bigcup_{i=1}^{n} O_{\alpha_i}$.

COMPACT
A subset K of $\Re$ is said to be **compact** if every open cover of K has a finite subcover.

EXAMPLE 3.6
(a) Show that no unbounded subset of $\Re$ is compact.

(b) Show that $(-1, 1)$ is not compact.

SOLUTION:

(a) Let B be an unbounded subset of $\Re$. Consider the open cover $\{I_n\}_{n=1}^{\infty}$ of B, where $I_n = (-n, n)$. Since $n_i < n_j$ implies that $I_{n_i} \subset I_{n_j}$, for any N: $\bigcup_{i=1}^{N} I_{n_i} = I_{n_N}$. Being unbounded, $B \not\subset I_{n_N}$ for any N, and is therefore not contained in $\bigcup_{i=1}^{N} I_{n_i}$ for any N.

(b) For each integer $n \geq 2$, let $I_n = \left(-1 + \frac{1}{n}, 1 - \frac{1}{n}\right)$. In the exercises you are asked to show that $\{I_n\}_{n=2}^{\infty}$ is an open cover of $(-1, 1)$. That cover has no finite subcover since, for any given N, $1 - \frac{1}{N} \in (-1, 1)$ and $1 - \frac{1}{N} \notin \bigcup_{n=2}^{N} I_n$.

The following result tells us that to challenge the compactness of $S \subseteq \Re$ one need only consider **countable** open covers of S.

> This theorem does not hold in a general topological space.

THEOREM 3.16 $S \subseteq \Re$ is compact if and only if every countable open cover of S has a finite subcover.

PROOF: If S is compact, then every open cover of S has a finite subcover. In particular, every countable open cover of S has a finite subcover.

For the converse, assume that every countable open cover of S has a finite subcover, and let $\{O_\alpha\}_{\alpha \in A}$ be an **arbitrary** open cover of S. We show that $\{O_\alpha\}_{\alpha \in A}$ contains a finite subcover:

For each $x \in S$, choose $O_{\alpha_x} \in \{O_\alpha\}_{\alpha \in A}$ such that $x \in O_{\alpha_x}$. Since O_{α_x} is open, and since the rationals are dense in $\Re$, we can find an open interval I_x with rational endpoints such that $x \in I_x \subseteq O_{\alpha_x}$ (see margin). Since there are but a countable number of intervals with rational endpoints (Exercise 27, page 86), $\{I_x\}_{x \in S}$ is a countable open cover of S. By our assumption, a finite number of those intervals, $I_{x_1}, I_{x_2}, \ldots, I_{x_n}$ cover S. It follows, since $I_{x_i} \subseteq O_{\alpha_{x_i}}$, that $O_{\alpha_{x_1}}, O_{\alpha_{x_2}}, \ldots, O_{\alpha_{x_n}}$ also covers S.

For $S_\varepsilon(x) \subseteq O_{\alpha_x}$, choose rational numbers r_1 and r_2 such that:
$x - \varepsilon < r_1 < x < r_2 < x + \varepsilon$
Then:
$x \in (r_1, r_2) \subseteq O_{\alpha_x}$

Heinrich Heine (1821-1881). Emile Borel (1871-1958)

THEOREM 3.17 Every closed bounded subset of $\Re$ is compact.
HEINE-BOREL

PROOF: Let S be a closed bounded subset of $\Re$. Employing Theorem 3.16 we establish compactness by showing that every countable open cover $\{O_n\}_{n=1}^\infty$ of S has a finite subcover.

Assume there exists a countable open cover $\{O_n\}_{n=1}^\infty$ of S containing no finite subcover. To arrive at a contradiction, consider the nested sequence $\{H_i\}_{i=1}^\infty$ of nonempty closed bounded sets, where (see margin):

$$(*)\quad H_i = S \cap \left(\bigcup_{n=1}^{i} O_n\right)^c \quad(*)$$

the intersection of two closed sets ← → the complement of an open set

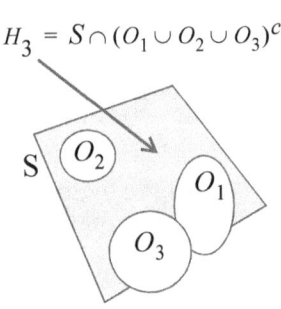

$H_3 = S \cap (O_1 \cup O_2 \cup O_3)^c$

Choose $x \in \bigcap H_i$ (see Theorem 3.15). Since each $H_i \subseteq S$, $x \in S$. Moreover, since x is contained in each H_i, x cannot be contained in any O_i [note the complement operator in (*)], contradicting the assumption that $\{O_n\}_{n=1}^\infty$ is an open cover of S.

The converse of Theorem 3.17 also holds:

THEOREM 3.18 Every compact subset of $\Re$ is closed and bounded.

PROOF: Let $K \subseteq \Re$ be compact. We observed, in Example 3.6(a), that K must be bounded, and now establish that K is closed by showing that its complement is open:

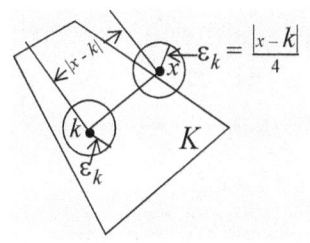

Let $x \in K^c$. For each $k \in K$ let $\varepsilon_k = \dfrac{|x-k|}{4}$ (see margin). Since K is compact, the open cover $\{S_{\varepsilon_k}(k)\}_{k \in K}$ has a finite subcover, say: $\{S_{\varepsilon_{k_1}}(k_1), S_{\varepsilon_{k_2}}(k_2), \ldots, S_{\varepsilon_{k_n}}(k_n)\}$. Noting that every element of K is contained in some $S_{\varepsilon_{k_i}}(k_i)$, and that $S_{\varepsilon_{k_i}}(k_i) \cap S_{\varepsilon_{k_i}}(x) = \varnothing$, we see that $x \in \underset{i=1}{\bigcap} S_{\varepsilon_{k_i}}(x) \subseteq K^c$.

↑— an open set

Combining Theorems 3.17 and 3.18 we have:

THEOREM 3.19 A subset of $\Re$ is compact if and only if it is both closed and bounded.

CHECK YOUR UNDERSTANDING 3.19

Determine if the given subset of $\Re$ is compact. Justify your answer.

(a) $\left\{ \dfrac{1}{n} \right\}_{n=1}^{\infty}$ (b) $\{0\} \cup \left\{ \dfrac{1}{n} \right\}_{n=1}^{\infty}$

(a) No   (b) Yes

In the real number system, one can also take a sequential approach to compactness, namely:

THEOREM 3.20 $K \subseteq \Re$ is compact if and only if every sequence in K contains a subsequence that converges to a point in K.

PROOF: Assume that K is compact and let $(k_n)_{n=1}^{\infty}$ be a sequence with each $k_n \in K$. Since K is bounded (Theorem 3.18) the sequence contains a convergent subsequence $(k_{n_i})_{i=1}^{\infty}$ (Theorem 3.10, page 129). Can the limit of that subsequence lie outside of K? No:

Let $a \in K^c$. Since K is closed (Theorem 3.18), K^c is an open set containing a which contains no element of the $(k_{n_i})_{i=1}^{\infty}$. It follows that a is **not** the limit of $(k_{n_i})_{i=1}^{\infty}$.

Conversely, assume that every sequence in K contains a subsequence that converges to a point in K. We show that K must be bounded and closed, and therefore compact (Theorem 3.18).

Assume that K is not bounded (we will arrive at a contradiction).

Choose $k_1 \in K$. Since K is not bounded, we can choose $k_2 \in K$ such that $|k_2 - k_1| > 1$, and $k_3 \in K$ such that $|k_3 - k_1| > 1$ and $|k_3 - k_2| > 1$, etc. Having chosen $k_1, k_2, k_3, \ldots, k_n$ with no two of the numbers within one unit of each other, we can still find a number k_{n+1} that is more than one unit from any of its predecessors (the finite set $\{k_1, k_2, k_3, \ldots, k_n\}$ is bounded, and K is not). The constructed sequence $(k_n)_{n=1}^{\infty}$ has no convergent subsequence since, for any $a \in \Re$, the open sphere $S_1(a)$ can contain at most one element of the sequence.

Assume that K is not closed (we will arrive at a contradiction).

Since K^c is not open, we can choose $a \in K^c$ such that $S_\varepsilon(a) \cap K \neq \varnothing$ for all $\varepsilon > 0$. In particular, for any $n \in Z^+$ we can choose $k_n \in K$ such that $|k_n - a| < \frac{1}{n}$. The sequence $(k_n)_{n=1}^{\infty}$ cannot contain a subsequence converging to a point in K, as by construction it converges to a which lies outside of K.

CHECK YOUR UNDERSTANDING 3.20

Prove that $K \subseteq \Re$ is compact if and only if every Cauchy sequence in K converges to a point in K.

Answer: See page A-17.

	EXERCISES	

Exercises 1-9. Determine if the given subset of $\Re$ is:

(a) Open, closed, or neither.   (b) Bounded above, below, or neither.   (c) Compact.

1. $(0, 1) \cup [1, 2)$ 2. $[0, 2) \cup (2, 3)$ 3. $[0, 2) \cap (2, 3)$

4. Z^+ 5. $\left\{ \dfrac{1}{n} \,\middle|\, n \in Z^+ \right\}$ 6. $\{0\} \cup \left\{ \dfrac{1}{n} \,\middle|\, n \in Z^+ \right\}$

7. $[0, 2] \cup \{3\}$ 8. Q (the set of rational numbers)   9. Q^c (the set of irrational numbers)

10. Prove that for any two distinct real numbers, x and y, there exist open sets O_x and O_y such that $x \in O_x$, $y \in O_y$, and $O_x \cap O_y = \emptyset$.

11. Show, by means of an example, that the intersection of a nested sequence of nonempty bounded sets may turn out to be empty.

12. Show, by means of an example, that the intersection of a nested sequence of nonempty closed sets may be empty.

13. (a) Prove that every finite subset of $\Re$ is compact.

    (b) Exhibit a countable subset of $\Re$ that is not compact.

14. Give an example of a nested sequence of nonempty open sets $O_1 \supseteq O_2 \supseteq O_3 \supseteq \cdots$ for which:

    (a) $\bigcap\limits_{n=1}^{\infty} O_n = \emptyset$.   (b) $\bigcap\limits_{n=1}^{\infty} O_n$ is closed and nonempty.

15. Show that $\left\{ \left(-1 + \dfrac{1}{n}, 1 - \dfrac{1}{n} \right) \right\}_{n=2}^{\infty}$ is an open cover of $(-1, 1)$

16. Construct a cover (necessarily not open) of the compact set $[0, 1]$ which does not contain a countable subcover.

17. Prove that if $K_1 \supseteq K_2 \supseteq 3 \supseteq \ldots$ is a nested sequence of nonempty compact sets, then

    $\bigcap\limits_{n=1}^{\infty} K_n \neq \emptyset$.

18. Prove that if K is compact, then lub K and glb K exist and are contained in K.

19. (a) Prove that if K is compact, and if C is closed, then $K \cap C$ is compact.
    (b) Show, by means of an example, that (a) need not hold if K is not compact.
    (b) Show, by means of an example, that (a) need not hold if C is not closed.

20. (a) Let B be a closed subset of $\Re$. Let x be such that $S_\varepsilon(x) \cap B \neq \emptyset$ for every $\varepsilon > 0$. Prove that $x \in B$.
    (b) Show, by means of an example, that (a) need not hold if B is not closed.

21. (a) Let O be an open subset of $\Re$. Prove that if $\lim_{n \to \infty} a_n = \alpha \in O$, then all but a finite number of the terms in $(a_n)_{n=1}^{\infty}$ are contained in O.
    (b) Show, by means of an example, that (a) need not hold if O is not open.

22. Let S be a subset of $\Re$, bounded below. Let glb $S \in O$, with O open. Show that $O \cap S \neq \emptyset$.

23. Let S be a subset of $\Re$, bounded above. Let lub $S \in O$, with O open. Show that $O \cap S \neq \emptyset$.

24. Prove that if $\lim_{n \to \infty} a_n = \alpha$, then $\{a_n\}_{n=1}^{\infty} \cup \{\alpha\}$ is compact.

25. Let K_1 and K_2 be compact subsets of $\Re$ with $K_1 \cap K_2 = \emptyset$. Prove that there exist open sets O_1 and O_2 with $K_1 \subset O_1$, $K_2 \subset O_2$, and $O_1 \cap O_2 = \emptyset$.

26. Prove that $\emptyset$ and $\Re$ are the only subsets of $\Re$ that are both open and closed.

PROVE OR GIVE A COUNTEREXAMPLE

27. Every open cover of any subset S of $\Re$ contains a countable subcover.

28. If O is an open subset of $\Re$ which contains every rational number, then $O = \Re$.

29. If O is an open subset of $\Re$ which contains every integer, then $O = \Re$.

30. If B is a closed subset of $\Re$ which contains every rational number, then $B = \Re$.

31. If the sequence $(a_n)_{n=1}^{\infty}$ converges, then the set $\{a_n\}_{n=1}^{\infty}$ is compact.

32. If the sequence $(a_n)_{n=1}^{\infty}$ converges to α, then the set $\{\alpha\} \cup \{a_n\}_{n=1}^{\infty}$ is compact.

33. If the sequence $(a_n)_{n=1}^{\infty}$ diverges, then the set $\{a_n\}_{n=1}^{\infty}$ cannot be compact.

34. Let B be a closed subset of $\Re$. Prove that if the sequence $\{a_n\}$ has a subsequence converging to α, then $\alpha \in B$.

35. If $A \cap B$ is compact, than either A or B is compact.

36. A finite intersection of compact sets is again compact.

37. An arbitrary intersection of compact sets is again compact.

38. If $A \cup B$ is compact, than either A or B is compact.

39. If $A \cup B$ is compact, than both A are B are compact.

40. A finite union of compact sets is again compact.

41. An arbitrary union of compact sets is again compact.

42. Let B be a closed subset of $\Re$. Prove that if the sequence $\{a_n\}$ has a subsequence converging to α, then $\alpha \in B$.

43. If S is bounded below and if glb $S \in K$, with K compact, then $O \cap S \neq \emptyset$.

44. If S_1 and S_2 are subsets of $\Re$ for which $S_1 \cup S_2$ is open, then S_1 and S_2 must both be open.

45. If S_1 and S_2 are subsets of $\Re$ for which $S_1 \cup S_2$ is open, then S_1 or S_2 must be open.

46. If S_1 and S_2 are subsets of $\Re$ for which $S_1 \cap S_2$ is open, then S_1 and S_2 must both be open.

47. If S_1 and S_2 are subsets of $\Re$ for which $S_1 \cap S_2$ is open, then S_1 or S_2 must be open.

48. If S_1 and S_2 are subsets of $\Re$ for which $S_1 \cup S_2$ is closed, then S_1 and S_2 must both be closed.

49. If S_1 and S_2 are subsets of $\Re$ for which $S_1 \cup S_2$ is closed, then S_1 or S_2 must be closed.

50. If S_1 and S_2 are subsets of $\Re$ for which $S_1 \cap S_2$ is closed, then S_1 and S_2 must both be closed.

51. If S_1 and S_2 are subsets of $\Re$ for which $S_1 \cap S_2$ is closed, then S_1 or S_2 must be closed.

52. If O_1 and O_2 are open subsets of $\Re$, then $O_1 \cup O_2$ is also open.

53. If H_1 and H_2 are closed subsets of $\Re$, then $H_1 \cup H_2$ is also closed.

54. For any $n \in Z^+$, if $\{O_i\}_{i=1}^{n}$ is a collection of open subsets of $\Re$, then $\bigcup_{i=1}^{n} O_i$ is also open.

55. For any $n \in Z^+$, if $\{H_i\}_{i=1}^n$ is a collection of closed subsets of $\Re$, then $\bigcup_{i=1}^n H_i$ is also closed.

56. If $\{O_i\}_{i=1}^\infty$ is a collection of open subsets of $\Re$, then $\bigcup_{i=1}^\infty O_i$ is also open.

57. If $\{H_i\}_{i=1}^\infty$ is a collection of closed subsets of $\Re$, then $\bigcup_{i=1}^\infty H_i$ is also closed.

58. If O_1 and O_2 are open subsets of $\Re$, then $O_1 \cap O_2$ is also open.

59. If H_1 and H_2 are closed subsets of $\Re$, then $H_1 \cap H_2$ is also closed.

60. For any $n \in Z^+$, if $\{O_i\}_{i=1}^n$ is a collection of open subsets of $\Re$, then $\bigcap_{i=1}^n O_i$ is also open.

61. For any $n \in Z^+$, if $\{H_i\}_{i=1}^n$ is a collection of closed subsets of $\Re$, then $\bigcap_{i=1}^n H_i$ is also closed.

62. If $\{O_i\}_{i=1}^\infty$ is a collection of open subsets of $\Re$, then $\bigcap_{i=1}^\infty O_i$ is also open.

63. If $\{H_i\}_{i=1}^\infty$ is a collection of closed subsets of $\Re$, then $\bigcap_{i=1}^\infty H_i$ is also closed.

§4. CONTINUITY

Let's begin with a formal definition:

DEFINITION 3.12
CONTINUITY AT A POINT

A function $f: D \to \Re$ is **continuous at** $c \in D$ if for any given $\varepsilon > 0$ there exists a $\delta > 0$ such that:

$$|x - c| < \delta \Rightarrow |f(x) - f(c)| < \varepsilon$$
(with $x \in D$)

> The Greek letter δ is pronounced "delta." Note the similarities between this definition and Definition 3.5, page 123.

> In a calculus course continuity is typically defined in terms of the limit concept. Specifically:
> f is continuous at c if
> $$\lim_{x \to c} f(x) = g(c)$$

What a beautiful definition! Here is what it is saying:

The distance between $f(x)$ and $f(c)$ can be made as **small** as you wish

$$\underbrace{|x - c| < \delta}_{\text{providing } x \text{ is sufficiently close to } c} \Rightarrow \underbrace{|f(x) - f(c)| < \varepsilon}$$

From a geometrical point of view:

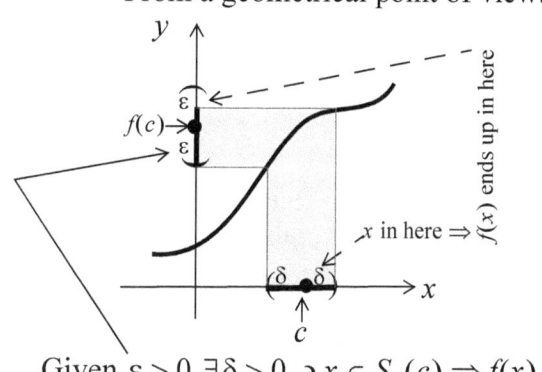

Given $\varepsilon > 0 \ \exists \delta > 0 \ \ni x \in S_\delta(c) \Rightarrow f(x) \in S_\varepsilon(f(c))$

> If a particular δ "works" for a given ε, then any smaller δ will also work for that ε. However, a smaller ε, may call for a smaller δ.

EXAMPLE 3.7 Show that the function $f(x) = 2x + 5$ is continuous at $x = 3$.

SOLUTION: For a given $\varepsilon > 0$ we are to find $\delta > 0$ such that:

$$|x - 3| < \delta \Rightarrow |f(x) - f(3)| < \varepsilon$$
i.e: $|x - 3| < \delta \Rightarrow |(2x + 5) - 11| < \varepsilon$
$|x - 3| < \delta \Rightarrow |2x - 6| < \varepsilon$
$|x - 3| < \delta \Rightarrow 2|x - 3| < \varepsilon$
$|x - 3| < \delta \Rightarrow |x - 3| < \frac{\varepsilon}{2}$

(same)

> Compare with Example 3.3(a), page 124.

While a choice of δ for which $|x - 3| \leq \delta \Rightarrow |(2x + 5) - f(3)| < \varepsilon$ may not be so apparent, it is trivial to find a δ that works if the task's rewritten form $|x - 3| < \delta \Rightarrow |x - 3| < \frac{\varepsilon}{2}$, namely: $\delta = \frac{\varepsilon}{2}$.

Answer: See page A-17.

CHECK YOUR UNDERSTANDING 3.21

Prove that the function $f(x) = 5x + 1$ is continuous at $x = 2$.

EXAMPLE 3.8 Show that the function $f(x) = x^2$ is continuous at $x = 3$.

SOLUTION: For a given $\varepsilon > 0$ we are to find $\delta > 0$ such that:

$$|x - 3| < \delta \Rightarrow |x^2 - 9| < \varepsilon$$
$$|x - 3| < \delta \Rightarrow |(x + 3)(x - 3)| < \varepsilon$$
$$|x - 3| < \delta \Rightarrow |x + 3||x - 3| < \varepsilon$$

The proof will be complete once we find a $\delta > 0$ for which:

$$|x - 3| < \delta \Rightarrow |x + 3||x - 3| < \varepsilon \quad (*)$$

While it is tempting to choose $\delta = \dfrac{\varepsilon}{|x + 3|}$, that temptation must be suppressed, for δ has to be a positive number and not a function of x. Since we are interested in what happens near $x = 3$, we decide to focus on the interval:

$$(2, 4) = \{x \mid |x - 3| < 1\}$$

Within that interval $|x + 3| < 7$ (see margin). Consequently, within that interval:

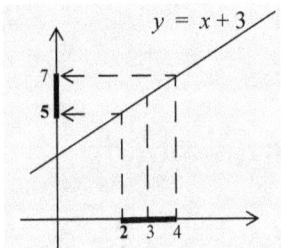

$$|x + 3||x - 3| < 7|x - 3|$$

Taking δ to be the smaller of the two numbers 1 and $\dfrac{\varepsilon}{7}$ [written: $\delta = \min\left(1, \dfrac{\varepsilon}{7}\right)$], we are assured that both $|x + 3| < 7$ and that $\delta \le \dfrac{\varepsilon}{7}$, and this enables us to meet our goal:

$$|x - 3| < \delta \Rightarrow |x^2 - 9| = |x + 3||x - 3| < 7 \cdot \dfrac{\varepsilon}{7} = \varepsilon$$

CHECK YOUR UNDERSTANDING 3.22

Answer: See page A-18.

Show that the function $f(x) = x^2$ is continuous at $x = 2$

DEFINITION 3.13 A **continuous function** is a function that is
CONTINUOUS FUNCTION continuous at every point in its domain.

EXAMPLE 3.9
Show that the function $f(x) = \sqrt{x}$ is continuous.

SOLUTION: Let $x_0 \in D_f$ (which is to say: $x_0 \geq 0$), and let $\varepsilon > 0$ be given. We are to find $\delta > 0$ such that:
$$|x - x_0| < \delta \Rightarrow |\sqrt{x} - \sqrt{x_0}| < \varepsilon$$
i.e: $\quad |\sqrt{x} + \sqrt{x_0}||\sqrt{x} - \sqrt{x_0}| < \delta \Rightarrow |\sqrt{x} - \sqrt{x_0}| < \varepsilon$ (*)

Assume that $x_0 > 0$. Since we are interested in what happens near x_0, we decide to focus on the interval $\left(x_0 - \frac{x_0}{2}, x_0 + \frac{x_0}{2}\right) = \left(\frac{x_0}{2}, \frac{3x_0}{2}\right)$.

Within that interval (see margin):
$$|\sqrt{x} + \sqrt{x_0}| > \sqrt{\frac{x_0}{2}} + \sqrt{x_0} > \frac{\sqrt{x_0}}{2} + \sqrt{x_0} = \frac{3}{2}\sqrt{x_0}$$

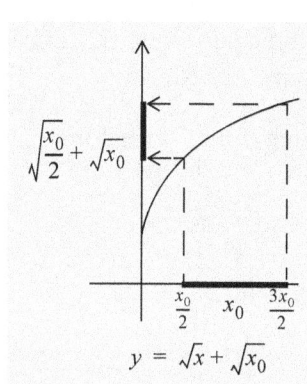
$y = \sqrt{x} + \sqrt{x_0}$

Letting $\delta = \min\left[\frac{x_0}{2}, \frac{3}{2}\sqrt{x_0}\varepsilon\right]$, we see that (*) is satisfied:
$$|\sqrt{x} + \sqrt{x_0}||\sqrt{x} - \sqrt{x_0}| < \delta \Rightarrow |\sqrt{x} + \sqrt{x_0}||\sqrt{x} - \sqrt{x_0}| < \frac{3}{2}\sqrt{x_0}\varepsilon$$
$$\Rightarrow |\sqrt{x} - \sqrt{x_0}| < \frac{\frac{3}{2}\sqrt{x_0}\varepsilon}{|\sqrt{x} + \sqrt{x_0}|} < \frac{\frac{3}{2}\sqrt{x_0}\varepsilon}{\frac{3}{2}\sqrt{x_0}} = \varepsilon$$

As for the $x_0 = 0$ case:

CHECK YOUR UNDERSTANDING 3.23

(a) Show that $f(x) = \sqrt{x}$ is continuous at $x = 0$.

(b) Show that $f(x) = x^2$ is a continuous function.

Answer: See page A-18.

Roughly speaking, a function is continuous if you can sketch its graph without lifting your writing utensil. In particular, we now show that the function $f(x) = \begin{cases} 2x & \text{if } x \leq 1 \\ x^2 + 3 & \text{if } x > 1 \end{cases}$ (margin) is not continuous at $x = 1$:

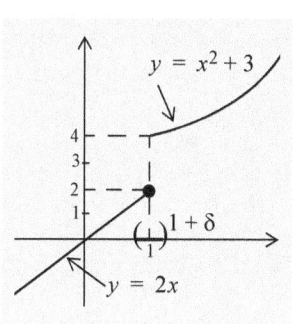

Let $\varepsilon = 1$ and let δ be ANY positive number whatsoever. Take any $x \in (1, 1 + \delta)$. Clearly $|x - 1| < \delta$, but:
$$|f(x) - f(1)| = |(x^2 + 3) - (2 \cdot 1)| = x^2 + 1 > 1 = \varepsilon$$

The above argument shows that there does not exist a $\delta > 0$ such that $|x - 1| < \delta \Rightarrow |f(x) - f(1)| < 1$. Since the continuity challenge fails for $\varepsilon = 1$ (as it would for any $0 < \varepsilon \leq 2$), f is not continuous at 1, and is therefore not a continuous function.

> **CHECK YOUR UNDERSTANDING 3.24**
>
> Prove that the function $f(x) = \begin{cases} x & \text{if } x \neq 1 \\ \dfrac{999}{1000} & \text{if } x = 1 \end{cases}$ is not continuous.

Answer: See page A-18.

THE ALGEBRA OF FUNCTIONS

Just as you can perform algebraic operations on numbers to arrive at other numbers, so then can one perform algebraic operations on functions to arrive at other functions, and in a most natural fashion:

DEFINITION 3.14 For real-valued functions f and g with domains D_f and D_g respectively, we define the functions $f+g$, $f-g$, fg, and $\dfrac{f}{g}$ as follows:

SUM OF FUNCTIONS

$f + g \colon D_f \cap D_g \to \Re$ with:
$$(f+g)(x) = f(x) + g(x)$$

DIFFERENCE OF FUNCTIONS

$f - g \colon D_f \cap D_g \to \Re$ with:
$$(f-g)(x) = f(x) - g(x)$$

PRODUCT OF FUNCTIONS

$fg \colon D_f \cap D_g \to \Re$ with:
$$(fg)(x) = f(x)g(x)$$

QUOTIENT OF FUNCTIONS

$\dfrac{f}{g} \colon D_f \cap D_g - \{x \mid g(x) \neq 0\}$ with:
$$\left(\frac{f}{g}\right)(x) = \frac{f(x)}{g(x)}$$

Continuity at a point is preserved under the above operations:

THEOREM 3.21 If f and g are continuous at c then so are the functions:

(a) $f + g$ (b) $f - g$

(c) fg (d) $\dfrac{f}{g}$ (if $g(c) \neq 0$)

Consequently: If f and g are continuous, then so are the functions
$f+g$, $f-g$, fg, and $\dfrac{f}{g}$
for $(g \neq 0)$

PROOF: We establish (a) and (c). You are invited to establish (b) in CYU 3.25 below, and (d) in the exercises.

Throughout this development we are assuming that the variable x is contained in the domain of both f and g.

(a) For a given $\varepsilon > 0$ we are to find $\delta > 0$ such that:
$$|x-c| < \delta \Rightarrow |(f+g)(x) - (f+g)(c)| < \varepsilon$$

We begin by manipulating $|(f+g)(x) - (f+g)(c)| < \varepsilon$ into a form which displays the expressions $|f(x) - f(c)|$ and $|g(x) - g(c)|$, as the continuity of f and g at c enables us to make those two expressions as small as we wish:

$$|x-c| < \delta \Rightarrow |(f+g)(x) - (f+g)(c)| < \varepsilon$$
$$|x-c| < \delta \Rightarrow |f(x) + g(x) - f(c) - g(c)| < \varepsilon$$
$$|x-c| < \delta \Rightarrow |[f(x) - f(c)] + [g(x) - g(c)]| < \varepsilon \quad (*)$$

Since $|[f(x) - f(c)] + [g(x) - g(c)]| \leq |f(x) - f(c)| + |g(x) - g(c)|$, We now set our sights on finding a $\delta > 0$ for which $|x-c| < \delta$ implies that both $|f(x) - f(c)| < \frac{\varepsilon}{2}$ and $|g(x) - g(c)| < \frac{\varepsilon}{2}$:

Since f and g are continuous at c, there exist $\delta_1 > 0$ such that $|x-c| < \delta_1 \Rightarrow |f(x) - f(c)| < \frac{\varepsilon}{2}$ and $\delta_2 > 0$ such that $|x-c| < \delta_2 \Rightarrow |g(x) - g(c)| < \frac{\varepsilon}{2}$. It follows that (*) holds for $\delta = \min(\delta_1, \delta_2)$:

$$|x-c| < \delta \Rightarrow |[f(x) - f(c)] + [g(x) - g(c)]|$$
$$\leq |f(x) - f(c)| + |g(x) - g(c)| < \frac{\varepsilon}{2} + \frac{\varepsilon}{2} = \varepsilon$$

(c) For a given $\varepsilon > 0$ we are to find $\delta > 0$ such that
$$|x-c| < \delta \Rightarrow |(fg)(x) - (fg)(c)| < \varepsilon$$

Note the similarity between this proof and that of the sequence version of Theorem 3.9(c), page 126.

Again, we somehow need to work $|f(x) - f(c)|$ and $|g(x) - g(c)|$ into the picture, and do so by introducing the **clever zero**, $-f(x)g(c) + f(x)g(c)$ within the expression $|(fg)(x) - (fg)(c)|$:

$$|x-c| < \delta \Rightarrow |(fg)(x) - (fg)(c)| < \varepsilon$$
$$|x-c| < \delta \Rightarrow |f(x)g(x) - f(c)g(c)| < \varepsilon$$
$$|x-c| < \delta \Rightarrow |f(x)g(x) - \boldsymbol{f(x)g(c)} + \boldsymbol{f(x)g(c)} - f(c)g(c)| < \varepsilon$$
$$|x-c| < \delta \Rightarrow |f(x)[g(x) - g(c)] + g(c)[f(x) - f(c)]| < \varepsilon \quad (*)$$

Noting that: $|f(x)[g(x) - g(c)] + g(c)[f(x) - f(c)]|$
$$\leq |f(x)||g(x) - g(c)| + |g(c)||f(x) - f(c)|$$

we set our sights on finding a $\delta > 0$ such that $|x-c| < \delta$ implies that:

(i) $|f(x)||g(x) - g(c)| < \frac{\varepsilon}{2}$ and (ii) $|g(c)||f(x) - f(c)| < \frac{\varepsilon}{2}$.

For (i): Since f is continuous at c, we can choose δ_f such that:

$$|x-c| < \delta_f \Rightarrow |f(x)-f(c)| < 1$$

could choose any positive number

Since $|f(x)-f(c)| \geq |f(x)| - |f(c)|$ (CYU 3.14, page 135):

$$|f(x)-f(c)| < 1 \Rightarrow |f(x)| - |f(c)| < 1 \Rightarrow |f(x)| < 1 + |f(c)| \quad (**)$$

Since g is continuous at c, we can choose $\delta_g > 0$ such that:

$$|x-c| \leq \delta_g \Rightarrow |g(x)-g(c)| < \frac{\varepsilon}{2(|f(c)|+1)} \quad (***)$$

can't be zero

Letting $\bar{\delta} = \min(\delta_f, \delta_g)$, we see that (i) is satisfied:

$$|x-c| \leq \bar{\delta} \Rightarrow |f(x)||g(x)-g(c)| < [1+|f(c)|] \cdot \frac{\varepsilon}{2(|f(c)|+1)} = \frac{\varepsilon}{2}$$

For (ii): $|g(c)||f(x)-f(c)| < \frac{\varepsilon}{2}$.

> We use $|g(c)|+1$ instead of simply $|g(c)|$ in the denominator since $|g(c)|$ might be zero.

Choose $\underline{\delta} > 0$ such that:

$$|x-c| < \underline{\delta} \Rightarrow |f(x)-f(c)| < \frac{\varepsilon}{2(|g(c)|+1)} \leftarrow \text{(margin)}$$

Then:

$$|x-c| < \underline{\delta} \Rightarrow |g(c)||f(x)-f(c)| < |g(c)| \cdot \frac{\varepsilon}{2(|g(c)|+1)} < \frac{\varepsilon}{2}$$

$|g(c)| < |g(c)| + 1$

End result: For $\delta = \min(\bar{\delta}, \underline{\delta})$, (*) is satisfied.

CHECK YOUR UNDERSTANDING 3.25

Prove Theorem 3.19(b)

Answer: See page A-19.

CONTINUITY, FROM A TOPOLOGICAL POINT OF VIEW

The abstract setting for the concept of continuity resides in the field of Topology, which is introduced in the next chapter. To provide a bridge to that field we now focus our attention on real-valued functions with domain all of $\Re$.

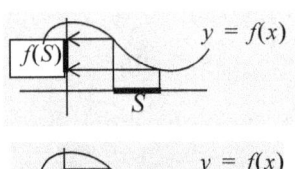

DEFINITION 3.15 For $f: \Re \to \Re$, and $S \subseteq \Re$:

IMAGE OF S The **image of S under f**, denoted by $f(S)$, is the set $f(S) = \{f(x) | x \in S\}$.

PRE-IMAGE OF S The **pre-image of S under f**, denoted by $f^{-1}(S)$, is the set $f^{-1}(S) = \{x \in X | f(x) \in S\}$.

Using the above notation we can restate Definition 3.12 as follows:

> A function $f\colon \Re \to \Re$ is **continuous at** c if for any given $\varepsilon > 0$ there exists a $\delta > 0$ such that $f[S_\delta(c)] \subseteq S_\varepsilon[f(c)]$

The following result will enable us, in the next chapter, to extend the concept of continuity to topological spaces:

THEOREM 3.22 A function $f\colon \Re \to \Re$ is continuous if and only if:
$$f^{-1}(O) \text{ is open } \forall \text{ open set } O$$

PROOF: Assume that $f\colon \Re \to \Re$ is continuous, and let O be open. We establish the fact that $f^{-1}(O)$ is open by showing that:

> For any given $c \in f^{-1}(O)$ there exists $\varepsilon > 0$ such that $S_\varepsilon(c) \subseteq f^{-1}(O)$. (*)

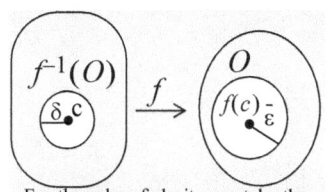

For the sake of clarity, we take the liberty of representing subsets of $\Re$ as two-dimensional objects.

Since $f(c) \in O$ and since O is open, there exists $\bar\varepsilon > 0$ such that $S_{\bar\varepsilon}[f(c)] \subseteq O$. Since f is continuous, we can choose $\delta > 0$ such that: $f[S_\delta(c)] \subseteq S_{\bar\varepsilon}[f(c)] \subseteq O$ (see margin). It follows that $S_\delta(c) \subseteq f^{-1}(O)$.

We found a $\delta > 0$ for which $S_\delta(c) \subseteq f^{-1}(O)$. Replacing δ with the symbol "ε" (just for the sake of appearance), we see that (*) is established.

Conversely, assume that for every open set O, $f^{-1}(O)$ is open. We show that f is continuous at $c \in \Re$ by showing that for given $\varepsilon > 0$ there exists a $\delta > 0$ such that $f[S_\delta(c)] \subseteq S_\varepsilon[f(c)]$:

Since $S_\varepsilon[f(c)]$ is open, $f^{-1}(S_\varepsilon[f(c)])$ is also open, and it clearly contains c. That being the case, we can choose $\delta > 0$ such that $S_\delta(c) \subseteq f^{-1}(S_\varepsilon[f(c)])$. It follows that: $f[S_\delta(c)] \subseteq S_\varepsilon[f(c)]$ (see Exercise 24).

We take advantage of Theorem 3.22 to establish the following important result:.

See Definition 2.9, page 64.

THEOREM 3.23 If $f\colon \Re \to \Re$ and $g\colon \Re \to \Re$ are continuous, then the composite function $g \circ f\colon \Re \to \Re$ is also continuous.

PROOF: Let O be open in $\Re$. Since g is continuous, $g^{-1}(O)$ is open. Since f is continuous, $f^{-1}[g^{-1}(O)]$ is also open. The desired result now follows from Theorem 2.6, page 73 which asserts that $(g \circ f)^{-1}(O) = f^{-1}[g^{-1}(O)]$.

CHECK YOUR UNDERSTANDING 3.26

Prove Theorem 3.23 using Definition 3.12 directly.

Answer: See page A-19.

EXERCISES

Exercises 1-8. Prove that the given function is continuous at the given point.

1. $f(x) = -5x + 3$ at $x = 4$
2. $f(x) = \frac{1}{2}x + 3$ at $x = -1$
3. $f(x) = 3x^2 + 1$ at $x = 2$
4. $f(x) = 2x^2 - x$ at $x = 1$
5. $f(x) = \frac{1}{2x}$ at $x = 5$
6. $f(x) = \frac{3}{2x + 1}$ at $x = 0$
7. $f(x) = \frac{x}{2x + 1}$ at $x = 0$
8. $f(x) = 3\sqrt{2x - 2}$ at $x = 8$

Exercises 9-12. Prove that the given function is not continuous at the given point.

9. $f(x) = \begin{cases} 2x + 3 & \text{if } x \leq 0 \\ -x + 4 & \text{if } x < 0 \end{cases}$ at $x = 0$

10. $f(x) = \begin{cases} x^2 & \text{if } x \leq 1 \\ x + 1 & \text{if } x < 1 \end{cases}$ at $x = 1$

11. $f(x) = \begin{cases} 2x + 3 & \text{if } x < 0 \\ -x + 3 & \text{if } x \geq 0 \end{cases}$ at $x = -1$

12. $f(x) = \begin{cases} \frac{1}{x+1} & \text{if } x \leq 0 \\ x & \text{if } x > 0 \end{cases}$ at $x = 1$

Exercises 13-18. Prove that the given function is continuous. (Recall that a function is continuous if it is continuous at each point in its domain.)

13. $f(x) = 3x + 5$
14. $f(x) = -\frac{1}{4}x + 1$
15. $f(x) = x^2 + 2x$
16. $f(x) = -x^2 - 4$
17. $f(x) = \frac{1}{x}$
18. $f(x) = \begin{cases} 2x + 3 & \text{if } x < 0 \\ -x + 3 & \text{if } x \geq 0 \end{cases}$

Exercises 19-22. Display a function with domain $\Re$ which fails to be continuous only at the numbers in S.

19. $S = \{-1\}$
20. $S = \{-1, 0, 1\}$
21. $S = Z^+$
22. $S = Z$

23. Prove Theorem 3.21(d).

24. Let $f: \Re \to \Re$. Prove that if $A \subseteq f^{-1}(B)$, then $f(A) \subseteq B$ (see Definition 3.15).

25. Use the Principle of Mathematical Induction to prove that if f is continuous at c, then so is the function f^n given by $f^n(x) = [f(x)]^n$.

26. Use the Principle of Mathematical Induction to prove that, for all $n \in Z^+$, if the functions f_n are continuous at c, then so is the function:
    (a) $f_1 + f_2 + \cdots + f_n$
    (b) $f_1 \cdot f_2 \cdot \cdots \cdot f_n$

27. Prove that every polynomial function, $p(x) = a_n x^n + a_{n-1} x^{n-1} + \ldots + a_1 x + x_0$, is continuous.

28. Use the Principle of Mathematical Induction to prove that if $f_i: \Re \to \Re$ is continuous for $1 \le i \le n$, then so is the function $f_n \circ f_{n-1} \circ \cdots \circ f_2 \circ f_1$.

29. Prove that every rational function, $\frac{p(x)}{q(x)}$ where $p(x)$ and $q(x)$ are polynomials with $q(x) \ne 0$, is continuous. (Recall that a function is continuous if it is continuous at each point in its domain.)

30. Prove that every function $f: Z \to \Re$ is continuous.

31. Prove that every function $f: \left\{\frac{1}{n}\right\}_{n=1}^{\infty} \to \Re$ is continuous.

32. Give an example of a function $f: \{0\} \cup \left\{\frac{1}{n}\right\}_{n=1}^{\infty} \to \Re$ that is not continuous.

33. Let $S \subseteq \Re$. Prove that a function $f: S \to \Re$ is continuous at c if and only if $\lim_{n \to \infty} (x_n) = c$, with each $x_n \in S$ implies $\lim_{n \to \infty} f(x_n) = f(c)$.

34. (a) Let $H \subseteq \Re$ be closed and let $f: H \to \Re$ be continuous. Prove that if (x_n) is a convergent sequence with each $x_n \in H$, then the sequence $(f(x_n))$ must converge.
    (b) Show, by means of an example, that (a) need not hold if the set H is not closed.

35. Prove that $H \subseteq \Re$ is closed if and only if every convergent sequence (x_n), with each $x_n \in H$, converges to a point in H.

36. Let $f: \Re \to \Re$ and $g: \Re \to \Re$ be continuous. Prove that the set $\{x | f(x) = g(x)\}$ is closed.

37. Let $f: \Re \to \Re$ and $g: \Re \to \Re$ be continuous. Prove that if $f(x) = g(x)$ for every rational number x, then $f = g$.

38. Display a function with domain $\Re$ which fails to be continuous everywhere.
    Suggestion: Consider the dense sets of the rational numbers and of the irrational numbers.

PROVE OR GIVE A COUNTEREXAMPLE

39. For given functions f and g, if $f + g$ is continuous at c, then both f and g must be continuous at c.

40. For given functions f and g, if $f + g$ is continuous at c, then f or g must be continuous at c.

41. For given functions f and g, if $f + g$ and f are continuous at c, then g is continuous at c.

42. For given functions f and g, if fg is continuous at c, then both f and g must be continuous at c.

43. For given functions f and g, if fg is continuous at c, then f or g must be continuous at c.

44. For given functions f and g, if fg and f are continuous at c, and g must be continuous at c.

45. For given functions f and g, if $\frac{f}{g}$ is continuous at c, then both f and g must be continuous at c.

46. For given functions f and g, if $\frac{f}{g}$ is continuous at c, then f or g must be continuous at c.

47. For given functions f and g, if $\frac{f}{g}$ and f are continuous at c, and g is continuous at c.

48. For given functions f and g, if $f \circ g$ is continuous at c, then g must be continuous at c, and f at $g(c)$.

49. For given functions f and g, if $f \circ g$ and g are continuous at c, then f must be continuous at $g(c)$.

50. For given functions f and g, if $f \circ g$ is continuous at c, and f is continuous at $g(c)$, then g must be continuous at c.

51. $f: \Re \to \Re$ is continuous if and only if $f^{-1}(H)$ is closed for every closed set H.

CHAPTER 4
A Touch of Topology

While a number of references to Chapter 3 appear in this chapter, they are only included to underline the fact that an abstract metric space stems from the standard Euclidean space $\Re$.

The distance concept of the previous chapter will lead us to the abstract setting of a metric space in Section 1. A further abstraction, from metric spaces to topological spaces takes place in Section 2. The all-important notion of continuity blossoms to that of a continuous function between topological spaces in Section 3. Cartesian products and quotient spaces of topological spaces are introduced in Section 4.

§1. METRIC SPACES

As you know, the absolute value of $a \in \Re$, denoted by $|a|$, is given by:

$$|a| = \begin{cases} a & \text{if } a \geq 0 \\ -a & \text{if } a < 0 \end{cases}$$

You can, and should, interpret $|a|$ as representing the distance (number of units) between a and 0 on the number line. For example, both 5 and -5 are 5 units from the origin, and we have:

$$|5| = 5 \text{ and } |-5| = 5$$

When you subtract one number from another the result is either **plus or minus** the distance (number of units) between those numbers on the number line. For example, $7 - 2 = 5$ while $2 - 7 = -5$. In either case, the absolute value of the difference is 5, the distance between the two numbers:

$$|7 - 2| = |5| = 5 \text{ and } |2 - 7| = |-5| = 5$$

In general: $|a - b|$ represents the distance (number of units) between a and b.

Roughly speaking, a metric space is a set with an imposed notion of distance — one inspired by the following result:

This theorem appears in Chapter 3, page 135. It is reproduced here for the sake of "chapter-independence."

THEOREM 4.1 (i) $\forall x, y \in \Re: |x - y| \geq 0$, with $|x - y| = 0$ if and only if $x = y$.
   The distance between two numbers is never negative, and is 0 only if the two numbers are one and the same.

(ii) $\forall x, y \in \Re: |x - y| = |y - x|$
   The distance between x and y is the same as that from y to x.

(iii) $\forall x, y, z \in \Re: |x - y| \leq |x - z| + |z - y|$
   The triangle inequality.

PROOF: The first two properties are direct consequences of the definition of the absolute value function. We establish (iii) by showing that $|x - y|^2 \leq (|x - z| + |z - y|)^2$ (this will do the trick, since neither side of the inequality can be negative):

$$\begin{aligned}
(|x-z|+|z-y|)^2 &= |x-z|^2 + 2|x-z||z-y| + |z-y|^2 \\
|a|^2 = a^2\colon &= (x-z)^2 + 2|x-z||z-y| + (z-y)^2 \\
a \le |a|\colon &\ge (x-z)^2 + 2(x-z)(z-y) + (z-y)^2 \\
\text{expanding:} &= x^2 - 2xz + z^2 + 2xz - 2xy - 2z^2 + 2zy + z^2 - 2zy + y^2 \\
&= x^2 - 2xy + y^2 = (x-y)^2 = |x-y|^2
\end{aligned}$$

CHECK YOUR UNDERSTANDING 4.1

Answer: See page A-19.

Prove that for any $x, y \in \Re$, $|x-y| \ge |x| - |y|$.

As you can see, the three properties of Theorem 4.1 morph into the defining axioms of a metric space:

DEFINITION 4.1
METRIC

A **metric** on a set X is a function $d\colon X \times X \to \Re$ which satisfies the following three properties:

(i) $\forall x, y \in X\colon d(x,y) \ge 0$, and $d(x,y) = 0$ if and only if $x = y$.

(ii) $\forall x, y \in X\colon d(x,y) = d(y,x)$

(iii) $\forall x, y, z \in X\colon d(x,y) \le d(x,z) + d(z,y)$

A pair (X, d) consisting of a set X and a metric d is said to be a **metric space**, and may simply be denoted by X.

Theorem 4.1 tells us that $(\Re, d)$, with $d(x,y) = |x-y|$ is a metric space (called the one-dimensional Euclidean space). In the exercises you are asked to verify that $(\Re^2, d)$ and $(\Re^3, d)$, with

$\Re^2 = \{(x,y) | x, y \in \Re\}$
and
$\Re^3 = \{(x,y,z) | x, y, z \in \Re\}$

$$d[(x_1, y_1), (x_2, y_2)] = \sqrt{(x_1 - x_2)^2 + (y_1 - y_2)^2} \text{ and}$$

$$d[(x_1, y_1, z_1), (x_2, y_2, z_2)] = \sqrt{(x_1 - x_2)^2 + (y_1 - y_2)^2 + (z_1 - z_2)^2}$$

are also metric spaces (called the two- and three-dimensional Euclidean spaces, respectively).

EXAMPLE 4.1 Verify that the function $\bar{d}$, given by:

$$\bar{d}[(x_1, y_1), (x_2, y_2)] = |x_1 - x_2| + |y_1 - y_2|$$

is a metric on the set $\Re^2 = \{(x,y) | x, y \in \Re\}$.

SOLUTION: We show that $\bar{d}$ satisfies each of the three defining axioms of Definition 4.1.

(i) For every $(x_1, y_1), (x_2, y_2) \in \Re^2$

$$\bar{d}[(x_1, y_1), (x_2, y_2)] = |x_1 - x_2| + |y_1 - y_2| \ge 0$$

since $|a| \ge 0$ for every $a \in \Re$

Moreover, $\bar{d}[(x_1,y_1),(x_2,y_2)] = 0$ only if both $|x_1 - x_2| = 0$ and $|y_1 - y_2| = 0$, and this can only happen if $x_1 = x_2$ and $y_1 = y_2$; which is to say, if $(x_1, y_1) = (x_2, y_2)$.

(ii) For every $(x_1, y_1), (x_2, y_2) \in \Re^2$:

$$\bar{d}[(x_1,y_1),(x_2,y_2)] = |x_1 - x_2| + |y_1 - y_2|$$
$$= |x_2 - x_1| + |y_2 - y_1| = \bar{d}[(x_2,y_2),(x_1,y_1)]$$

(iii) Let $A = (x_1, y_1)$, $B = (x_2, y_2)$, and $C = (x_3, y_3)$. We show that $\bar{d}(A, C) \leq \bar{d}(A, B) + \bar{d}(B, C)$:

$$\bar{d}(A, C) = \bar{d}[(x_1,y_1),(x_3,y_3)]$$
$$= |x_1 - x_3| + |y_1 - y_3|$$

Theorem 4.1(iii):
$$\leq |x_1 - x_2| + |x_2 - x_3| + |y_1 - y_2| + |y_2 - y_3|$$

regroup:
$$= |x_1 - x_2| + |y_1 - y_2| + |x_2 - x_3| + |y_2 - y_3|$$
$$= \bar{d}[(x_1,y_1),(x_2,y_2)] + \bar{d}[(x_2,y_2),(x_3,y_3)]$$
$$= \bar{d}(A, B) + \bar{d}(B, C)$$

Note how the definition of $\bar{d}$; in used in both directions in this development.
One direction:
$\bar{d}[(x_1,y_1),(x_3,y_3)]$
$= |x_1 - x_3| + |y_1 - y_3|$
Other direction:
$|x_1 - x_2| + |y_1 - y_2|$
$= \bar{d}[(x_1,y_1),(x_2,y_2)]$
Yes:
DEFINITIONS RULE

CHECK YOUR UNDERSTANDING 4.2

(a) Let X be **any** set. Show that the function $d: X \times X \to \Re$ given by:

$$d(x, y) = \begin{cases} 1 & \text{if } x \neq y \\ 0 & \text{if } x = y \end{cases}$$

is a metric on X. The metric d is said to be the **discrete metric** on X, and (X, d) is said to be a **discrete space**.

Answer: See page A-19.

DEFINITION 4.2
OPEN SPHERE

Let (X, d) be a metric space. For $x_0 \in X$ and $r > 0$, the **open sphere** of radius r about x_0 is the set:

$$S_r(x_0) = \{x \in X | d(x, x_0) < r\}$$

Compare with the notation introduced at the bottom of page 123.

Open spheres in metric spaces need not even be round. Consider, for example the adjacent sphere of radius 1, centered at the origin, in the metric space of Example 4.1, wherein:

$$\bar{d}[(x_1,y_1),(x_2,y_2)] = |x_1 - x_2| + |y_1 - y_2|$$

so that: $S_1(0, 0) = \{(x, y) | (|x| + |y| < 1)\}$

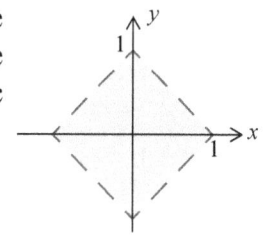

Answer: See page A-20.

> **CHECK YOUR UNDERSTANDING 4.3**
>
> Determine the open spheres $S_1(5)$ and $S_5(1)$ in the discrete metric space Z^+, wherein:
>
> $$d(n, m) = \begin{cases} 1 & \text{if } n \neq m \\ 0 & \text{if } n = m \end{cases} \quad \text{(see CYU 4.2)},$$

Open and Closed subsets of the Euclidean space $\Re$ were introduced in the previous chapter (Definition 3.9, and Definition 3.10).

DEFINITION 4.3 Let (X, d) be a metric space.

OPEN SUBSET OF A METRIC SPACE $O \subseteq X$ is **open** in X if for every $x_0 \in O$ there exists $r > 0$ such that $S_r(x_0) \subseteq O$.

CLOSED SUBSET OF A METRIC SPACE $H \subseteq X$ is **closed** in X if its complement $H^c = \{x \in X | x \notin H\}$ is open in X.

EXAMPLE 4.2 (a) Show that the interval $(1, 5)$ is open in the Euclidean space $\Re$.
(b) Show that the square
$$D = \{(x, y) | (0 \leq x \leq 1), (0 \leq y \leq 1)\}$$
is closed in the Euclidean space $\Re^2$.
(c) Show that every subset of a discrete space X is both open and closed. (See CYU 4.2.)

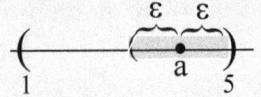

SOLUTION: (a) For any given $a \in (1, 5)$ we are to find $\varepsilon > 0$ such that $S_\varepsilon(a) \subseteq (1, 5)$. This is easy to do: take ε to be the smaller of the two numbers $a - 1$ and $5 - a$ (see margin).

Any smaller $\varepsilon > 0$ will do just as well; but no larger ε will work.

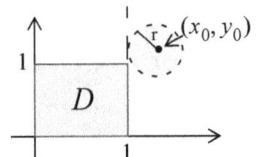

(b) We show D^c is open:

If $(x_0, y_0) \notin D$, then either $x_0 \notin [0, 1]$ or $y_0 \notin [0, 1]$. For definiteness, assume that $x_0 \notin [0, 1]$. Then, for $r = \min(|x_0|, |x_0 - 1|)$: $S_r(x_0, y_0) \cap D = \emptyset$ (indeed $S_r(x_0, y_0) \cap ([0, 1] \times \Re) = \emptyset$).

(c) Let $X = (X, d)$ be discrete, which is to say:
$$d(x, y) = \begin{cases} 1 & \text{if } x \neq y \\ 0 & \text{if } x = y \end{cases}$$

For any $D \subseteq X$ and any $x \in D$: $S_1(x) \cap D = \{x\} \subseteq D$. It follows that every subset D of X is open. That being the case, every subset D must also be closed, and, for the simple reason that D^c is open.

CHECK YOUR UNDERSTANDING 4.4

(a) See page A-20.
(b) 2^n

(a) Show that the interval $(1, 5]$ is neither open nor closed in the Euclidean space $\Re$.
(b) Let X be a discrete metric space consisting of n elements. Determine the number of open subsets of X.

The following theorem will lead us to the definition of a topological space in the next section.

THEOREM 4.2 For any metric space X:
(i) X and $\varnothing$ are open subsets of X.
(ii) Arbitrary unions of open sets are open.
(iii) Finite intersections of open sets are open.

This is the identical proof offered for Theorem 3.13, page 136.

PROOF:

(i) X is open: For any $x \in X$, $S_1(x) \subseteq X$.

$\varnothing$ is open (by default): For any $a \in \varnothing$ there exists an $\varepsilon > 0$ such that $S_\varepsilon(a) \subseteq \varnothing$, for the simple reason that no such a exists.

(ii) Let $\{O_\alpha\}_{\alpha \in A}$ be a collection of open sets, and let $x \in \bigcup_{\alpha \in A} O_\alpha$. Since x is in the union of the O_α's, there must exist some $\alpha_0 \in A$ such that $x \in O_{\alpha_0}$. Since O_{α_0} is open, we can choose $\varepsilon > 0$ such that $S_\varepsilon(x) \subseteq O_{\alpha_0}$. Then:
$$x \in S_\varepsilon(x) \subseteq O_{\alpha_0} \subseteq \bigcup_{\alpha \in A} O_\alpha$$

(iii) Let $\{O_i\}_{i=1}^n$ be a collection of open sets, and let $x \in \bigcap_{i=1}^n O_i$. For each i, choose ε_i such that $S_{\varepsilon_i}(x) \subseteq O_i$, and let $\varepsilon = \min\{\varepsilon_1, \varepsilon_2, \ldots, \varepsilon_n\}$. Since $S_\varepsilon(x) \subseteq O_i$ for each i,
$$x \in S_\varepsilon(x) \subseteq \bigcap_{i=1}^n O_i.$$

CHECK YOUR UNDERSTANDING 4.5

Let X be a metric space. Prove that:
(i) X and $\varnothing$ are both closed in X.
(ii) The arbitrary intersection of closed sets is closed.
(iii) The finite union of closed sets is closed.

Answer: See page A-20.

An order relation ($a < b$) need not reside in an abstract metric space. There is, however, a sense of finite containment:

DEFINITION 4.4
BOUNDED SUBSETS OF A METRIC SPACE

A subset S of a metric space (X, d) is **bounded** if there exists $M \in Z^+$ such that $d(x_1, x_2) \leq M$ for every $x_1, x_2 \in S$. Any such M is said to be a bound for S.

In particular, if there exists $M \in Z^+$ such that $d(x_1, x_2) \leq M$ for every $x_1, x_2 \in X$, then the space X is said to be bounded.

CHECK YOUR UNDERSTANDING 4.6

Prove or give a counterexample. In a metric space X
(a) The union of two bounded subsets is bounded.
(b) The arbitrary union of bounded subsets is bounded.
(c) The arbitrary intersection of bounded subsets is bounded.

Answer: See page A-20.

COMPACT SPACES

A collection $\{O_\alpha\}_{\alpha \in A}$ of open subsets of a metric space X is said to be an **open cover** of a subset $S \subseteq X$ if $S \subseteq \bigcup_{\alpha \in A} O_\alpha$. The open cover is said to have a **finite subcover** if there exists $\{\alpha_1, \alpha_2, \ldots, \alpha_n\} \subseteq A$ such that $S \subseteq \bigcup_{i=1}^{n} O_{\alpha_i}$.

See the indexing remarks that follow CYU 2.3, page 58.

DEFINITION 4.5
COMPACT

A subset K of a metric space X is **compact** if every open cover of K has a finite subcover.

In particular, a metric space X is said to be compact if every open cover of X has a finite subcover.

Compare with Definition 3.11, page 139

THEOREM 4.3 Compact subsets of metric spaces are closed and bounded.

Compare with Theorem 3.18, page 140.

PROOF: Let K be compact. Assume that K is **not** closed (we will arrive at a contradiction).

Choose $x_0 \in X$ such that for every $\varepsilon > 0$, $S_\varepsilon(x_0) \cap K \neq \emptyset$ (if this were not the case, then the complement of K would be open, and K would therefore be closed). For each $x \in K$, let $\varepsilon_x = \dfrac{d(x, x_0)}{2}$, and consider the open cover $\{S_{\varepsilon_x}(x)\}_{x \in K}$ of K. Being compact that open cover has a finite subcover $\{S_{\varepsilon_{x_i}}(x_i)\}_{i=1}^{n}$.

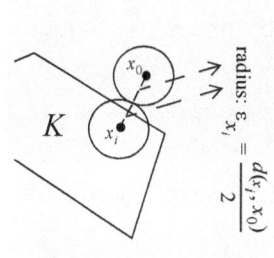

Let $\varepsilon = \min(\varepsilon_{x_1}, \varepsilon_{x_2}, \ldots, \varepsilon_{x_n})$. Since $S_\varepsilon(x_0) \cap S_{\varepsilon_{x_i}}(x_i) = \varnothing$ for $1 \leq i \leq n$ (see margin), and since $K \subseteq \bigcup_{i=1}^{n} S_{\varepsilon_{x_i}}(x_i)$:

$S_\varepsilon(x_0) \cap K = \varnothing$ — a contradiction.

Assume that K is not bounded (we will arrive at a contradiction). Choose $x_0 \in K$. Since K is not bounded, for any $n \in Z^+$ we can choose $x_n \in K$ such that $d(x_n, x_0) > n$. Since every element of K is a finite distance from x_0, $\{S_n(x_0)\}_{n \in Z^+}$ is an open cover of K, which has no finite subcover — contradicting the assumption that K is compact.

CHECK YOUR UNDERSTANDING 4.7

Answer: See page A-21.

Give an example of a closed bounded subset of a metric space X that is not compact. (Note Theorem 3.17, page 140.)

CONTINUITY

We begin by introducing some useful set-theory concepts:

This is a generalization of Definition 3.15, page 152.

DEFINITION 4.6
IMAGE OF S

For $f: X \to Y$, and $S \subseteq X$ we define the **image of S under f**, denoted by $f(S)$, to be the set $f(S) = \{f(x) | x \in S\}$ (see margin).

PRE-IMAGE OF S

For $f: X \to Y$, and $S \subseteq Y$ we define the **pre-image of S under f**, denoted by $f^{-1}(S)$, to be the set $f^{-1}(S) = \{x \in X | f(x) \in S\}$ (see margin).

Your turn:

CHECK YOUR UNDERSTANDING 4.8

(a) Let $f: X \to Y$. Prove that
  (i) $f(A \cup B) = f(A) \cup f(B)$ for any subsets A, B of X.
  (ii) $f^{-1}(A \cup B) = f^{-1}(A) \cup f^{-1}(B)$ for any subsets A, B of Y.
  (iii) $f^{-1}(A \cap B) = f^{-1}(A) \cap f^{-1}(B)$ for any subsets A, B of Y.
  (iv) $f^{-1}(A^c) = [f^{-1}(A)]^c$ for any $A \subseteq Y$.

(b) Show that if $f: X \to Y$ is a bijection, then $f(A) = [f(A^c)]^c$ for every $A \subseteq X$.

Answer: See page A-21.

Since the distance notion resides in metric spaces, the dreaded "ε/δ-method" could be used to define the concept of continuity in the current metric space environment. We take a different approach:

166 Chapter 4 A Touch of Topology

Motivated by Theorem 3.22, page 153.

DEFINITION 4.7
CONTINUOUS FUNCTION
Let X and Y be metric spaces. A function $f: X \to Y$ is continuous if $f^{-1}(O)$ is open in X for every set O open in Y.

EXAMPLE 4.3 Let d be the standard Euclidean metric on $\Re^2 = \{(x, y) | x, y \in \Re\}$:

$$d[(x_1, y_1), (x_2, y_2)] = \sqrt{(x_1 - x_2)^2 + (y_1 - y_2)^2}$$

and let $\bar{d}$ be the metric of Example 4.1:

$$\bar{d}[(x_1, y_1), (x_2, y_2)] = |x_1 - x_2| + |y_1 - y_2|$$

Show that the identity function:

$$I: [(\Re^2, d) \to (\Re^2, \bar{d})]$$

given by $I(x, y) = (x, y)$ is continuous.

SOLUTION: Let O be open in $(\Re^2, \bar{d})$. Since $I^{-1}(O) = O$, the continuity of I will be established once we show that the set O, open in $(\Re^2, \bar{d})$, is also open in the Euclidean space $(\Re^2, d)$. Let's do it:

For given $(x_0, y_0) \in O$ choose $\varepsilon > 0$ be such that

$$\{(x, y) | |x - x_0| + |y - y_0| < \varepsilon\} \subseteq O$$

open sphere of radius ε centered at (x_0, y_0) in $(\Re^2, \bar{d})$

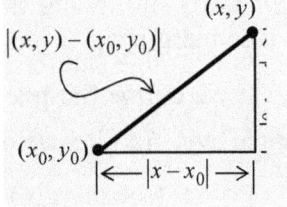

Since $|(x, y) - (x_0, y_0)| \leq |x - x_0| + |y - y_0|$ (see margin), the open sphere

$$\{(x, y) | |(x, y) - (x_0, y_0)| < \varepsilon\}$$
$$\uparrow$$
$$d[(x, y), (x_0, y_0)] = \sqrt{(x - x_0)^2 + (y - y_0)^2}$$

in the Euclidean space $(\Re^2, d)$ is contained in O. It follows that O is open in $(\Re^2, d)$.

CHECK YOUR UNDERSTANDING 4.9

Referring to the notation of the above example, show that the identity function that "goes the other way:" $I: (\Re^2, \bar{d}) \to (\Re^2, d)$ is also continuous.

Answer: See page A-21.

If you worked your way through the previous chapter you discovered that if $f: \Re \to \Re$ and $g: \Re \to \Re$ are continuous, then the composite function $g \circ f: \Re \to \Re$ is also continuous. That result extends to arbitrary metric spaces:

THEOREM 4.4 Let X, Y, and Z be metric spaces. If $f: X \to Y$ and $g: Y \to Z$ are continuous, then the composite function $g \circ f: X \to Z$ is also continuous.

This is the proof of Theorem 3.23, page 153.

PROOF: Let O be open in Z. Since g is continuous, $g^{-1}(O)$ is open in Y. Since f is continuous, $f^{-1}[g^{-1}(O)]$ is open in X. The desired result now follows from Theorem 2.6, page 73 which asserts that $(g \circ f)^{-1}(O) = f^{-1}[g^{-1}(O)]$.

CHECK YOUR UNDERSTANDING 4.10

Let X, Y, and Z be metric spaces. For given functions $f: X \to Y$ and $g: Y \to Z$, if $g \circ f: X \to Z$ is continuous must both f and g be continuous? Justify your answer.

Answer: See page A-21.

ISOMETRIES

In any abstract setting, one attempts to define a notion of "sameness" for its denizens. Bringing us to:

DEFINITION 4.8
ISOMETRIC SPACES

A metric space (X, d) is **isometric** to a space $(Y, \bar{d})$ if there exists a bijection $f: X \to Y$ such that $d(x_1, x_2) = \bar{d}(f(x_1), f(x_2))$ for every $x_1, x_2 \in X$. Such a function f is said to be an **isometry** from X to Y.

If two spaces are isometric, then they are the "same, up to appearances" (the naming of elements, for example).

EXAMPLE 4.4 Show that the metric space (Z, d) with $d(a, b) = |a - b|$ is isometric to the space $(2Z, \bar{d})$, where $2Z = \{2n | n \in Z\}$ and $\bar{d}(2a, 2b) = \frac{1}{2}|2a - 2b|$.

SOLUTION: The function $f: Z \to 2Z$ given by $f(n) = 2n$ is easily seen to be a bijection:

f is one-to-one: $f(a) = f(b) \Rightarrow 2a = 2b \Rightarrow a = b$.

f is onto: For any given $2a \in 2Z$, we have $f(a) = 2a$.

f also preserves the distance between any two points $a, b \in Z$:

$$d(a, b) = |a - b| = \left|\frac{1}{2}(2a - 2b)\right|$$
$$= \frac{1}{2}|2a - 2b| = \bar{d}(2a, 2b) = \bar{d}[f(a) - f(b)]$$

CHECK YOUR UNDERSTANDING 4.11

Show that *isometry* is an equivalence relation on any set of metric spaces. (See Definition 2.20, page 88.)

Answer: See page A-22.

EXERCISES

Exercise 1-5. Verify that each of the following is a metric space.

1. $(\mathfrak{R}^2, d)$, where $\mathfrak{R}^2 = \{(x, y) | x, y \in \mathfrak{R}\}$ and:
$$d[(x_1, y_1), (x_2, y_2)] = \sqrt{(x_1 - x_2)^2 + (y_1 - y_2)^2}$$

2. $(\mathfrak{R}^3, d)$, where $\mathfrak{R}^3 = \{(x, y, z) | x, y, z \in \mathfrak{R}\}$ and:
$$d[(x_1, y_1, z_1), (x_2, y_2, z_2)] = \sqrt{(x_1 - x_2)^2 + (y_1 - y_2)^2 + (z_1 - z_2)^2}$$

3. $(\mathfrak{R}^n, d)$, where $\mathfrak{R}^n = \{(x_1, x_2, ..., x_n) | x_i \in \mathfrak{R}, 1 \le i \le n\}$ and:
$$d[(x_1, x_2, ..., x_n), (y_1, y_2, ..., y_n)] = \sqrt{(x_1 - y_1)^2 + (x_2 - y_2)^2 + \cdots + (x_n - y_n)^2}$$

4. $(\mathfrak{R}^2, d)$, where $\mathfrak{R}^2 = \{(x, y) | x, y \in \mathfrak{R}\}$ and:
$$d[(x_1, y_1), (x_2, y_2)] = \max\{|x_1 - x_2|, |y_1 - y_2|\}$$

   Sketch the unit sphere $S_1(0, 0) = \{(x, y) | d[(x, y), (0, 0)]\} = 1$ in $(\mathfrak{R}^2, d)$.

5. (Calculus Dependent). (C, d), where $C = \{f: [0, 1] \to \mathfrak{R} | f \text{ is continuous}\}$ and:
$$d(f, g) = \int_0^1 |f(x) - g(x)| dx$$

Exercise 6-11. Explain why the given function is **NOT** a metric on $\mathfrak{R}$.

6. $d(x, y) = |x + y|$
7. $d(x, y) = x^2 - y^2$
8. $d(x, y) = x^2 + y^2$
9. $d(x, y) = xy$
10. $d(x, y) = |x| - |y|$
11. $d(x, y) = x - y$

12. Let x, y be any two distinct elements of a metric space X. Show that there are disjoint open sets O_x, O_y with $x \in O_x$ and $y \in O_y$.

13. Prove that if a metric space contains at least two points, then it must contain at least four distinct open sets.

14. Let S be an unbounded subset of a metric space (X, d). Prove that for any $x_0 \in S$ and any given $n \in Z^+$ there exists $x_n \in S$ such that $d(x, x_0) > n$.

15. Let X be a discrete space. Prove that every function $f: X \to Y$ is continuous for any metric space Y.

16. Let X, and Y be metric spaces. Prove that for any $y_0 \in Y$ the constant function $f_{y_0}: X \to Y$ given by $f_{y_0}(x) = y_0$ for every $x \in X$ is continuous.

17. Prove that the continuous image of a compact metric space is compact.

18. Let (X, d_X) and (Y, d_Y) be two metric spaces. Prove that $(X \times Y, d)$, where $X \times Y = \{(x, y) | x \in X \text{ and } y \in Y\}$ and $d[(x_1, y_1), (x_2, y_2)] = d_X(x_1, x_2) + d_Y(y_1, y_2)$ is a metric space.

19. Prove that a space X is compact if and only for any collection $\{H_\alpha\}_{\alpha \in A}$ of closed subsets of X with $\bigcap_{\alpha \in A} H_\alpha = \emptyset$ one can choose a finite subcollection $\{H_{\alpha_i}\}_{i=1}^n$ such that $\bigcap_{i=1}^n H_{\alpha_i} = \emptyset$.

20. Let H be a compact subset of a metric space X. Show that for any given $x \notin H$ there exist disjoint open sets O_x, O_H with $x \in O_x$ and $H \subseteq O_H$.

21. Let H_1 and H_2 be two disjoint compact subsets of a space X. Show that there exist two disjoint open sets O_1, O_2 with $H_1 \subseteq O_1$ and $H_2 \subseteq O_2$.

22. Let X and Y be isometric spaces. Prove that X is bounded if and only if Y is bounded.

23. Let X and Y be isometric spaces. Prove that X is compact if and only if Y is compact.

24. (Set theory). Let $f: X \to Y$. Prove:

   (a) For $S_i \subseteq X$, $1 \leq i \leq n$: $f\left(\bigcup_{i=1}^n S_i\right) = \bigcup_{i=1}^n f(S_i)$. (b) For $S \subseteq X$: $f[f^{-1}(S)] \subseteq S$.

25. (Set theory). Let $f: X \to Y$. Prove that for $S_\alpha \subseteq X$, $\alpha \in A$:

   (a) $f^{-1}\left(\bigcup_{\alpha \in A} S_\alpha\right) = \bigcup_{\alpha \in A} f^{-1}(S_\alpha)$ (b) $f^{-1}\left(\bigcap_{\alpha \in A} S_\alpha\right) = \bigcap_{\alpha \in A} f^{-1}(S_\alpha)$.

Exercise 26-32. (Closure) For A a subset of a metric space X, let $C_A = \{H \subseteq X | A \subseteq H \text{ and } H \text{ is closed in } X\}$. The set $\bar{A} = \bigcap_{H \in C_A} H$ is called the **closure** of A.

26. Determine $\bar{A}$, for the given subset A of the Euclidean space $\Re$.

   (a) $A = (1, 3)$ (b) $A = Z^+$ (c) $A = Q$ (the set of rational numbers)

   (d) $A = (1, 3) \cup \{5\}$ (e) $A = \left\{\frac{1}{n} \middle| n \in Z^+\right\}$

27. Prove that for $A \subseteq X$, $\bar{A}$ is the smallest closed subset of X that contains A, in that $\bar{A}$ is a subset of every closed set containing A.

28. Prove that $\bar{A}$ is closed for every $A \subseteq X$.

29. Prove that if A is closed if and only if $\bar{A} = A$.

30. Prove that $x \in \bar{A}$ if and only if $O \cap A \neq \emptyset$ for every open set O containing x.

31. Let x_1, x_2 be any two distinct elements of a metric space X. Show that there exists $r \in \Re$ such that $\overline{S_r(x_1)} \cap \overline{S_r(x_2)} = \emptyset$.

32. Let x be an element of an open subset O of a metric space X. Show that there exists $r \in \Re$ such that $\overline{S_r(x)} \subseteq O$.

PROVE OR GIVE A COUNTEREXAMPLE

33. $(\Re, d)$, where $d(x, y) = r|x - y|$ is a metric space for any $r \in \Re$.

34. $(\Re^2, d)$, where $\Re^2 = \{(x, y) | x, y \in \Re\}$ and $d[(x_1, y_1), (x_2, y_2)] = |x_1 - x_2| - |y_1 - y_2|$ is a metric space.

35. A function f from a metric space X to a metric space Y is continuous if and only if $f^{-1}(H)$ is closed in X for every closed subsets H of Y.

36. Let $f: X \to Y$ be a bijection from the metric space X to the metric space Y. If f is continuous, then f is an isometry.

37. Let $f: X \to Y$ be a bijection from the metric space X to the metric space Y. If f and f^{-1} are continuous, then f is an isometry.

38. (**Closure**: See Exercises 26-32) If S_i is a subset of a metric space X, for $i \in Z^+$, then:

   (a) $\overline{S_1} \cap \overline{S_2} = \overline{S_1 \cap S_2}$ (b) $\overline{S_1} \cup \overline{S_2} = \overline{S_1 \cup S_2}$ (c) $(\overline{S_1} \cup \overline{S_2})^c = \overline{S_1^c} \cap \overline{S_2^c}$

   (d) $(\overline{S_1} \cup \overline{S_2})^c = \overline{(S_1 \cup S_2)^c}$ (e) $\bigcap_{i=1}^{\infty} \overline{S_i} = \overline{\bigcap_{i=1}^{\infty} S_i}$ (f) $\bigcup_{i=1}^{\infty} \overline{S_i} = \overline{\bigcup_{i=1}^{\infty} S_i}$

   (g) $\left(\bigcap_{i=1}^{\infty} \overline{S_i}\right)^c = \overline{\left(\bigcap_{i=1}^{\infty} S_i\right)^c}$

§2 TOPOLOGICAL SPACES

Theorem 4.2, page 163, inspires the definition of a topological space:

DEFINITION 4.9
TOPOLOGY

A **topology** on a set X is a collection τ of subsets of X, called **open** subsets of X, satisfying the following properties:

(i) X and $\varnothing$ are in τ.

(ii) Any union of elements of τ is again in τ.
(τ is closed under arbitrary unions)

(iii) Any finite intersection of elements of τ is again in τ.
(τ is closed under finite intersections)

TOPOLOGICAL SPACE
A pair (X, τ) consisting of a set X and a topology τ is called a **topological space**, and may at times simply be denoted by X.

The Greek letter τ, spelled "tau" — rhymes with "cow."

CHECK YOUR UNDERSTANDING 4.12

For any set X show that:

(a) $\tau_0 = \{X, \varnothing\}$ is a topology on X [it is called the **indiscrete topology** on X].

(b) $\tau_1 = \{S | S \subseteq X\}$ is a topology on X [it is called the **discrete topology** on X].

Answer: See page A-22.

EXAMPLE 4.5
FINITE-COMPLEMENT SPACE

For any set X, let
$$\tau = \{\varnothing\} \cup \{S \subseteq X | S^c \text{ is finite}\}$$
Show that (X, τ) is a topological space.

SOLUTION: We verify that the three axioms of Definition 4.9 hold for the family τ.

(i) We are given that $\varnothing \in \tau$. In addition, since $X^c = \varnothing$ is a finite subset of X, $X \in \tau$.

(ii) Consider the set $\{O_\alpha\}_{\alpha \in A}$ with each $O_\alpha \in \tau$. Since
$$\left(\bigcup_{\alpha \in A} O_\alpha\right)^c = \bigcap_{\alpha \in A} O_\alpha^c$$
(Theorem 2.3, page 58), the number of ele-

ments in $\left(\bigcup_{\alpha \in A} O_\alpha\right)^c$ cannot exceed the number of elements in any of the O_α^c, all of which are finite. It follows that $\bigcup_{\alpha \in A} O_\alpha \in \tau$.

(iii) Consider the finite set $\{O_i\}_{i=1}^n$ with each $O_i \in \tau$. Since $\left(\bigcap_{i=1}^n O_i\right)^c = \bigcup_{i=1}^n O_i^c$ (Theorem 2.3, page 58), the number of elements in $\left(\bigcap_{i=1}^n O_i\right)^c$ cannot exceed the sum of the finite number of elements in the sets O_i, $1 \leq i \leq n$; and is therefore finite. It follows that $\bigcap_{i=1}^n O_i \in \tau$.

CHECK YOUR UNDERSTANDING 4.13

Let $X = \{a, b\}$ and $\tau = \{\emptyset, X, \{a\}\}$. Show that (X, τ) is a topological space (called the **Sierpinski space**).

Answer: See page A-22.

Every metric space gives rise to a topological space. Specifically:

THEOREM 4.5 For any metric space X:
$$\tau = \{O \subseteq X | O \text{ is open in the metric space } X\}$$
is a topology on X.

In the Euclidean space $\Re$, $\tau = \{O \subseteq \Re | O \text{ is open in } \Re\}$ is called the standard or Euclidean topology on $\Re$).

PROOF: Theorem 4.2, page 163.

DEFINITION 4.10 A topological space (X, τ) is said to be **metrizable** if there exists a metric d on X such that
$$\tau = \{O \subseteq X | O \text{ is open in the metric space } X\}$$

Not every topological space is metrizable:

CHECK YOUR UNDERSTANDING 4.14

Show that the Sierpinski space [CYU 4.13] is not metrizable. Suggestion: consider Exercise 13, page 168.

Answer: See page A-22.

The complement of open sets in a metric space were called closed sets. Following suit:

DEFINITION 4.11
CLOSED SUBSET OF A TOPOLOGICAL SPACE
Let $(X, \mathcal{T})$ be a topological space. $H \subseteq X$ is **closed** if its complement H^c is open (that is: $H^c \in \mathcal{T}$).

As is the case in any metric space (see CYU 4.5, page 163):

THEOREM 4.6 In any topological space X:
(i) X and $\varnothing$ are closed in X (they are also open).
(ii) Any intersection of closed sets is again a closed set.
(iii) Any finite union of closed sets is again a closed set.

PROOF: (i) Since $X^c = \varnothing$ and $\varnothing^c = X$ are open (Definition 4.9), both X and $\varnothing$ are closed.

(ii) Let $\{H_\alpha\}_{\alpha \in A}$ be a collection of closed sets. We show $\bigcap_{\alpha \in A} H_\alpha$ is closed by showing that its complement $\left(\bigcap_{\alpha \in A} H_\alpha\right)^c = \bigcup_{\alpha \in A} (H_\alpha)^c$ is open:

Since each H_α is closed, $\{(H_\alpha)^c\}_{\alpha \in A}$ is a collection of open sets. Since unions of open sets are open, $\bigcup_{\alpha \in A} (H_\alpha)^c$ is open.

(iii) Let $\{H_i\}_{i=1}^n$ be a collection of closed sets. Since $\left(\bigcup_{i=1}^n H_i\right)^c = \bigcap_{i=1}^n (H_i)^c$, and since finite intersections of open sets are open, $\left(\bigcup_{i=1}^n H_i\right)^c$ is open; which is to say: $\bigcup_{i=1}^n H_i$ is closed.

CHECK YOUR UNDERSTANDING 4.15

Characterize the closed subsets of the given topological space.
(a) A discrete space $(X, \mathcal{T}_1)$. (See CYU 4.12.)
(b) An indiscrete space $(X, \mathcal{T}_0)$. (See CYU 4.12)
(c) The Sierpinski space of CYU 4.13.

Answer: See page A-22.

SUBSPACES OF A TOPOLOGICAL SPACE

DEFINITION 4.12
SUBSPACE

For (X, τ) a topological space and $S \subseteq X$ let
$$\tau_S = \{O \cap S \mid O \in \tau\}$$
The ordered pair (S, τ_S) is said to be a **subspace** of the space (X, τ) and the elements of τ_S are said to be open in S.

As might be expected:

THEOREM 4.7 Let (X, τ) be a topological space, $S \subseteq X$. The subspace (S, τ_S) is also a topological space.

PROOF: We verify that the three axioms of Definition 4.9 hold for the family τ_S.

(i) Since $\emptyset$ and X are open in (X, τ), both $\emptyset \cap S = \emptyset$ and $X \cap S = S$ are open in (S, τ_S).

(ii) Let $\{U_\alpha\}_{\alpha \in A}$ be a family of open sets in (S, τ_S). We show that $\bigcup_{\alpha \in A} U_\alpha$ is again in τ_S:

For each $\alpha \in A$ choose $O_\alpha \in \tau$ such that $O_\alpha \cap S = U_\alpha$. Since τ is a topology, $\bigcup_{\alpha \in A} O_\alpha \in \tau$. The desired result now follows from:

$$\bigcup_{\alpha \in A} U_\alpha = \bigcup_{\alpha \in A} (O_\alpha \cap S) \underset{\text{Exercise 82(a), page 61}}{=} \left(\bigcup_{\alpha \in A} O_\alpha\right) \cap S$$

As for (iii):

CHECK YOUR UNDERSTANDING 4.16

Let (X, τ) be a topological space and $S \subseteq X$. Show that $\tau_S = \{O \cap S \mid O \in \tau\}$ is closed under finite intersections.

Answer: See page A-23.

BASES AND SUBBASES

When analyzing a topological space it is often sufficient to consider the following subsets of its topology:

DEFINITION 4.13
BASE

A **base** for a space (X, τ) is a subset β of τ which satisfies the following property:

$$\forall x \in O \in \tau \; \exists B \in \beta \ni x \in B \subseteq O$$

(In other words, every open set is a union of elements from β)

SUBBASE

A **subbase** for (X, τ) is a subset Γ of τ such that the set of finite intersections of elements of Γ is a base for the topology.

Γ (gamma) is the Greek letter for C.

EXAMPLE 4.6 (a) Show that $\beta = \{S_r(x) | x \in X, r > 0\}$ is a basis for any metric space X.

(b) Show that
$$\Gamma = \{(-\infty, b) | b \in \Re\} \cup \{(a, \infty) | a \in \Re\}$$
is a subbase for the Euclidean space $\Re$

*$O \subseteq X$ is **open** in X if for every $x_0 \in O$ there exists $r > 0$ such that $S_r(x_0) \subseteq O$.*

SOLUTION: (a) A direct consequence of Definition 4.3, page 162 (see margin).

(b) Since every open interval in $\Re$ is the intersection of two elements in $\Gamma = \{(-\infty, b) | b \in \Re\} \cup \{(a, \infty) | a \in \Re\}$, and since $S_r(x)$ is the open interval $(x - r, x + r)$, Γ is a subbase for the Euclidean topology on $\Re$.

CHECK YOUR UNDERSTANDING 4.17

Let X be a metric space. Show that $\beta = \{S_r(x) | x \in X, r \in Q^+\}$, where Q^+ denotes the set of positive rational numbers, is a base for the topology of X.

Answer: See page A-23.

In the exercises you are invited to show that for any collection $S = \{S_\alpha\}_{\alpha \in A}$ of subsets of X, the set:

$$\Gamma = \left\{ \bigcap_{i=1}^{n} S_{\alpha_i} \middle| S_{\alpha_i} \in S, n \in Z^+ \right\}$$

is a subbase for a topology τ_S on X, called the **topology generated by S**. In other words:

> If you start with any collection S of subsets of X, and then take all unions of the finite intersections of the sets in S, you will end up with a topology τ_S on X.

In particular, consider the set $S = \{[n,m] | n, m \in Z, n < m\}$ in $\Re$. Since $\{r\} = [r-1, r] \cap [r, r+1]$, every subset of $\Re$, being the union of the open sets $\{\{r\} | r \in \Re\}$, is open in τ_S. In other words:

τ_S is the discrete topology on $\Re$ [see CYU 4.12(b)]

CHECK YOUR UNDERSTANDING 4.18

Let τ denote the standard Euclidean topology on $\Re$, and let τ_S denote the topology on $\Re$ generated by $S = \{[x,y) | x < y\}$. Show that $\tau \subset \tau_S$.

($H = (\Re, \tau_S)$ is called the **half-open-interval space**)

Answer: See page A-23.

COMPACT SPACES

Reminiscence of Definition 4.5, page 164:

DEFINITION 4.14
COMPACT
A subset K of a topological space X is **compact** if every open cover of K has a finite subcover.

In particular, the space X is said to be compact if every open cover of X has a finite subcover.

Theorem 4.3, page 164, asserts that compact subsets of metric spaces are closed and bounded. This result cannot carry over to general topological spaces, for the notion of "bounded" involves the concept of distance which does not reside in the general setting of topological spaces. Moreover, as it turns out, compact subsets of a general topological space need not even be closed:

EXAMPLE 4.7 Let τ denote the finite complement topology on the set Z of integers of Example 4.5:

$$\tau = \{\varnothing\} \cup \{S \subseteq Z | S^c \text{ is finite}\}$$

Exhibit a compact subset of (Z, τ) that is not closed.

SOLUTION: As it turns out, every single subset $H \subseteq Z$ of (Z, τ) is, in fact, compact:

Let $\{O_\alpha\}_{\alpha \in A}$ be an open cover of H. Since the empty set is certainly compact, we can assume that $H \neq \emptyset$ and choose an arbitrary element $h_0 \in H$. Some element of the open cover, say O_{α_0} contains h_0. By the very definition of τ, only finitely many elements of Z lie outside of O_{α_0}. A fortiori, there are only a finite number of elements of H that are not contained in O_{α_0}, say $\{h_i\}_{i=1}^n$. For each $1 \le i \le n$ choose an element O_{α_i} of the given cover which contains h_i. Clearly $\{O_i\}_{i=0}^n$ covers H.

We next show that the compact subset Z^+ of Z is not closed by showing that its complement is not open:

Let O be any open set containing $0 \in (Z^+)^c$. Since O^c is finite, $O \cap Z^+ \neq \emptyset$. Thus $(Z^+)^c$ is not open.

CHECK YOUR UNDERSTANDING 4.19

Answer: See page A-23.

Prove that every closed subset of a compact space is compact.

In the exercises, you are asked to show that a space X is compact if every open cover of X by sets taken from a base β of X has a finite subcover. We now state, without proof, a stronger result:

James W. Alexander (1888-1971).

THEOREM 4.8
ALEXANDER'S SUBBASE THEOREM

Let X be a topological space and let Γ be a subbase for the topology of X. If every open cover of X by sets in Γ has a finite subcover, then X is compact.

> The above result can be shown to be equivalent to the Axiom of Choice (see page 109).

Felix Hausdorff (1868-1942).

HAUSDORFF SPACES

In the previous section, analytical properties of the real number system directed us to the definition of a metric space. The concept of a metric space was then generalized further to that of a topological space. This generalization may, in fact, be a bit too general in that there are some rather uninteresting topological spaces hanging around. For example, a set with indiscrete topology has very little "topological personality." What distinguishes a topological space from a plain old set is its collection of open sets, and while one does not like to degrade any structure, it is nonetheless difficult to think of an indiscrete space (X, τ), with $\tau = \{\emptyset, X\}$, as having evolved far from its underlying set X.

The following important class of spaces contain sufficiently many open sets to "separate points:"

DEFINITION 4.15 HAUSDORFF SPACE (X, τ) is a **Hausdorff space** if for any two distinct points x, y in X there exists disjoint open sets O_x, O_y with $x \in O_x$ and $y \in O_y$.

> Distinct points in a Hausdorff space can be separated by disjoint open sets. Additional separation properties are introduced in the exercises.

An open set in a space X that contains a point $x \in X$ is said to be an (open) **neighborhood of x**. With this terminology at hand, we can say that a space is Hausdorff if any two points in the space reside in disjoint neighborhoods.

CHECK YOUR UNDERSTANDING 4.20

Prove that every metrizable space is Hausdorff.

> Answer: See page A-23.

We previously observed that compact subsets of a general topological space need not be closed. However:

THEOREM 4.9 Any compact subset of a Hausdorff space is closed.

> This result along with CYU 4.19 tell us that: A subset of a compact Hausdorff space is compact if and only if it is closed.

PROOF: Suppose K is a compact subset of a Hausdorff space (X, τ) that is not closed (we will arrive at a contradiction):

Since K is not closed, its complement is not open, and we can choose $x_0 \in X$ such that $O(x_0) \cap K \neq \emptyset$ for every open neighborhood $O(x_0)$ of x_0 (if this were not the case, then the complement of K would be open). For each $x \in K$, choose disjoint open sets O_x and $O_x(x_0)$ containing x and x_0, respectively (margin), and consider the open cover $\{O_x\}_{x \in K}$ of K. Being compact that open cover has a finite subcover $\{O_{x_i}\}_{i=1}^{n}$. It follows that the neighborhood $\bigcap_{i=1}^{n} O_{x_i}(x_0)$ of x_0 contains no element of K — a contradiction.

CHECK YOUR UNDERSTANDING 4.21

Let K be a compact set in a Hausdorff space X. Show that for any given $x_0 \notin K$ there exists disjoint open sets O_{x_0} and O_K containing x_0 and K, respectively.

> Answer: See page A-23.

EXERCISES

1. Let $X = \{a, b, c, d\}$. Determine whether or not the given collection of subsets of X is a topology on X.

   (a) $\{\emptyset, X\}$
   (b) $\{\emptyset, X, \{a, b\}\}$
   (c) $\{\emptyset, X, \{a\}, \{a, b\}, \{a, c\}\}$
   (d) $\{\{a\}, \{b\}, \{c\}, \{d\}\}$
   (e) $\{\emptyset, X, \{a\}\}$
   (f) $\{\emptyset, X, \{a\}, \{b\}\}$
   (g) $\{\emptyset, X, \{a, b, c\}, \{a, b, d\}, \{a, b, c\}\}$
   (h) $\{\emptyset, X, \{a\}, \{a, b\}, \{a, c, d\}\}$

2. Show that the topology of Example 4.6 is a proper subset of the standard Euclidean topology on $\Re$.

3. Prove that if β is a base for a discrete space X, then $\{x\} \in \beta$ for every $x \in X$.

4. Prove that if β is a base for a topological space $(X, \mathcal{T})$, and if $\beta \subseteq \beta' \subseteq \mathcal{T}$, then β' is also a base for $\mathcal{T}$.

5. Let $(S, \mathcal{T}_S)$ be a subspace of a topological space $(X, \mathcal{T})$. Prove that $D \subseteq S$ is closed in $(S, \mathcal{T}_S)$ if and only if there exists a closed subset H of X such that $H \cap S = D$.

6. Exhibit three topologies $\mathcal{T}_1, \mathcal{T}_2, \mathcal{T}_3$ on the set $X = \{a, b, c\}$ with $\mathcal{T}_1 \subset \mathcal{T}_2 \subset \mathcal{T}_3$.

7. (a) Let $\mathcal{T}_1, \mathcal{T}_2$ be topologies on a set X. Prove that $\mathcal{T}_1 \cap \mathcal{T}_2$ is also a topology on X.

   (b) Let $\{\mathcal{T}_i\}_{i=1}^{n}$ be a collection of topologies on a set X. Prove that $\bigcap_{i=1}^{n} \mathcal{T}_i$ is also a topology on X.

   (b) Let $\{\mathcal{T}_\alpha\}_{\alpha \in A}$ be a collection of topologies on a set X. Prove that $\bigcap_{\alpha \in A} \mathcal{T}_\alpha$ is also a topology on X.

8. Let S be open in the topological space X. Prove that $O \subseteq S$ is open in the subspace S if and only if O is open in X.

9. Show that $\mathcal{T} = \{\emptyset\} \cup \{O_n\}_{n \in Z^+}$ where $O_n = \{i \in Z^+ | i \geq n\}$ is a topology on Z^+.

10. Let X be an uncountable set. Prove that the collection $\mathcal{T} = \{O \subseteq X | O^c \text{ is countable}\}$ is a topology on X.

11. Prove, without appealing to Theorem 4.8, that a space X is compact if every open cover of X by sets taken from a base β of X has a finite subcover.

12. Prove that $\Gamma = \{(-\infty, b) | b \in Q\} \cup \{(a, \infty) | a \in Q\}$ is a subbase for $\Re$.

13. Prove that a subbase of a topological space is a base for the space if an only if it is closed under finite intersections.

14. Show that for any collection $S = \{S_\alpha\}_{\alpha \in A}$ of subsets of a set X, the set:
$$\Gamma = \left\{ \bigcap_{i=1}^{n} S_{\alpha_i} \,\middle|\, S_{\alpha_i} \in S, n \in Z^+ \right\}$$
is a subbase for a topology on X.

15. **(Rational-Real Topology)** Let τ denote the standard Euclidean topology on $\Re$ and Q the set of rational numbers. Let τ' denote the topology on $\Re$ generated by $\tau \cup Q$. Prove that:

    (a) $\tau \subset \tau'$.

    (b) $\beta = \{(a, b) | a < b\} \cup \{(a, b) \cap Q | a < b\}$ is a base for τ'.

    (b) $\tau = \{(a, \infty) | a \in \Re\} \cup \{(-\infty, a) | a \in \Re\} \cup Q$ is a subbase for τ'.

16. **(Half-Open Interval Topology)**

    (a) Show that $H = \{[a, b) | a < b\}$ is closed under finite intersections, and is therefore a base for a topology τ' on $\Re$.

    (b) Show that $\tau' \subset \tau$, where τ denotes the Euclidean topology on $\Re$.

17. **(Tangent Disc Topology)** Consider the upper closed plane $T = \{(x, y) \in R^2 | y \geq 0\}$. For any $p = (x, y)$ with $y > 0$ let $S(p) = \{S_r(p) | r < y\}$, and let $S(q)$ for any $q = (x, 0)$ consist of those sets of the form $\{q\} \cup D$ where D is an open disk in T tangent to q (see adjacent figure).

    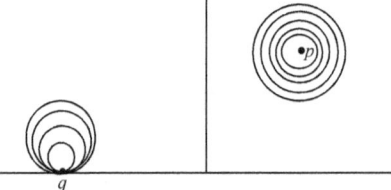

    (a) Show that $S = \{S_{(x,y)}\}_{(x,y) \in T}$ is closed under finite intersections, and is therefore a base for a topology τ' on T. [(T, τ') is called the Moor plane.]

    (b) Is the point q open in τ'? Justify your answer.

    (c) Is the topology τ' the same as the subspace topology τ_T, where τ denotes the standard Euclidean topology on $\Re^2$? Justify your answer.

18. Let $f: X \to Y$ be a function from a non-empty set X to a topological space (Y, τ). Show that $\{f^{-1}(O) | (O \in \tau)\}$ is a topology on X. (See Definition 3.15, page 152.)

19. (a) Let (X_1, τ_1), (X_2, τ_2) be two topological spaces. Prove that $(X_1 \times X_2, \tau)$, where $\tau = \{O_1 \times O_2 | O_1 \in \tau_1 \text{ and } O_2 \in \tau_2\}$ is also a topological space.

    (b) If β_1 and β_2 are base for (X_1, τ_1) and (X_2, τ_2), is $\beta_1 \times \beta_2$ a base for $(X_1 \times X_2, \tau)$? Justify your answer.

    (c) State and establish a generalization of part (a) to accommodate a collection of n topological spaces $\{(X_i, \tau_i)\}_{i=1}^{n}$.

20. Let S be a subset of a topological space X. Prove that the subspace S is compact if and only if every cover of S by sets **open in X** contains a finite subcover.

21. (a) Show that every infinite subset S of a compact space X contains a point whose every neighborhood contains infinitely many elements of S.

    (b) Show, by means of an example, that (a) need not hold if X is not compact.

    (c) Show that the converse of (a) does not hold.

22. Prove that a topological space X is compact if and only if for any given collection $\{H_\alpha\}_{\alpha \in A}$ of closed sets such that $\bigcap_{\alpha \in A} H_\alpha = \emptyset$, there exists a finite subcollection $\{H_{\alpha_i}\}_{i=1}^{n}$ with $\bigcap_{i=1}^{n} H_{\alpha_i} = \emptyset$.

Exercise 23-24. (Sequences) A sequence $(x_i)_{i=1}^{\infty}$ in a topological space X is said to be a convergent sequence which converges to $x_0 \in X$ if for any neighborhood O of x_0 there exists N such that $n > N \Rightarrow x_n \in O$. In the event that the sequence converges to x_0 we write $\lim_{n \to \infty} x_n = x_0$, and say that x_0 is a limit point of the sequence.

23. (a) Give an example of a convergent sequence in a topological space which has more than one limit point.

    (b) Prove that in a Hausdorff space a convergent sequence has a unique limit point.

24. (a) Let H be a closed subset of a topological space. Prove that if a sequence $(x_i)_{i=1}^{\infty}$ with each $x_i \in H$ converges to x_0, then $x_0 \in H$.

    (b) Prove that (a) need not hold if H is not closed in X. Suggestion: Think indiscreetly.

    (c) Show, by means of an example, that the condition in (a) can hold without H being closed.

Exercise 25-33. (Closure) For S a subset of a topological space X, let
$$C_S = \{H \subseteq X \mid S \subseteq H \text{ and } H \text{ is closed in } X\}. \text{ The set } \bar{S} = \bigcap_{H \in C_S} H \text{ is called the \textbf{closure of } } A.$$

25. Prove that $x \in \bar{S}$ if and only if $O \cap A \neq \emptyset$ for every open set O containing x.

26. For the topology $\mathcal{T} = \{\emptyset, Z\} \cup \{n, n+1, n+2, \ldots\}_{n \in Z^+}$ on Z^+:
    (a) List the closed subsets of Z^+.
    (b) Determine the closure of the sets $\{5, 11, 13\}$ and $\{3, 5, 7, 9, \ldots\}$.

27. For the topology $\mathcal{T} = \{\emptyset, X, \{a\}, \{a, b\}, \{a, c, d\}, \{a, b, e\}, \{a, b, c, d\}\}$ on $X = \{a, b, c, d, e\}$:
    (a) List the closed subsets of X.
    (b) Determine the closure of the sets $\{a\}$, $\{b\}$, $\{c\}$, $\{d\}$, $\{e\}$, $\{c, e\}$, and $\{b, e\}$.

28. Prove that $\bar{S}$ is closed for every $S \subseteq X$.

29. Prove that $\bar{\bar{S}} = \bar{S}$ for every $S \subseteq X$.

30. Prove that if $S \subseteq H$ with H closed, then $\bar{S} \subseteq H$.

31. Prove that if S is closed if and only if $\bar{S} = S$.

32. (a) Prove that for any finite collection $\{S_i\}_{i=1}^n$ of subsets of a topological space X:
    $$\overline{\bigcup_{i=1}^n S_i} = \bigcup_{i=1}^n \bar{S}_i$$
    (b) Prove that for any collection $\{S_\alpha\}_{\alpha \in A}$, $\bigcup \bar{S}_\alpha \subseteq \overline{\bigcup S_\alpha}$.
    (c) Show, by means of an example, that (a) need not hold if the collection is not finite.
    (d) Prove that for any collection $\{S_\alpha\}_{\alpha \in A}$, $\overline{\bigcap_{\alpha \in A} S_\alpha} \subseteq \bigcap_{\alpha \in A} \bar{S}_\alpha$.
    (e) Show, by means of an example, that even for a finite collection $\{S_i\}_{i=1}^n$ of subsets, equality in (d) need not hold.

33. Determine $\bar{A}$, for the given subset A of the topological space $\mathfrak{R}$.
    (a) $A = (1, 3)$
    (b) $A = Z^+$
    (c) $A = Q$ (the set of rational numbers)
    (d) $A = (1, 3) \cup \{5\}$
    (e) $A = \left\{\dfrac{1}{n} \,\middle|\, n \in Z^+\right\}$

Exercise 34-37. (**Dense**) A subset A of a topological space X is said to be dense in X if $\bar{A} = X$ (see Exercise 24-32).

34. Prove that $A \subseteq X$ is dense if and given any $x \in X$ and any neighborhood O of x, $O \cap A \neq \emptyset$.

35. Prove that the set Q of rational numbers is dense in $\Re$.

36. For the topology $\tau = \{\emptyset, X, \{a\}, \{a, b\}, \{a, c, d\}, \{a, b, e\}, \{a, b, c, d\}\}$ on $X = \{a, b, c, d, e\}$, determine the dense subsets of X.

37. For the topology $\tau = \{\emptyset, Z\} \cup \{n, n+1, n+2, \ldots\}_{n \in Z^+}$ on Z^+, determine the dense subset of Z^+.

Exercise 38-43. (**Separation Properties**)

38. A T_0-**space** is a topological space in which for any two distinct points there is an open set containing one of the points and not the other.
    (a) Show that $\Re$ is a T_0-space.
    (b) Give an example of a topological space that is not a T_0-space.

39. A T_1-**space** is a topological space in which for any two distinct points there is a neighborhood of each point not containing the other point.
    (a) Show that $\Re$ is a T_1-space.
    (b) Give an example of a T_0-space that is not a T_1-space (see Exercise 38).

40. T_2-**space** or **Hausdorff** space (see Definition 4.14).
    (a) Show that $\Re$ is a T_2-space.
    (b) Give an example of a T_1-space that is not a T_2-space (see Exercise 39).

41. A T_3-**space** or **regular space** is a T_1-space (see Exercise 40) in which for any closed set H and any point $x \notin H$ there exist disjoint open sets U, V with $x \in U$ and $H \subseteq V$
    (a) Show that $\Re$ is a T_3-space.
    (b) Give an example of a T_2-space that is not a T_3-space (see Exercise 40).

42. A T_4-**space** or **normal space** is a T_1-space (see Exercise 40) in which for any two disjoint closed sets H_1, H_2 there exist disjoint open sets O_1, O_2 with $H_1 \subseteq O_1$ and $H_2 \subseteq O_2$.
    (a) Show that $\Re$ is a T_4-space.
    (b) Give an example of a T_4-space that is not a T_3-space (see Exercise 41).

43. Prove that any compact Hausdorff space is normal (see Exercise 42).

Exercise 44-49. (**Connected Spaces**) A topological space is **connected** if it is not the union of two nonempty disjoint open sets.

44. Prove that a topological space X is connected if and only if the only subsets of X that are both open and closed are $\emptyset$ and X.

45. Prove that any discrete space consisting of 2 or more elements is not connected.

46. Show that the space (X, τ), where $X = \{a, b, c, d\}$ and $\tau = \{\emptyset, X, \{a, b\}\}$ is connected.

47. Show that $\Re$ with the standard Euclidean topology is connected. (You will need to invoke the completion axiom of Definition 3.1, page 111.)

48. Show that the subspace Q (rational numbers) of the Euclidean space $\Re$ is not connected.

49. Prove that if A is a connected subspace of a space X, then the closure of A in X is also connected. (See Exercises 25-33).

	PROVE OR GIVE A COUNTEREXAMPLE	

50. If τ_1, τ_2 are topologies on a set X, then so is $\tau_1 \cup \tau_2$ a topology on X.

51. If Γ is a subbase for a topological space (X, τ), and if $\Gamma \subseteq \Gamma' \subseteq \tau$, then Γ' is also a subbase for τ.

52. If β is a base for a topological space (X, τ), and if $\beta' \subseteq \beta \subseteq \tau$, then β' is also a base for τ.

53. If X is an indiscrete space (CYU 4.11), then any base for X must contain two elements.

54. If X is an indiscrete space (CYU 4.11), then any base for X cannot contain more than two elements.

55. (See Exercises 26-34) For subsets S and T of a topological space X:

    (a) $(\bar{S} \cup \bar{T})^c = \bar{S}^c \cap \bar{T}^c$ 　　(b) $(\bar{S} \cap \bar{T})^c = (\bar{S})^c \cup (\bar{T})^c$ 　　(c) $(\bar{S^c})^c = \bar{S}$

    (d) If $S \subset T$, then $\bar{S} \subset \bar{T}$ 　　(e) If $\bar{S} = \bar{T}$, then $S = T$

56. (See Exercises 44-49) If A and B are connected subspaces of a space X, then:

    (a) $A \cap B$ is also connected. 　　(b) $A \cup B$ is also connected.

    (c) If $A \cap B \neq \emptyset$, then $A \cup B$ is also connected.

57. The set of open subsets of a topological space and that of the closed subsets of the space are of the same cardinality (see Definition 2.15, page 77).

§3 CONTINUOUS FUNCTIONS AND HOMEOMORPHISMS

We extend Definition 4.7 of page 166 to accommodate general topological spaces:

DEFINITION 4.16
CONTINUITY
Let X and Y be topological spaces. A function $f: X \to Y$ is **continuous** if $f^{-1}(O)$ is open in X for every O open in Y.

EXAMPLE 4.8 Show that:

(a) If X is a discrete topological space (CYU 4.12, page 171), then every function $f: X \to Y$ is continuous for every topological space Y.

(b) Let X and Y be topological spaces. For any $y_0 \in Y$ the constant function $f_{y_0}: X \to Y$ given by $f_{y_0}(x) = y_0$ for every $x \in X$ is continuous.

(c) For any subspace S of a topological space X, the inclusion function $I_S: S \to X$ given by $I_S(s) = s$ for every $s \in S$ is continuous.

SOLUTION: (a) Since the discrete topology consists of all subsets of X, $f^{-1}(O)$ is open in X for every open set O in Y.

(b) Let O be open in Y. We consider two cases.
Case 1: $y_0 \in O$. Since, by the definition of f, every $x \in X$ maps to y_0, $f_{y_0}^{-1}(O) = X$ — an open subset of X.

Case 2: $y_0 \notin O$. Since no element in X maps to y_0, $f_{y_0}^{-1}(O) = \emptyset$ — an open subset of X.

(c) Let O be open in X. By Definition 4.12, page 174, $O \cap S$ is open in S. Noting that $I_S^{-1}(O) = O \cap S$, we conclude that $I_S: S \to X$ is continuous.

CHECK YOUR UNDERSTANDING 4.22

Exhibit a continuous function f from the Sierpinski space of CYU 4.13 (page 172) to $\Re$, and a non-continuous function g from the Sierpinski space to $\Re$.

Answer: See page A-24.

Here are several equivalent characterization of the continuity concept:

> **THEOREM 4.10** Let f be a function from a topological space X to a topological space Y. The following are equivalent:
>
> (i) f is continuous.
>
> (ii) $f^{-1}(H)$ is closed in X for every H closed in Y.
>
> (iii) $f^{-1}(S)$ is open in X for every S in a subbase for the topology of Y.
>
> (iv) $f^{-1}(B)$ is open in X for every B in a base for the topology of Y.

PROOF: We show (i) $\to$ (ii) $\to$ (iii) $\to$ (iv) $\to$ (i).

(i) $\to$ (ii). Let H be closed in Y. Since H^c is open, $f^{-1}(H^c)$ is open in X (by continuity). But $[f^{-1}(H)]^c = f^{-1}(H^c)$ [CYU 4.8(a-iv), page 165]. It follows that $f^{-1}(H)$ is closed in X.

(ii) $\to$ (iii). Let Γ be a subbase for Y, with $S \in \Gamma$. Since S is open in Y, S^c is closed. By (ii), $f^{-1}(S^c)$ is closed in X. But $[f^{-1}(S)]^c = f^{-1}(S^c)$ [CYU 4.8(a-iv), page 165]. Taking the complement of both sides we have $f^{-1}(S) = [f^{-1}(S^c)]^c$. Being the complement of a closed set, $f^{-1}(S)$ is open in X.

(iii) $\to$ (iv). Follows directly from the observation that every base for a topology is also a subbase for the topology.

(iv) $\to$ (i). Let β be a base for Y, and let O be open in Y. If $f^{-1}(O) = \varnothing$, then it is open in X.

For $f^{-1}(O) \neq \varnothing$ and $x_0 \in f^{-1}(O)$ choose $y_0 \in O$ such that $f(x_0) = y_0$. Let $B \in \beta$ be such that $y_0 \in B \subseteq O$. We then have $x_0 \in f^{-1}(B) \subseteq f^{-1}(O)$. Since, by assumption, $f^{-1}(B)$ is open: $f^{-1}(O)$ is open.

CHECK YOUR UNDERSTANDING 4.23

Let X be an arbitrary topological space and Y an indiscrete topological space (see CYU 4.12, page 171). Prove that every function $f: X \to Y$ is continuous.

Answer: See page A-24.

Theorem 4.4, page 166, extends to general topological spaces:

THEOREM 4.11 Let X, Y, and Z be any topological spaces.

If $f: X \to Y$ and $g: Y \to Z$ are continuous, then the composite function $g \circ f: X \to Z$ is also continuous.

We just copied the proof of Theorem 4.4.

PROOF: Let O be open in Z. Since g is continuous, $g^{-1}(O)$ is open in Y. Since f is continuous, $f^{-1}[g^{-1}(O)]$ is open in X. The desired result now follows from Theorem 2.6 of page 73 which asserts that $(g \circ f)^{-1}(O) = f^{-1}[g^{-1}(O)]$.

CHECK YOUR UNDERSTANDING 4.24

Construct functions $f: X \to Y$ and $g: Y \to Z$ (where X, Y, and Z are topological spaces) such that:

(a) f is continuous and $g \circ f: X \to Z$ is not continuous.

(b) g is continuous and $g \circ f: X \to Z$ is not continuous.

(c) $g \circ f: X \to Z$ is continuous with neither f nor g continuous.

Answer: See page A-24.

The following result is of particular importance in analysis:

THEOREM 4.12 Let $f: X \to Y$ be continuous. If K is a compact subset of X, then $f(K)$ is a compact subset of Y.

PROOF: Let K be a compact subset of X and let $\{V_\alpha\}_{\alpha \in A}$ be an open cover of $f(K)$. Since f is continuous, $\{f^{-1}(V_\alpha)\}_{\alpha \in A}$ is an open cover of K. Since K is compact, that cover contains a finite subcover $\{f^{-1}(V_{\alpha_i})\}_{i=1}^{n}$. We then have:

$$K \subseteq \bigcup_{i=1}^{n} f^{-1}(V_{\alpha_i}) \Rightarrow f(K) \subseteq f\left[\bigcup_{i=1}^{n} f^{-1}(V_{\alpha_i})\right] = \bigcup_{i=1}^{n} f[f^{-1}(V_{\alpha_i})] \subseteq \bigcup_{i=1}^{n} V_{\alpha_i}$$

Exercise 24, page 169

CHECK YOUR UNDERSTANDING 4.25

PROVE OR GIVE A COUNTEREXAMPLE:

Let $f: X \to Y$ be continuous and let $S \subseteq X$. If $f(S)$ is a compact subset of Y, then S is a compact subset of X.

Answer: See page A-24.

Open and Closed Functions

The definition of a continuous function $f: X \to Y$ kind of "works in reverse" in that it hinges on the behavior of $f^{-1}: Y \to X$:

$f^{-1}(O)$ is open in X for every O open in Y.

Here are a couple of "forward looking" concepts for your consideration:

DEFINITION 4.17
OPEN AND CLOSED FUNCTIONS

Let X and Y be topological spaces. A function $f: X \to Y$ is said to be:

Open if $f(O)$ is open in Y for every O open in X.

Closed if $f(C)$ is closed in Y for every C closed in X.

The following example shows the independent nature of continuous, open, and closed functions.

EXAMPLE 4.9

(a) Give an example of a continuous function that is neither open nor closed.

(b) Give an example of an open function that is neither continuous nor closed.

(c) Give an example of a closed function that is neither continuous nor open.

SOLUTION: (a) Let $X = \{a, b\}$. Let τ_1 denote the discrete topology on X and τ_0 the indiscrete topology (CYU 4.12, page 171). It is easy to see that the identity function $I_X: (X, \tau_1) \to (X, \tau_0)$, given by $I_X(x) = x$ for every $x \in X$, is a **continuous** function which is **neither open nor closed**.

(b) Let X be the subspace $(0, 1)$ of $\Re$, and let Y be the subspace $[0, 1]$ of the half-open interval space H of CYU 4.18, page 176. Consider the inclusion function $J: X \to Y$, given by $J(x) = x$ for every $x \in X$.

J is open: Let O be open in X. Since $X = (0, 1)$ is open in $\Re$, O is open in $\Re$ (Exercise 8, page 179). Since the topology of H contains the Euclidean topology (CYU 4.18), O is open in H. It follows that $J(O) = O \subseteq Y$ is open in the subspace $Y = [0, 1]$ of H.

J is not continuous: The set $[\frac{1}{2}, 1)$ is open in Y but $J^{-1}([\frac{1}{2}, 1)) = [\frac{1}{2}, 1)$ is not open in X.

J is not closed: $X = (0, 1)$ is certainly closed in X. However, $J[(0, 1)] = (0, 1)$ is not closed in Y, since every open neighborhood of 0 in the space Y meets $(0, 1)$.

(c) Let X be the following subspace of $\Re$

$$X = \{0\} \cup \left\{\frac{1}{n} \,\middle|\, n \in Z^+\right\} \cup \left\{2 - \frac{1}{n} \,\middle|\, n \in Z^+\right\} \cup \{2\}$$

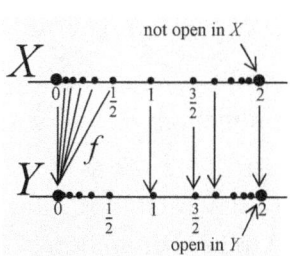

not open in X

open in Y

Let Y be the same as the topological space X, but with $\{2\}$ added to its topology [$\{2\}$ is open in Y (see margin)]. Consider the function $f: X \to Y$ given by:

$$f(x) = \begin{cases} x & \text{if } x \geq 1 \\ 0 & \text{if } x < 1 \end{cases}$$

f is closed: For C closed in X, we show that $[f(C)]^c$ is open in Y. Let $y \notin f(C)$.

Case 1: $y \neq 0$. Since $Y - \{0\}$ is discrete (recall that $\{2\}$, is open in Y) we have that $\{y\}$ is an open set in Y containing no element of $f(C)$.

Case 2: $y = 0$. Since $f(0) = 0$ and $y \notin f(C)$, $0 \notin C$. Since C is closed in X, we can choose an integer N such that

$$\left[\{0\} \cup \left\{\frac{1}{n} \,\middle|\, n > N\right\}\right] \cap C = \varnothing$$

The desired result now follows from the observation that $\{0\} \cup \left\{\frac{1}{n} \,\middle|\, n > N\right\}$ is also open in Y.

f is not continuous: The set $\{2\}$, is open in Y but not in X, and $f^{-1}(\{2\}) = \{2\}$.

f is not open: The set $\left\{\frac{1}{2}\right\}$ is open in X, but $f\left(\left\{\frac{1}{2}\right\}\right) = \{0\}$ is not open in Y.

CHECK YOUR UNDERSTANDING 4.26

Answer: See page A-24.

Give an example of a function $f: X \to Y$ that is open and closed, but not continuous.

HOMEOMORPHIC SPACES

Two topological spaces are considered to be "the same" if there exists a bijection from one to the other which preserves the topological structure of those spaces. To be more precise:

DEFINITION 4.18
HOMEOMORPHIC SPACES
A topological space (X, τ_X) is **homeomorphic** to a topological space (Y, τ_Y), written $X \cong Y$, if there exists a bijection $f: X \to Y$ such that:
$$f(\tau_X) = \{f(O) | O \in \tau_X\} = \tau_Y$$
Such a function f is said to be a **homeomorphism** from X to Y.

> Recall that $f: X \to Y$ is a bijection if it is both one-to-one and onto (Definition 2.13, page 70).

CHECK YOUR UNDERSTANDING 4.27

Prove that a bijection $f: X \to Y$ is a homeomorphism if and only if both f and $f^{-1}: Y \to X$ are continuous.

> Answer: See page A-24.

THEOREM 4.13 Any continuous closed bijection is a homeomorphism.

PROOF: Let $f: X \to Y$ be a continuous closed bijection. Appealing to CYU 4.26, we show that $f^{-1}: Y \to X$ is also continuous, and do so by showing that for any O, open in X, $[f^{-1}]^{-1}(O)$ is open in Y:

Noting that the inverse of the inverse function f^{-1} is the function f, we have $[f^{-1}]^{-1}(O) = f(O) \underset{\text{CYU 4.8(b), page 165}}{=} [f(O^c)]^c$.

And why is $[f(O^c)]^c$ open? Because:

O open $\Rightarrow O^c$ is closed $\underset{f \text{ is closed}}{\Rightarrow} f(O^c)$ is closed $\Rightarrow [f(O^c)]^c$ is open

---

CHECK YOUR UNDERSTANDING 4.28

(a) Prove that any continuous open bijection is a homeomorphism.

(b) Give an example of a continuous bijection $f: X \to Y$ which is not a homeomorphism.

> Answer: See page A-25.

THEOREM 4.14 Any continuous bijection from a compact space to a Hausdorff space is a homeomorphism.

PROOF: Let $f: X \to Y$ be a continuous bijection from a compact space X to a Hausdorff space Y. We complete the proof by showing that f is closed (see Theorem 4.13):

Let C be closed in X. Since X is compact, C is compact (see CYU 4.19, page 177). By Theorem 4.12, $f(C)$ is compact, and therefore closed [Theorem 4.9 (page 178)].

CHECK YOUR UNDERSTANDING 4.29

Answer: See page A-25.

Give an example of a bijection that is both open and closed from a compact space to a Hausdorff space that is not a homeomorphism.

TOPOLOGICAL INVARIANT PROPERTIES

A major goal in topology is to determine when two spaces are homeomorphic. Finding an explicit homeomorphism may not be easy. There is, however, a useful technique that can at times be used to show that two spaces are not homeomorphic:

> Properties of a topological space which are preserved under homeomorphisms are said to be **topological invariant properties.** Exhibiting a topologically invariant property possessed by one space and not by another can serve to show that the two spaces are not homeomorphic.

In particular, if two topological spaces X and Y are not of the same cardinality then they cannot be homeomorphic. Why not? Because any homomorphism $f: X \to Y$ must be a bijection.

Here is a less obvious topological invariant property:

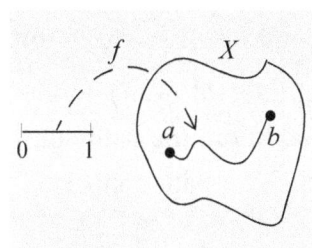

DEFINITION 4.19
PATH

A **path** in a topological space X is the image of a continuous function $f: ([0, 1] \to X)$.

PATH CONNECTED

A space X is **path connected** if for any two given points $a, b \in X$ there exists a path: $f: [0, 1] \to X$ with $f(0) = a$ and $f(1) = b$.

THEOREM 4.15 Path connectedness is a topological invariant property.

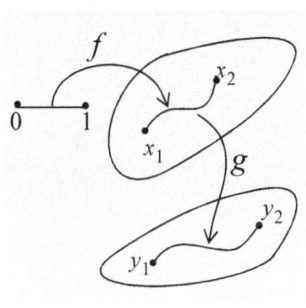

PROOF: Let X be a path connected space, and let Y be homeomorphic to X. We show that Y must also be path connected:

Since $X \cong Y$, there exists a homeomorphism $g: X \to Y$. For any two given points $y_1, y_2 \in Y$, let $x_1, x_2 \in X$ be such that $f(x_1) = y_1$ and $f(x_2) = y_2$. Since X is path connected, we know that there exists a continuous function $f: [0, 1] \to X$ such that $f(0) = x_1$ and $f(1) = x_2$. The continuous function $g \circ f: [0, 1] \to Y$ (Theorem 4.11), produces a path joining y_1 to y_2:

$$(f \circ g)(0) = f[g(0)] = f(x_1) = y_1 \text{ and } (f \circ g)(1) = y_2.$$

In the exercises you are asked to verify that the subspace $(1, 3)$ of $\Re$ is path connected, and that $(0, 1) \cup (2, 3)$ is not. It follows, from the previous theorem, that the two spaces are not homeomorphic.

CHECK YOUR UNDERSTANDING 4.30

(a) Prove that compactness is a topological invariant property.

(b) Are the subspaces $(0, 1)$ and $[0, 1]$ of $\Re$ homeomorphic? Justify your answer.

Answer: See page A-25.

SOME CASUAL REMARKS

Roughly speaking two spaces are homeomorphic if by stretching one of them you can arrive at the other. Indeed, topology is often called the "rubber-sheet geometry." In particular, one can stretch the "rubber" interval $(1, 3)$ to get to the interval $(1, 5)$, and the interval $\left(-\frac{\pi}{2}, \frac{\pi}{2}\right)$ to the interval $(-\infty, \infty)$ (just keep on stretching):

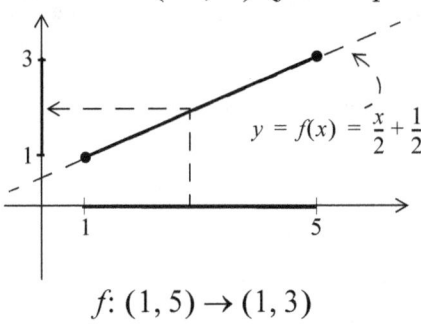

$f: (1, 5) \to (1, 3)$

A homeomorphism [Exercise 17(a)]

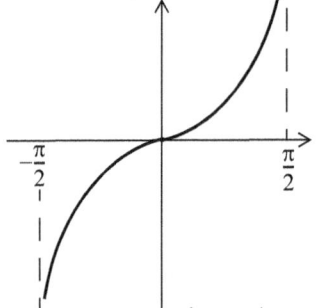

$f(x) = \tan x: \left(-\frac{\pi}{2}, \frac{\pi}{2}\right) \to \Re$

A homeomorphism [Exercise 17(b)]

One cannot, however, stretch the non-compact interval $(1, 3)$ to the compact interval $[1, 5]$ [see CYU 4.28(a)]. In addition, the path connected interval $(1, 3)$ is not homeomorphic to the non path connected $(1, 2) \cup (2, 3)$ (see Theorem 4.15).

CHECK YOUR UNDERSTANDING 4.31

Prove that *homeomorphic* is an equivalence relation ($\cong$) on any set of topological spaces (see Definition 2.20, page 88).

Answer: See page A-25.

	EXERCISES	

1. Let f be a function from a topological space X to $\Re$. Prove that f is continuous if and only if for any $a \in \Re$ both the sets $\{x | f(x) > a\}$ and $\{x | f(x) < a\}$ are open.

2. Let Y be any space that is not indiscrete. Show there exists a space X and a function $f: X \to Y$ that is not continuous.

3. Let X be any space that is not discrete. Show there exists a space Y and a function $f: X \to Y$ that is not continuous.

4. Let (X, τ) be a topological space and let (Y, d) be a metric space. Prove that a function $f: X \to Y$ is continuous at $x \in X$ if and only if for any $\varepsilon > 0$ there exists a neighborhood O of x such that $d([f(x), f(y)]) < \varepsilon$ for every $y \in O$.

5. Let f and g be continuous functions from a topological space X to $\Re$. Prove that:

    (a) $af + bg: X \to \Re$ is continuous for any $a, b \in \Re$.

    (a) $fg: X \to \Re$ is continuous.

    (a) $\frac{f}{g}: X \to \Re$ is continuous if $g(x) \neq 0$ for any $x \in X$.

6. Give an example of a function $f: X \to Y$ that is continuous and open, but not closed.

7. Give an example of a function $f: X \to Y$ that is continuous and closed, but not open.

8. Let $f: X \to Y$ be a bijection. Prove that f is open if and only if $f^{-1}: Y \to X$ is continuous.

9. (a) Show that a function $f: X \to Y$ is open in Y if $f(B)$ is open in Y for every B in a base for the topology of X.

    (b) Show, by means of an example, that (a) need not hold when the word "base" is replaced with "subbase."

    (c) Show that in the event that f is a bijection, then (a) will hold when the word "base" is replaced with "subbase."

10. (a) Prove that a bijection $f: X \to Y$ is a closed function if and only if it is an open function.

    (b) Give and example of an open onto function that is neither closed nor continuous.

    (c) Give and example of an open one-to-one function that is neither closed nor continuous.

    (d) Give and example of a closed onto function that is neither open nor continuous.

    (e) Give and example of a closed one-to-one function that is neither open nor continuous.

11. Let $\{f_i: X_i \to X_{i+1}\}_{i=1}^{n}$ be a collection of continuous functions. Use the Principle of Mathematical Induction to show that $f_n \circ f_{n-1} \circ \cdots \circ f_2 \circ f_1: X_1 \to X_{n+1}$ is also continuous.

12. **Closure** (See Exercises 25-33, page 182)]. Establish the equivalence of the following three properties:

    (i) $f: X \to Y$ is continuous.

    (ii) $f(\bar{A}) \subseteq \overline{f(A)}$ for every $A \subseteq X$.

    (iii) $\overline{f(\bar{A})} \subseteq \overline{f(A)}$ for every $A \subseteq X$.

13. [**Closure** (See Exercises 25-33, page 182)]. Prove that $f: X \to Y$ is closed if and only if $\overline{f(A)} \subseteq f(\bar{A})$ for every $A \subseteq X$.

14. Let $f: X \to Y$ be closed. Show that for any $A \subseteq Y$ and any open set U containing $f^{-1}(A)$ there exists an open set V containing A such that $f^{-1}(V) \subseteq U$.

15. Let $f: X \to Y$ be open. Show that for any $A \subseteq Y$ and any closed set H containing $f^{-1}(A)$ there exists a closed set V containing A such that $f^{-1}(V) \subseteq H$.

16. Give an example of a set X, and two topologies τ and τ' such that the identity function $I_X: (X, \tau) \to (Y, \tau')$ is:

    (a) Continuous but not open.
    (b) Continuous but not closed.
    (c) Continuous but not a homeomorphism.
    (d) Open but not closed.
    (e) Open but not continuous.
    (f) Open but not a homeomorphism.
    (g) Closed but not open.
    (h) Closed but not continuous.
    (i) Closed but not a homeomorphism.

17. (a) Show that $f: (1, 5) \to (1, 3)$ given by $f(x) = \dfrac{x}{2} + \dfrac{1}{2}$ is a homeomorphism.

    (b) Show that $f: \left(-\dfrac{\pi}{2}, \dfrac{\pi}{2}\right) \to \Re$ given by $f(x) = \tan x$ is a homeomorphism.

18. Is the closed unit interval $[0, 1]$ homeomorphic to the open interval $(-1, 1)$? Justify your answer.

19. Give an example of an open bijection from a compact Hausdorff space to a Hausdorff space that is not a homeomorphism.

20. Give an example of a closed bijection from a compact Hausdorff space to a Hausdorff space that is not a homeomorphism.

21. [Let $f: X \to Y$ be a bijection. Show that the following properties are equivalent:

    (i) f is a homeomorphism.
    (ii) f is continuous and open.
    (ii) f is continuous and closed.
    (iii) $f(\bar{A}) = \overline{f(A)}$ for every $A \subseteq X$.

22. Show that the Sierpinski space (CYU 4.13, page 172) is not homeomorphic to the discrete space of two points (CYU 4.12, page 171).

23. Show that the subspace $(1, 3)$ of $\Re$ is path connected, and that $(0, 1) \cup (2, 3)$ is not.

24. **(Fixed Point Property).** A nonempty space X satisfies the **fixed point property** if for any continuous function $f: X \to X$ there exists $x \in X$ such that $f(x) = x$. Prove that the fixed point property is a topological invariant.

25. (a) Prove that the existence of a proper subset of a topological space X that is both open and closed is a topological invariant property.

    (b) Use (a) to show that the subspaces $(0, 1)$ and $(0, 1) \cup \{2\}$ of $\Re$ are not homeomorphic.

26. Prove that connectedness is a topological invariant property (see Exercise 44-49, page 184).

27. Prove that the cardinality of the set of subsets of a topological space that are both open and closed is a topological invariant property.

28. Prove that metrizable is a topological invariant property.

29. Prove that each of the separation properties: T_0, T_1, T_2, T_3, T_4 of Exercises 38-42, page 183, is a topological invariant property.

	PROVE OR GIVE A COUNTEREXAMPLE	

30. If Y is an indiscrete space, then every function $f: X \to Y$ is continuous for every space X.

31. Any continuous open bijection is a homeomorphism.

32. Two topological spaces, X and Y, are homeomorphic if and only if $\mathrm{Card}(X) = \mathrm{Card}(Y)$.

33. If the space X is homeomorphic to a space Y, then $\mathrm{Card}(X) = \mathrm{Card}(Y)$.

34. (See Exercises 34-37, page 183.) If $f: X \to Y$ is a homeomorphism, and if A is dense in Y, then $f(A)$ is dense in Y.

35. (See Exercises 34-37, page 183.) If $f: X \to Y$ is continuous, and if A is dense in X, then $f(A)$ is dense in Y.

36. (See Exercises 34-37, page 183.) If $f: X \to Y$ is onto and continuous, and if A is dense in X, then $f(A)$ is dense in Y.

37. (See Exercises 34-37, page 183.) If $f: X \to Y$ is continuous, and if A is dense in X, then $f(A)$ is dense in Y.

§4 PRODUCT AND QUOTIENT SPACES

See Definition 2.7, page 63.

Let's impose a topology on the Cartesian product of two topological spaces:

DEFINITION 4.20

PRODUCT TOPOLOGY ON TWO SPACES

Let (X, τ_X) and (Y, τ_Y) be topological spaces. The **product topology** on
$$X \times Y = \{(x, y) | x \in X, y \in Y\}$$
is that topology with basis:
$$\beta = \{U \times V | U \in \tau_X, V \in \tau_Y\}$$

Note that while the above collection β is closed under finite intersections (Exercise 2), it need not be closed under unions. For example, while the intersection of the two rectangles $(1, 3) \times (1, 2)$ and $(2, 4) \times (0, 3)$ in Figure 4.1 is again of the form $U \times V$ with U and V open in $\Re$, their union is not. It follows that β is not a topology on $\Re \times \Re$; but then again, it does not profess to be a topology, but rather a basis for the product topology on $\Re \times \Re$: the collection of all arbitrary unions of elements from β.

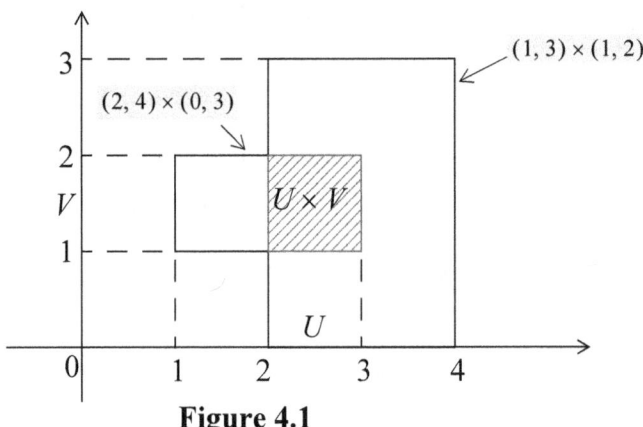

Figure 4.1

CHECK YOUR UNDERSTANDING 4.32

(a) Show that if β_X and β_Y are bases for the topological spaces X and Y, respectively, then
$$\gamma = \{D \times E | D \in \beta_X \text{ and } E \in \beta_Y\}$$
is a basis for the product topology on $X \times Y$.

(b) Show that the product topology on $\Re \times \Re$ coincides with the Euclidean topology on $\Re^2$.

Answer: See page A-26.

As you will see, the following result continues to hold for any collection of compact spaces.

THEOREM 4.16 If X and Y are compact spaces, then the product space $X \times Y$ is also compact.

PROOF: If either X or Y is empty, then so is $X \times Y$, and, as such, is compact. That being the case, we need only consider the case where neither X nor Y is empty.

Appealing to Exercise 11, page 179, we show that $X \times Y$ is compact by showing that every open cover of $X \times Y$, taken from the **basis** $U \times V$, where U is open in X and V is open in Y, has a finite subcover:

Let $\{U_\alpha \times V_\alpha\}_{\alpha \in A}$ be a cover of $X \times Y$, where each U_α is open in X, and each V_α is open in Y. For any $x \in X$ and $y \in Y$, the element (x, y) is contained in some $U_\alpha \times V_\alpha$. It follows that $\{U_\alpha\}_{\alpha \in A}$ and $\{V_\alpha\}_{\alpha \in A}$ are open covers of X and Y, respectively. By compactness, we can choose finite subcovers $\{U_{\alpha_i}\}_{i=1}^{n}$ and $\{V_{\alpha_i}\}_{i=1}^{m}$ of X and Y, respectively. Noting that for any $(x, y) \in X \times Y$ there exist elements U_{α_i} and V_{α_j} of those subcovers containing x and y respectively, we conclude that the finite collection $U_{\alpha_i} \times V_{\alpha_j} | 1 \leq i \leq n, 1 \leq j \leq m$ covers $X \times Y$.

CHECK YOUR UNDERSTANDING 4.33

Establish the converse of Theorem 4.16: If $X \times Y$ is compact, then X and Y are compact.

Answer: See page A-26.

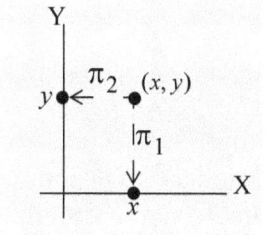

DEFINITION 4.21
PROJECTION FUNCTIONS

Let X and Y be nonempty sets. The functions $\pi_1 : X \times Y \to X$ and $\pi_2 : X \times Y \to Y$ given by $\pi_1(x, y) = x$ and $\pi_2(x, y) = y$ are called the **projection functions** onto X and Y, respectively.

THEOREM 4.17 If X and Y are nonempty topological spaces, the projection functions $\pi_1 : X \times Y \to X$ and $\pi_2 : X \times Y \to Y$ are continuous.

PROOF: To see that π_1 is continuous you need but note that if U is open in X, then $\pi_1^{-1}(U) = U \times Y$ is open in $X \times Y$ (see Definition 4.20). The same argument can be applied to $\pi_2 : X \times Y \to Y$.

CHECK YOUR UNDERSTANDING 4.34

Answer: See page A-26.

Are the projection functions of Theorem 4.17 necessarily open? Closed? Justify your answer.

In Exercise 14, page 180, you were asked to show that any collection Γ of subsets of a given set generates a topology on that set; namely: the set of arbitrary unions of finite intersections of elements taken from Γ. As it turns out:

THEOREM 4.18 Let X and Y be topological spaces. The product topology on $X \times Y$ is generated by:

$$\{\pi_1^{-1}(U) | U \text{ is open in } X\} \cup \{\pi_2^{-1}(V) | V \text{ is open in } Y\}$$

(In other words: the above collection constitutes a subbase for the product topology.)

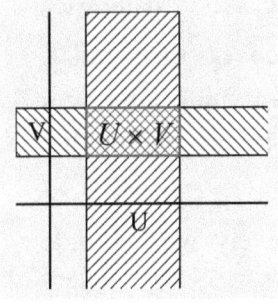

PROOF: The claim is a consequence of Definition 4.20 and the observation that for any open set U in X, and any open set V in Y:

$$U \times V = \pi_1^{-1}(U) \cap \pi_2^{-1}(V) \quad \text{(see margin)}$$

GENERAL PRODUCT SPACES

The Cartesian product of two sets can easily be extended to accommodate a finite number of sets $X_1, X_2, \ldots, X_n$:

$$\prod_{i=1}^{n} X_i = \{(x_1, x_2, \ldots, x_n) | x_i \in X_i\}$$

A countable number of sets:

$$\prod_{i=1}^{\infty} X_i = \{(x_1, x_2, \ldots, x_n, \ldots) | x_i \in X_i\}$$

Any collection of sets:

$$\prod_{\alpha \in A} X_\alpha = \{(x_\alpha) | x_\alpha \in X_\alpha\}$$

By the same token, Definition 4.29 can be generalized to accommodate any collection of topological spaces. Let's first turn to a finite collection:

DEFINITION 4.22
PRODUCT TOPOLOGY ON n SPACES

Let $\{(X_i, \tau_{X_i})\}_{i=1}^{n}$ be a collection of topological spaces. The **product topology** on $\prod_{i=1}^{n} X_i$ is that topology with basis:

$$\beta = \{U_1 \times U_2 \times \ldots \times U_n | U_i \in \tau_{X_i}\}$$

Which is to say: The set consisting of arbitrary unions of elements from β.

The following generalization of Definition 4.21 will enable us to extend the concept of a product topology to any collection of topological spaces:

DEFINITION 4.23
PROJECTION FUNCTION

Let $\{X_\alpha\}_{\alpha \in A}$ be a collection of nonempty sets. For $\bar{\alpha} \in A$, the function

$$\pi_{\bar{\alpha}} : \prod_{\alpha \in A} X_\alpha \to X_{\bar{\alpha}} \text{ given by } \pi_{\bar{\alpha}}(x_\alpha)_{\alpha \in A} = x_{\bar{\alpha}}$$

is called the **projection function** onto the $\bar{\alpha}^{\text{th}}$ component of $\prod_{\alpha \in A} X_\alpha$.

DEFINITION 4.24
PRODUCT TOPOLOGY

Let $\{X_\alpha, \tau_{X_\alpha}\}_{\alpha \in A}$ be a collection of topological spaces. The **product topology** on $\prod_{\alpha \in A} X_\alpha$ is that topology with **subbase**

$$\Gamma = \{\pi_\alpha^{-1}(U_a) \mid U_a \in \tau_{X_\alpha}, \alpha \in A\}$$

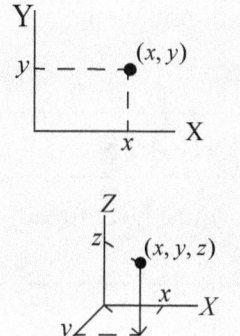

While one is able to geometrically depict the Cartesian product of two or three sets (see margin), the same cannot be said for four or more sets. We can, however, provide a visual sense of the product topology $\prod_{\alpha \in A} X_\alpha$ of an arbitrary collection of spaces by first representing each X_α as a vertical line, as is done in Figure 4.2. An open set O is also depicted (in blue) in the figure. Note that being a union of **finite intersections** of the subbase elements $\pi_\alpha^{-1}(U_a)$, only a **finite** number of the X_α lines can fail to be "entirely blue."

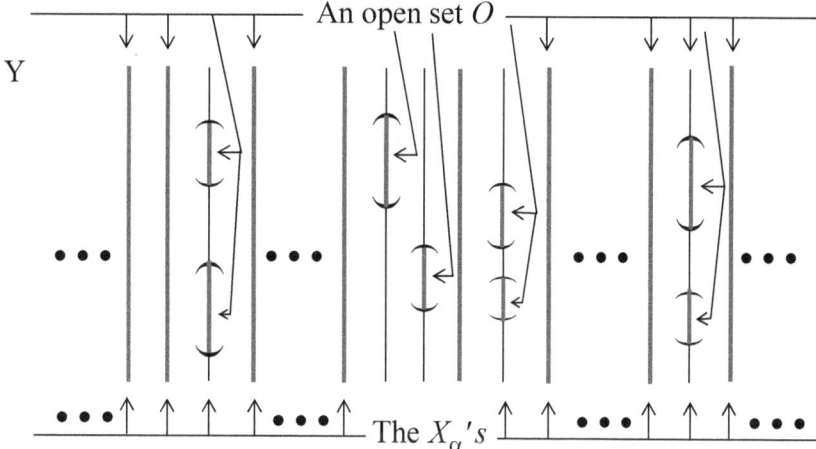

Figure 4.2

Theorem 4.16 assured us that the product of two compact spaces is again compact. Much more can be said:

Andrey Tychonoff (1906-1993)

THEOREM 4.19
TYCHONOFF'S PRODUCT THEOREM

Any product of compact spaces is compact.

PROOF: Appealing to Alexander's Subbase Theorem of page 177, we assume that $F \subseteq \{\pi_\alpha^{-1}(U_\alpha) | U_\alpha \in \tau_{X_\alpha}\}_{\alpha \in A}$ is a cover of $X = \prod_{\alpha \in A} X_\alpha$ that has no finite subcover, and arrive at a contradiction:

For any $\alpha \in A$, $\{U_\alpha | \pi_\alpha^{-1}(U_\alpha) \in F\}$ cannot cover the compact space X_α. For if it did, then there would exist a finite subfamily $\{U_{\alpha_i} | \pi_\alpha^{-1}(U_{\alpha_i}) \in F\}_{i=1}^n$ which covers X_α, resulting in a finite subfamily of F which covers X; namely: $\{\pi_\alpha^{-1}(U_{\alpha_i})\}_{i=1}^n$. Consequently, for every $\alpha \in A$, we can choose an element $x_\alpha \in X_\alpha$ which is not contained in any U_α with $\pi_\alpha^{-1}(U_\alpha) \in F$. Since the particular point $(x_\alpha)_{\alpha \in A} \in X$ is not contained in any element of F, F does not cover X — a contradiction.

CHECK YOUR UNDERSTANDING 4.35

Show that if $X = \prod_{\alpha \in A} X_\alpha$ is compact then each X_α must be compact.

Answer: See page A-27.

QUOTIENT SPACES

We remind you that:
- An **equivalence relation** on a set X is a relation, $\sim$, that is reflexive, symmetric, and transitive (Definition 2.20, page 88).
- The **equivalence class** of $x_0 \in X$, denoted by $[x_0]$, is given by $[x_0] = \{x \in X | x \sim x_0\}$ (Definition 2.21, page 91).

We will denote the set of equivalence classes associated with an equivalence relation $\sim$ on a topological space X by the symbol $X/\sim$:

$$X/\sim \, = \{[x]\}_{x \in X}$$

In addition, the function $\pi: X \to X/\sim$, given by $\pi(x) = [x]$, will be called the **projection function** of X onto $X/\sim$.

As it turns out, the topology on X imposes a topology on the set $X/\sim$:

THEOREM 4.20 If $\sim$ be an equivalence relation on a topological space $(X, \mathcal{T})$, then

QUOTIENT TOPOLOGY
$$\tilde{\mathcal{T}} = \{\pi^{-1}(O) \subseteq X \mid O \in \mathcal{T}\}$$

is a topology on $X/\sim$ (called a **quotient topology**).

QUOTIENT SPACE
The resulting topological space $\left(X/\sim, \tilde{\mathcal{T}}\right)$ is said to be a **quotient space** of X.

PROOF: We verify that the three defining axioms of a topological space (Definition 4.9, page 171) are satisfied:

(i) Since $\varnothing$ and X are open in X, $\pi^{-1}(\varnothing) = \varnothing$ and $\pi^{-1}(X) = X/\sim$ are open in $X/\sim$.

(ii) Arbitrary unions of open sets in $X/\sim$ are again open.

$$\bigcup_{\alpha \in A} \pi^{-1}(O_\alpha) \underset{\text{Exercise 25, page 169}}{=} \pi^{-1}\left(\underset{\text{open in } X}{\bigcup_{\alpha \in A} O_\alpha}\right),$$

(iii) Finite intersections of open sets in $X/\sim$ are again open.

$$\bigcap_{i=1}^{n} \pi^{-1}(O_i) \underset{\text{Exercise 25, page 169}}{=} \pi^{-1}\left(\underset{\text{open in } X}{\bigcap_{i=1}^{n} O_i}\right)$$

Here are a couple of specific quotient spaces for your consideration:

> While every point in the open interval $(0, 2\pi)$ is identified with itself only, the two end points of the closed interval are identified with each other. The visual effect is that of gluing one end point of the interval to the other.

(a) Let $\sim$ be the equivalence relation on the subspace $[0, 2\pi]$ of $\Re$ represented by the partition $[x] = \{x\}$ for every $x \in (0, 2\pi)$, and $[0] = [2\pi] = \{0, 2\pi\}$. As is suggested in Figure 4.3, the quotient space $[0, 2\pi]/\sim$ is homeomorphic to the unit circle $S^1 = \{(x, y) \in \Re^2 \mid x^2 + y^2 = 1\}$ — a fact that can be rigorously established by showing that the function $f: [0, 2\pi]/\sim \to S^1$ given by $f([x]) = (\cos x, \sin x)$ is a homeomorphism.

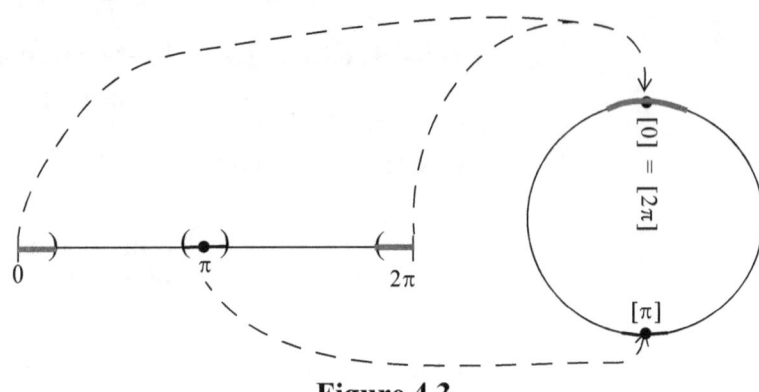

Figure 4.3

(b) Consider the space $X/\sim$ where $X = [0, 2\pi] \times [0, 1]$, and $\sim$ is the relation represented by the partition $[(x, y)] = \{(x, y)\}$ for $0 < x < 2\pi$, $0 \le y \le 1$, and $[(0, y)] = [1, y] = \{(0, y), (1, y)\}$ for $0 \le y \le 1$. As is suggested in Figure 4.4, the quotient space $X/\sim$ is homeomorphic to the cylinder $C = S^1 \times [0, 1]$ — a fact that can be rigorously established by showing that the function $f: (X/\sim) \to C$ given by

$$f([(x, y)]) = (\cos x, \sin x, y) \quad \text{is a homeomorphism.}$$

> Visually, we are gluing the left and right edges of the square together.

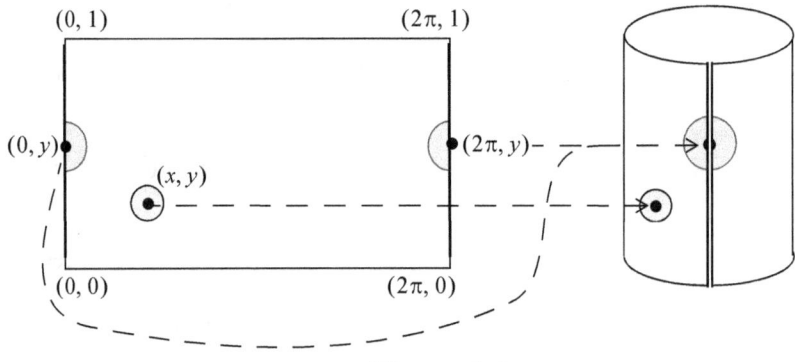

Figure 4.4

CHECK YOUR UNDERSTANDING 4.36

Let X denote the rectangle of Figure. 4.4. Define an equivalence relation $\sim$ on X in such a way so that $X/\sim$ is homeomorphic to a torus.

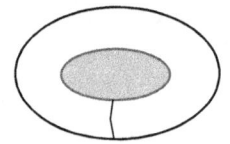

Answer: See page A-27.

4.4 Product and Quotient Spaces

EXERCISES

1. Let $\tau_1 = \{\emptyset, X, \{a, b\}\}$ and $\tau_2 = \{\emptyset, X, \{a\}, \{a, b\}, \{a, c\}\}$ be two topologies on the set $X = \{a, b, c\}$. Determine the topology on the product space:

   (a) $(X, \tau_1) \times (X, \tau_1)$ (b) $(X, \tau_1) \times (X, \tau_2)$ (c) $(X, \tau_2) \times (X, \tau_2)$

2. (a) Let (X, τ_X) and (Y, τ_Y) be topological spaces. Prove that $S = \{U \times V | U \in \tau_X, V \in \tau_Y\}$ is closed under finite intersections.

   (b) Let $\{(X_i, \tau_{X_i})\}_{i=1}^n$ be a collection of topological spaces. Prove that $\beta = \{U_1 \times U_2 \times \ldots \times U_n | U_i \in \tau_{X_i}\}$ is closed under finite intersections.

3. Prove that a space X is Hausdorff if and only if $\{(x, x) | x \in X\}$ is closed in $X \times X$.

4. (a) Prove that the product of two Hausdorff spaces is Hausdorff.
   (b) Prove that the product of Hausdorff spaces is Hausdorff.

5. (a) Prove that the product of two regular spaces is regular (see Exercise 41, page 183).
   (b) Prove that the product of regular spaces is regular (see Exercise 41, page 183).

6. Let f and g be functions from a topological space X to a topological space Y. Let $h: X \to Y \times Y$ be given by $h(x) = (f(x), g(x))$. Prove that h is continuous if and only if both f and g are continuous.

7. For given functions $f: X_1 \to Y_1$ and $g: X_2 \to Y_2$, let $(f \times g): X_1 \times X_2 \to Y_1 \times Y_2$ be given by $(f \times g)(x_1, x_2) = (f(x_1), g(x_2))$. Prove that $f \times g$ is continuous if and only if f and g are continuous.

8. (a) Let U and V be a subspaces of X and Y, respectively. Prove that $U \times V$ is a subspace of $X \times Y$.

   (b) Let S_i be a subspace of a space X_i, $1 \leq i \leq n$. Prove that $\prod_{i=1}^n S_i$ is a subspace of $\prod_{i=1}^n X_i$.

   (c) Let S_α be a subspace of a space X_α, $\alpha \in A$. Prove that $\prod_{\alpha \in A} S_\alpha$ is a subspace of $\prod_{\alpha \in A} X_\alpha$.

9. (a) Let U and V be subsets of the spaces X and Y, respectively. Prove that $U \times V$ is dense in $X \times Y$ if and only if U is dense in X, and V is dense in Y. (See Exercises 34-37, page 183.)

   (b) Let S_i be a subset of a space X_i, $1 \le i \le n$. Prove that $\prod_{i=1}^{n} S_i$ is dense in $\prod_{i=1}^{n} X_i$ if and only if each S_i is dense in X_i.

   (c) Let S_α be a subset of a space X_α, $\alpha \in A$. Prove that $\prod_{\alpha \in A} S_\alpha$ is dense in $\prod_{\alpha \in A} X_\alpha$ if and only if each S_α is dense in X_α.

10. (a) Let X and Y be topological spaces. Prove that the projection functions $\pi_1 : X \times Y \to X$ and $\pi_2 : X \times Y \to Y$ are open.

    (b) Prove that each projection function $\pi_\alpha : \prod_{\alpha \in A} X_\alpha \to X_\alpha$ is open.

11. (a) Prove that for any functions $f: X_1 \to Y_1$ and $g: X_2 \to Y_2$, the function $h: X_1 \times X_2 \to Y_1 \times Y_2$ given by $h(x_1, x_2) = [f(x_1), g(x_2)]$ is continuous if and only if f and g are both continuous.

    (b) Let $f_\alpha: X_\alpha \to Y_\alpha$ be given, for $\alpha \in A$. Prove that the function $h: \prod_{\alpha \in A} X_\alpha \to \prod_{\alpha \in A} Y_\alpha$ given by $h(x_\alpha)_{\alpha \in X} = [f_\alpha(x_\alpha)]_{\alpha \in A}$ is continuous if and only if each f_α is continuous.

12. (a) Let X, Y_1, and Y_2 be topological spaces. Prove that a function $f: X \to Y_1 \times Y_2$ is continuous if and only if $\pi_1 \circ f: X \to Y_1$ and $\pi_2 \circ f: X \to Y_2$ are continuous.

    (b) Prove that a function $f: X \to \prod_{\alpha \in A} Y_\alpha$ is continuous if and only if the functions $\pi_\alpha \circ f: X \to Y_\alpha$ are continuous for every $\alpha \in A$.

13. (a) Let S_1 and S_2 be closed subsets of the topological spaces X_1 and X_2, respectively. Prove that $S_1 \times S_2$ is a closed subset of $X_1 \times X_2$.

    (b) Let S_α be a closed subset of X_α, for $\alpha \in A$. Prove that $\prod_{\alpha \in A} S_\alpha = \prod_{\alpha \in A} X_\alpha$.

14. (a) Let S_1 and S_2 be subsets of the topological spaces X_1 and X_2, respectively. Prove that $\overline{S_1 \times S_2} = \overline{S_1} \times \overline{S_2}$. (See Exercises 25-33, page 182.)

    (b) Let S_α be a subset of X_α, for $\alpha \in A$. Prove that $\overline{\prod_{\alpha \in A} S_\alpha} = \prod_{\alpha \in A} \overline{S_\alpha}$.

15. Prove that for any two topological spaces X and Y, $X \times Y \cong Y \times X$.

16. (a) Prove that if X_1 is homeomorphic to Y_1, and if X_2 is homeomorphic to Y_2, then $X_1 \times X_2$ is homeomorphic to $Y_1 \times Y_2$.

    (b) Prove that if X_α is homeomorphic to Y_α, for $\alpha \in A$, then $\prod_{\alpha \in A} X_\alpha$ is homeomorphic to $\prod_{\alpha \in A} Y_\alpha$.

17. (a) Let ~ be an equivalence relation on a space X. Prove that $X/\sim$ is T_1 if and only if each equivalence class $[x]$ is closed in X. (See Exercise 39, page 183.)

    (b) Give an example of a T_1-space X and an equivalence relation ~ on X such that $X/\sim$ is not T_1.

18. Let ~ be the equivalence relation on the space $X = [0, 1] \times [0, 1]$ given by $(x_1, y_1) \sim (x_2, y_2)$ if and only if $(y_1 = y_2) > 0$. Describe the quotient space $X/\sim$ and show that it is not a Hausdorff space.

19. (a) Let X and Y be topological spaces. Let ~ be the equivalence relation given by $(x_1, y_1) \sim (x_2, y_2)$ if and only if $x_1 = x_2$. Prove that $(X \times Y)/\sim$ is homeomorphic to X.

    (b) Generalize (a) to accommodate a collection $\{X_i\}_{i=1}^n$ of spaces.

    (c) Generalize (a) to accommodate a collection $\{X_\alpha\}_{\alpha \in A}$ of spaces.

20. Let ~ be an equivalence relation on a compact space X. Prove that $X/\sim$ is compact.

21. Let ~ be the equivalence relation on the subspace $[0, 2\pi]$ of $\Re$ induced by the partition $[x] = \{x\}$ for every $x \in (0, 2\pi)$, and $[0] = [2\pi] = \{0, 2\pi\}$ (see Figure 4.4). Show that the function $f: ([0, 2\pi]/\sim) \to S_1$ given by $f([x]) = (\cos x, \sin x)$ is a homeomorphism.

22. Let ~ be the equivalence relation on the space $X = [0, 2\pi] \times [0, 1]$ induced by the partition $[(x, y)] = \{(x, y)\}$ for $0 < x < 2\pi$, $0 \le y \le 1$, and $[(0, y)] = [1, y] = \{(0, y), (1, y)\}$ for $0 \le y \le 1$ (see Figure 4.4). Let S^1 denote the unit circle: $S^1 = \{(x, y) \in \Re^2 | x^2 + y^2 = 1\}$. Show that the function $f: X/\sim \to S^1 \times [0, 1]$ given by $f([(x, y)]) = (\cos x, \sin x, y)$ is a homeomorphism.

	PROVE OR GIVE A COUNTEREXAMPLE	

23. The Cartesian Product of two metrizable spaces is again metrizable.

24. The Cartesian Product of any collection of metrizable spaces is again metrizable.

25. For any two topological spaces X and Y, the projection functions $\pi_1 : X \times Y \to X$ and $\pi_2 : X \times Y \to Y$ are closed.

26. For any three topological spaces X, Y_1, and Y_2, a function $f: X \to Y_1 \times Y_2$ is open if and only if $\pi_1 \circ f: X \to Y_1$ and $\pi_2 \circ f: X \to Y_2$ are open.

27. For any four nonempty topological spaces X_1, X_2, Y_1, and Y_2, if $X_1 \times Y_1$ is homeomorphic to $X_2 \times Y_2$, then X_1 is homeomorphic to X_2 and Y_1 is homeomorphic to Y_2.

28. For any three nonempty topological spaces X, Y_1, and Y_2, if $Y_1 \times X$ is homeomorphic to $Y_2 \times X$, then Y_1 is homeomorphic to Y_2.

CHAPTER 5
A Touch of Group Theory

§1. DEFINITIONS AND EXAMPLES

The following properties reside in the familiar set Z of integers:

	Property	Example:
Closure	$a + b \in Z \ \forall a, b \in Z$	$5 + 7 \in Z$
Associative	1. $a + (b + c) = (a + b) + c \ \forall a, b \in Z$	$5 + (4 + 1) = (5 + 4) + 1$
Identity	2. $a + 0 = a \ \forall a \in Z$	$4 + 0 = 4$
Inverse	3. $a + (-a) = 0 \ \forall a \in Z$	$5 + (-5) = 0$

A generalization of the above properties bring us to the definition of a group — an abstract structure upon which rests a rich theory, with numerous applications throughout mathematics, the sciences, architecture, music, the visual arts, and elsewhere:

A **binary operator** on a set X is a function that assigns to any **two** elements in X an element **in** X. Since the function value resides back in X, one says that the operator is **closed**.

Evariste Galois defined the concept of a group in 1831 at the age of 20. He was killed in a duel one year later, while attempting to defend the honor of a prostitute.

We show, in the next section, that both the **identity element** e and the **inverse element** a' of Axioms 2 and 3 are, in fact, both unique and "ambidextrous:"
$$a*e = e*a = a$$
$$a*a' = a'*a = e$$

DEFINITION 5.1
GROUP

A **group** $\langle G, *\rangle$, or simply G, is a nonempty set G together with a binary operator, $*$, (see margin) such that:

Associative Axiom: 1. $a*(b*c) = (a*b)*c$ for every $a, b, c \in G$.

Identity Axiom: 2. There exists an element in G, which we will label e, such that $a*e = a$ for every $a \in G$.

Inverse Axiom: 3. For every $a \in G$ there exists an element, $a' \in G$ such that $a*a' = e$.

In particular, $\langle Z, +\rangle$ is a group; with "+, 0, and $-a$" playing the role of "$*$, e, and a'" in the above definition.

Is the set of integers under multiplication a group? No:

While "regular" multiplications is an associative binary operator on Z, with 1 as identity, no integer other than ± 1 has a multiplicative inverse **in Z**.

Yes, there is a number whose product with 2 is 1:
$$2 \cdot \frac{1}{2} = 1, \text{ but } \frac{1}{2} \notin Z.$$

Bottom line: The set of integers under multiplication is **not** a group.

> **CHECK YOUR UNDERSTANDING 5.1**
>
> Determine if the given set is a group under the given operation. If not, specify which of the axioms of Definition 5.1 do not hold.
>
> (a) The set Q of rational numbers under addition.
>
> (b) The set $\Re$ of real numbers under addition.
>
> (c) The set $\Re$ of real numbers under multiplication.
>
> (d) The set $\Re^+ = \{r \in \Re | r > 0\}$ of positive real numbers under multiplication.

(a), (b), and (d) are groups. (c) is not a group.

We now move Theorem 2.18 of page 94 up a notch:

THEOREM 5.1 For given $n \in Z^+$, let $[Z]_n$ denote the set of equivalence classes associated with the equivalence relation $a \sim b$ if $n|(a-b)$; i.e:

$$[Z]_n = \{[0]_n, [1]_n, ..., [n-1]_n\}$$

Then: $\langle [Z]_n, [+] \rangle$ with $[a]_n [+] [b]_n = [a+b]_n$ is a group

PROOF: We already know that $[+]$ is a well defined associative operator. The identity and inverse axioms of Definition 5.1 are also met:

Identity: For any $[a]_n \in [Z]_n$: $[a]_n [+] [0]_n = [a+0]_n = [a]_n$.

Inverses: For any $[a]_n \in [Z]_n$: $[a]_n [+] [-a]_n = [a-a]_n = [0]_n$

Molding Theorem 5.1 into a more compact form by replacing each equivalent class $[a]_n$ with the smallest nonnegative integer in that class, we come to:

THEOREM 5.2 For given $n \in Z^+$, let $Z_n = \{0, 1, 2, ..., n-1\}$, and let $a +_n b = r$, where $a + b = dn + r$.

Then $\langle Z_n, +_n \rangle$ is a group.

You are invited to formally establish this result in Exercise 51.

The above sum is called **addition modulo n.** Note that $a +_n 0 = a$ for every $a \in Z_n$, and that for any $a \in Z_n$: $a + (n-a) = 0$

For example, if $n = 5$ then $Z_5 = \{0, 1, 2, 3, 4\}$, and:

$$1 +_5 2 = 3, \quad 4 +_5 4 = 3, \text{ and } 3 +_5 2 = 0$$

$\uparrow$
$3 \equiv (4+4) \bmod 5$

CHECK YOUR UNDERSTANDING 5.2

Complete the following (self-explanatory) **group table** for $\langle Z_4, +_4 \rangle$.

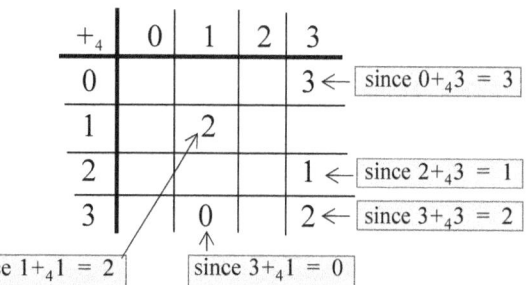

Answer: See page A-27.

Groups containing infinitely many elements, like $\langle Z, + \rangle$ and $\langle \Re, + \rangle$, are said to be **infinite groups**. Those containing finite may elements, like $\langle Z_n, +_n \rangle$ which contains n elements, are said to be **finite groups**.

DEFINITION 5.2
ORDER OF A GROUP
Let G be a finite group. The number of elements in G is called the order **of G**, and is denoted by $|G|$.

GROUP TABLES AND BEYOND

The group Z_4, with table depicted in Figure 5.1(a), has order 4. Another group of order 4, the so-called **Klein 4-group**, appears in Figure 5.1(b).

Z_4:

$+_4$	0	1	2	3
0	0	1	2	3
1	1	2	3	0
2	2	3	0	1
3	3	0	1	2

(a)

K:

*	e	a	b	c
e	e	a	b	c
a	a	e	c	b
b	b	c	e	a
c	c	b	a	e

(b)

Figure 5.1

Is K really a group? Well, the above table leaves no doubt that the closure and identity axioms are satisfied (e is the identity element). Moreover, each element has an inverse, namely itself: $ee = e, aa = e, bb = e,$ and $cc = e$. Finally, though a bit tedious, you can check directly that the associative property holds [for example: $(ab)a = ca = b$ and $a(ba) = ac = b$]. You can also see that K is an abelian group; where:

Abelian groups are also said to be commutative groups.

DEFINITION 5.3
ABELIAN GROUP
A group $\langle G, * \rangle$ is **abelian** if
$$a*b = b*a \text{ for every } a, b \in G$$

We will soon show that Z_4 and K are the only groups of order 4, but first:

THEOREM 5.3 Every element of a finite group G must appear once and only once in each row and each column of its group table.

PROOF: Let $G = \{e, a_1, a_2, \ldots, a_{n-1}\}$. By construction, the ith row of G's group table is precisely $a_i e, a_i a_1, a_i a_2, \ldots, a_i a_{n-1}$. The fact that every element of G appears exactly one time in that row is a consequence of Exercise 50, which asserts that the function $f_{a_i}: G \to G$ given by $f_{a_i}(g) = a_i g$ is a bijection. As for the columns:

CHECK YOUR UNDERSTANDING 5.3

Answer: See page A-27.

Complete the proof of Theorem 5.3.

We now show that the two groups in Figure 5.1 represent all groups of order four. To begin with, we note that any group table featuring the four elements $\{e, a, b, c\}$ must "start off" as in T in Figure 5.2, for e represents the identity element.

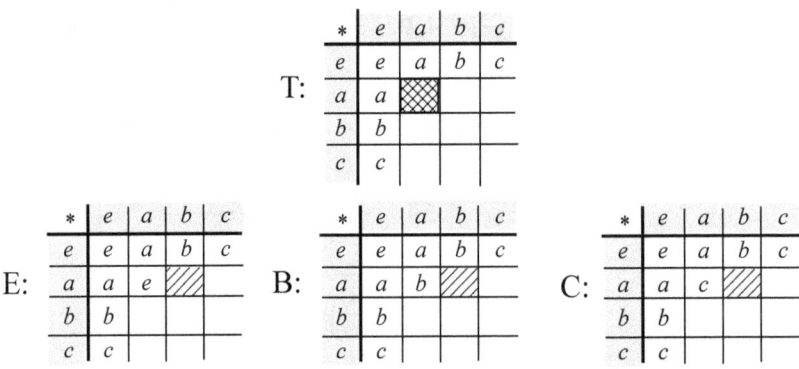

Figure 5.2

Since no element of a group can occur more than once in any row or column of the table, the ▧-box in T can only be inhabited by e, b or c, with each of those possibilities displayed as E, B, and C in Figure 5.2. Repeatedly reemploying Theorem 5.2, we observe that while E leads to two possible group tables, both B and C can only be completed in one way (see Figure 5.3)

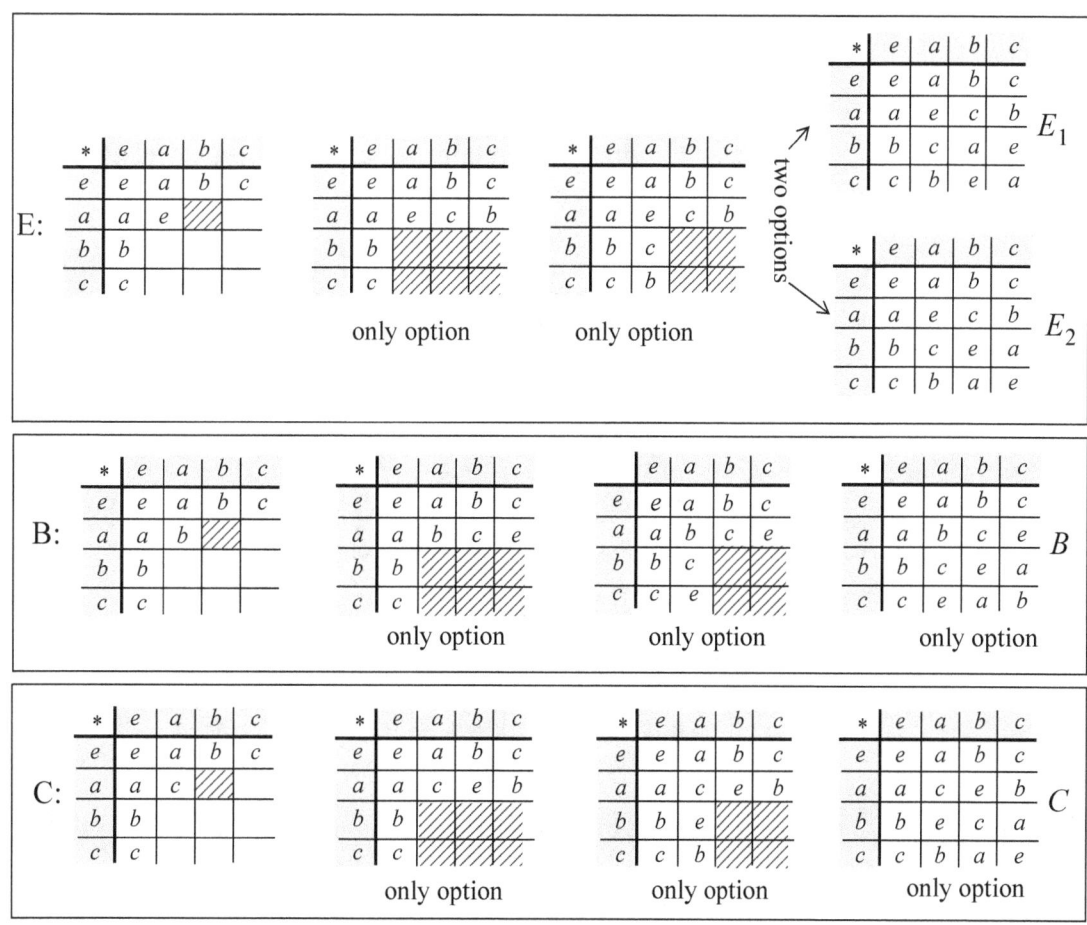

Figure 5.3

At this point we know that there can be at most four groups of order 4, and their corresponding group tables appear in Figure 5.4. The group tables for Z_4 and K of Figure 5.1 are also displayed at the bottom Figure 5.4.

Figure 5.4

While table E_2 and the Klein group table K are identical, those of the remaining four tables in Figure 5.4 look different.

But looks can be deceiving:

To show, for example, that Z_4 and E_1 only differ superficially, we begin by reordering the elements in the first row and first column of Z_4 in Figure 5.5(a) from "0, 1, 2, 3" to "0, 2, 1, 3" [see Figure 5.5 (b)]. We then transform Figure 5.5(b) to E_1 in (c) by replacing the symbols "0, 2, 1, 3" with the symbols "e, a, b, c," respectively, and the operator symbol "$+_4$" with "*".

Z_4:

$+_4$	0	1	2	3
0	0	1	2	3
1	1	2	3	0
2	2	3	0	1
3	3	0	1	2

$+_4$	0	2	1	3
0	0	2	1	3
2	2	0	3	1
1	1	3	2	0
3	3	1	0	2

E_1:

*	e	a	b	c
e	e	a	b	c
a	a	e	c	b
b	b	c	a	e
c	c	b	e	a

(a)   (b)   (c)

Figure 5.5

> This "appearances aside" concept is formalized in Section 4.

So, appearances aside, the group structure of E_1 coincides with that of Z_4. In a similar fashion you can verify that tables B and C of Figure 5.4 only differ from table Z_4 syntactically.

PERMUTATIONS AND SYMMETRIC GROUPS

For any non-empty set X, let $S_X = \{f: X \to X | f \text{ is a bijection}\}$. We then have:

> The composition operator "$\circ$" is defined on page 64.

THEOREM 5.4 For any non-empty set X, $\langle S_X, \circ \rangle$ is a group.

PROOF: Turning to Definition 5.1:

Operator. $\forall f, g \in S_X: g \circ f \in S_X$ [Theorem 2.5(c), page 72].

Associative. $\forall f, g, h \in S_X: h \circ (g \circ f) = (h \circ g) \circ f$ [Exercise 43, page 75].

Identity. $\forall f \in S_X: f \circ I_X = I_X \circ f = f$, where $I_X: X \to X$ is the identity function: $I_X(x) = x$ for every $x \in X$.

Inverse. $\forall f \in S_X: f \circ f^{-1} = I_X$ [Theorem 2.4(b), page 70].

> The elements (functions) in S_X are said to be **permutations** (on X), and $\langle S_X, \circ \rangle$ is said to be the **symmetric group** on X.

IN PARTICULAR:

> For $X = \{1, 2, ..., n\}$, $\langle S_X, \circ \rangle$ is called the **symmetric group of degree n**, and will be denoted by S_n.

Let's get our feet wet by considering the symmetric croup S_3, the set of permutations on $X = \{1, 2, 3\}$. Since there are $n!$ ways of ordering n objects (Exercise 35, page 42), the group S_3 consists of $3! = 1 \cdot 2 \cdot 3 = 6$ elements:

e	α_1	α_2	α_3	α_4	α_5
$1 \to 1$	$1 \to 2$	$1 \to 3$	$1 \to 1$	$1 \to 3$	$1 \to 2$
$2 \to 2$	$2 \to 3$	$2 \to 1$	$2 \to 3$	$2 \to 2$	$2 \to 1$
$3 \to 3$	$3 \to 1$	$3 \to 2$	$3 \to 2$	$3 \to 1$	$3 \to 3$

Directly below each elements of the first row appears its image under the permutations. The fact that 3 lies below 1 in α_4, for example, simply indicates that the permutation α_4 maps 1 to 3: $1 \to 3$.

In a more compact (and more standard) form (see margin), we write:

$$e = \begin{pmatrix} 1 & 2 & 3 \\ 1 & 2 & 3 \end{pmatrix}, \quad \alpha_1 = \begin{pmatrix} 1 & 2 & 3 \\ 2 & 3 & 1 \end{pmatrix}, \quad \alpha_2 = \begin{pmatrix} 1 & 2 & 3 \\ 3 & 1 & 2 \end{pmatrix}$$

$$\alpha_3 = \begin{pmatrix} 1 & 2 & 3 \\ 1 & 3 & 2 \end{pmatrix}, \quad \alpha_4 = \begin{pmatrix} 1 & 2 & 3 \\ 3 & 2 & 1 \end{pmatrix}, \quad \alpha_5 = \begin{pmatrix} 1 & 2 & 3 \\ 2 & 1 & 3 \end{pmatrix}$$

Note that α_0 is the identity function e: $\alpha_0(1) = 1$, $\alpha_0(2) = 2$, and $\alpha_0(3) = 3$

The symmetric group S_3

Figure 5.6

Generalizing the above observation we have:

THEOREM 5.5 The symmetric group S_n of degree n contains $n!$ elements.

EXAMPLE 5.1 Referring to the group S_3 featured in Figure 5.6, Determine:

(a) $\alpha_2 \circ \alpha_4$ (b) $\alpha_4 \circ \alpha_2$ (c) $(\alpha_2)^{-1}$

SOLUTION: (a) To find $\alpha_2 \circ \alpha_4$ we first perform α_4 and then apply α_2 to the resulting function values:

From Figure 5.6:

α_2	α_4
$1 \to 3$	$1 \to 3$
$2 \to 1$	$2 \to 2$
$3 \to 2$	$3 \to 1$

$$\begin{array}{cc} \alpha_4 & \alpha_2 \\ 1 \to 3 \to 2 & 1 \to 2 \\ 2 \to 2 \to 1 & \Rightarrow \alpha_2 \circ \alpha_4 = 2 \to 1 = \alpha_5 \\ 3 \to 1 \to 3 & 3 \to 3 \end{array}$$

(b) Using the standard form we show that $\alpha_4 \circ \alpha_2 = \alpha_3$:

$$\begin{pmatrix} 1 & 2 & 3 \\ 3 & 1 & 2 \\ 1 & 3 & 2 \end{pmatrix} \begin{array}{l} \nearrow \text{first } \sigma_2 : \begin{pmatrix} 1 & 2 & 3 \\ 3 & 1 & 2 \end{pmatrix} \\ \searrow \text{then } \sigma_4 : \begin{pmatrix} 1 & 2 & 3 \\ 3 & 2 & 1 \end{pmatrix} \end{array} \Rightarrow \alpha_4 \circ \alpha_2 = \begin{pmatrix} 1 & 2 & 3 \\ 1 & 3 & 2 \end{pmatrix} = \alpha_3$$

(c) Tor arrive at the inverse of the permutation $\alpha_2 = \begin{pmatrix} 1 & 2 & 3 \\ 3 & 1 & 2 \end{pmatrix}$, simply reverse its action:

$$\alpha_2^{-1} = \begin{pmatrix} 1 & 2 & 3 \\ 3 & 1 & 2 \end{pmatrix}^{-1} = \begin{pmatrix} 3 & 1 & 2 \\ 1 & 2 & 3 \end{pmatrix} = \begin{pmatrix} 1 & 2 & 3 \\ 2 & 3 & 1 \end{pmatrix} = \alpha_1$$

Answer:
$$\tau \circ \sigma: \begin{pmatrix} 1 & 2 & 3 & 4 & 5 \\ 5 & 4 & 3 & 2 & 1 \end{pmatrix}$$
$$\sigma \circ \tau: \begin{pmatrix} 1 & 2 & 3 & 4 & 5 \\ 4 & 3 & 2 & 5 & 1 \end{pmatrix}$$

CHECK YOUR UNDERSTANDING 5.4

With reference to the symmetric group S_5, determine $\tau \circ \sigma$ and $\tau \circ \sigma$, where:

$$\sigma = \begin{pmatrix} 1 & 2 & 3 & 4 & 5 \\ 1 & 5 & 2 & 3 & 4 \end{pmatrix} \text{ and } \tau = \begin{pmatrix} 1 & 2 & 3 & 4 & 5 \\ 5 & 3 & 2 & 1 & 4 \end{pmatrix}$$

Adhering to convention, we will start using ab (rather than $a*b$) to denote the binary operation in a generic group. Under this notation, the symbol a^{-1} (rather than a') is used to denote the inverse of a, while e continues to represent the identity element. In a generic abelian group, however, the symbol "+" is typically used to represent the binary operator, with 0 denoting the identity element, and $-a$ denoting the inverse of a. To summarize:

In Summery:

Original Form	Product Form	Sum Form (Reserved **only** for abelian groups)
1. $a*b \in G \;\; \forall a, b \in G$	1. $ab \in G$	1. $a + b \in G$
2. $a*(b*c) = (a*b)*c$	2. $a(bc) = (ab)c$	2. $a + (b+c) = (a+b) + c$
3. $a*e = a$	3. $ae = a$	3. $a + 0 = a$
4. $a*a' = e$	4. $aa^{-1} = e$	4. $a + (-a) = 0$

SOME ADDITIONAL NOTATION:

Referring to the product form, do not express a^{-n} in the form $\frac{1}{a^n}$ (there is no "division" in the group).
From its very definition we find that the following exponent rules hold in any group G:
For any $n, m \in Z$:
$$a^n a^m = a^{n+m}$$
$$(a^n)^m = a^{nm}$$
In the sum form, it is acceptable utilize the notation $a - b$. By definition:
$$a - b = a + (-b).$$

For any positive integer n:
a^n represents $\underbrace{aaa \cdots a}_{n \; a\text{'s}}$
and $a^{-n} = (a^{-1})^n$.
We also define a^0 to be e.

For any positive integer n:
na represents $\underbrace{a + a + a + \cdots + a}_{n \; a\text{'s}}$
and $(-n)a = n(-a)$.
We also define $0a$ to be 0.

Utilizing the above notation:

DEFINITION 5.4
CYCLIC GROUP

(**Product form**) A group G is **cyclic** if there exists $a \in G$ such that $G = \{a^n | n \in Z\}$.

(**Sum form**) An abelian group G is **cyclic** if there exists $a \in G$ such that $G = \{na | n \in Z\}$.

GENERATOR

In either case we say that the element a is a **generator** of G, and write $G = \langle a \rangle$.

EXAMPLE 5.2 Show that:
(a) Z_6 is cyclic   (b) S_3 is not cyclic.

SOLUTION: (a) Clearly $Z_6 = \langle 1 \rangle$. In fact, as we now show, **5** is also a generator of Z_6 (don't forget that we are summing modulo 6):

	Note that:
$1(\mathbf{5}) = 5$	$5 = 0 \cdot 5 + 5$
$2(\mathbf{5}) = 5 +_6 5 = 4$	$10 = 1 \cdot 6 + 4$
$3(\mathbf{5}) = 5 +_6 5 +_6 5 = 3$	$15 = 2 \cdot 6 + 3$
$4(\mathbf{5}) = 5 +_6 5 +_6 5 +_6 5 = 2$	$20 = 3 \cdot 6 + 2$
$5(\mathbf{5}) = 5 +_6 5 +_6 5 +_6 5 +_6 5 = 1$	$25 = 4 \cdot 6 + 1$
$6(\mathbf{5}) = 5 +_6 5 +_6 5 +_6 5 +_6 5 +_6 5 = 0$	$30 = 5 \cdot 6 + 0$

Since every element of $Z_6 = \{0, 1, 2, 3, 4, 5\}$ is a multiple of **5**, we conclude that $Z_6 = \langle 5 \rangle$.

(b) We could use a brute-force method to verify, directly, that no element of S_3 generates all of S_3. Instead, we appeal to the following theorem [and Example 5.2(b)] to draw the desired conclusion.

THEOREM 5.6 Every cyclic group is abelian.

PROOF: Let $G = \langle a \rangle = \{a^n | n \in Z\}$. For any two elements a^s and a^t in G (not necessarily distinct) we have:
$$a^s a^t = a^{s+t} = a^{t+s} = a^t a^s$$

CHECK YOUR UNDERSTANDING 5.5

(a) Show that 1 and 5 are the only generators of Z_6.

(b) Show that S_2 is cyclic.

(c) Show that S_n is not cyclic for any $n > 2$.

Answer: See page A-28.

At this point we have two groups of order 6 at our disposal:
$$\langle Z_6, +_n \rangle \text{ and } S_3$$

Do these groups differ only superficially, or are they really different in some algebraic sense? They do differ algebraically in that one is cyclic while the other is not, and also in that one is abelian while the other is not.

EXERCISES

Exercise 1-11. Determine if the given set is a group under the given operator. If not, specify why not. If it is, indicate whether or not the group is abelian, and whether or not it is cyclic. If it is cyclic, find a generator for the group.

1. The set $\{2n | n \in Z\}$ of even integers under addition.

2. The set $\{2n + 1 | n \in Z\}$ of odd integers under addition.

3. The set of integers Z, with $a*b = c$, where c is the smaller of the two integers a and b (the common value if $a = b$).

4. The set Q^+ of positive rational numbers, with $a*b = \dfrac{ab}{2}$.

5. The set $\{x \in \mathfrak{R} | x \neq 0\}$, with $a*b = \dfrac{a^2}{b}$.

6. The set $\{0, 2, 4, 6, 8\}$ under the operation of addition modulo 10.

7. The set $\{0, 1, 2, 3\}$ under multiplication modulo 4. (For example: $2*3 = 2$, since $2 \cdot 3 = 6 = 1 \cdot 4 + 2$; and $3*3 = 1$, since $3 \cdot 3 = 9 = 2 \cdot 4 + 1$.)

8. The set $\{0, 1, 2, 3, 4\}$ under multiplication modulo 5. (See Exercise 7.)

9. The set $\{a + b\sqrt{2} | a, b \in Z\}$ under addition.

10. The set $\{a + b\sqrt{2} | a, b \in Q$ with not both a and b equal to $0\}$ under the usual multiplication of real numbers.

11. The set $Z \times Z = \{(a, b) | a, b \in Z\}$, with $(a, b) + (c, d) = (a + c, b + d)$.

Exercise 12-23. Referring to the group S_3:

$$e = \begin{pmatrix} 1 & 2 & 3 \\ 1 & 2 & 3 \end{pmatrix}, \alpha_1 = \begin{pmatrix} 1 & 2 & 3 \\ 2 & 3 & 1 \end{pmatrix}, \alpha_2 = \begin{pmatrix} 1 & 2 & 3 \\ 3 & 1 & 2 \end{pmatrix}, \alpha_3 = \begin{pmatrix} 1 & 2 & 3 \\ 1 & 3 & 2 \end{pmatrix}, \alpha_4 = \begin{pmatrix} 1 & 2 & 3 \\ 3 & 2 & 1 \end{pmatrix}, \alpha_5 = \begin{pmatrix} 1 & 2 & 3 \\ 2 & 1 & 3 \end{pmatrix}$$

determine:

12. $\alpha_2 \alpha_4$ and $\alpha_4 \alpha_2$

13. α_3^2 and α_3^3

14. α_3^n for $n \in Z^+$.

15. α_1^n for $n \in Z^+$.

16. α_3^{-2} and α_3^{-3}

17. α_3^{-n} for $n \in Z^+$.

18. α_3^{-n} for $n \in Z^+$.

19. α_2^2 and α_2^3

20. α_2^n for $n \in Z^+$.

21. α_2^n for $n \in Z^+$.

22. α_2^{-2} and α_2^{-3}

23. α_2^{-n} for $n \in Z^+$.

5.1 Definitions and Examples

Exercise 24-33. For

$$\alpha = \begin{pmatrix} 1 & 2 & 3 & 4 & 5 & 6 \\ 2 & 3 & 4 & 5 & 6 & 1 \end{pmatrix} \qquad \beta = \begin{pmatrix} 1 & 2 & 3 & 4 & 5 & 6 \\ 2 & 1 & 4 & 3 & 6 & 5 \end{pmatrix} \qquad \gamma = \begin{pmatrix} 1 & 2 & 3 & 4 & 5 & 6 \\ 6 & 5 & 4 & 3 & 2 & 1 \end{pmatrix}$$

Determine:

24. $\alpha\beta$
25. $\beta\alpha$
26. $\beta\gamma$
27. $\gamma\beta$
28. $\alpha\beta\gamma$
29. α^5
30. α^{100}
31. α^{101}
32. β^{100}
33. β^{101}

34. Let $S = \{1\}$. Show that $\langle S, * \rangle$ with $1*1 = 1$ is a group. Is the group abelian? Cyclic?

35. Is $\langle M_{2 \times 2}, + \rangle$ with $\begin{bmatrix} a & b \\ c & d \end{bmatrix} + \begin{bmatrix} \bar{a} & \bar{b} \\ \bar{c} & \bar{d} \end{bmatrix} = \begin{bmatrix} a+\bar{a} & b+\bar{b} \\ c+\bar{c} & d+\bar{d} \end{bmatrix}$ a group? If so, is it abelian? Cyclic?

36. Is $\langle M_{2 \times 2}, * \rangle$ with $\begin{bmatrix} a & b \\ c & d \end{bmatrix} * \begin{bmatrix} \bar{a} & \bar{b} \\ \bar{c} & \bar{d} \end{bmatrix} = \begin{bmatrix} a\bar{a} & b\bar{b} \\ c\bar{c} & d\bar{d} \end{bmatrix}$ a group? If so, is it abelian? Cyclic?

37. Is $\langle M_{2 \times 2}, * \rangle$ with $\begin{bmatrix} a & b \\ c & d \end{bmatrix} * \begin{bmatrix} \bar{a} & \bar{b} \\ \bar{c} & \bar{d} \end{bmatrix} = \begin{bmatrix} a\bar{a}+b\bar{c} & a\bar{b}+b\bar{d} \\ c\bar{a}+d\bar{c} & c\bar{b}+d\bar{d} \end{bmatrix}$ a group? If so, is it abelian? Cyclic?

38. Let $S = \{a, b, c\}$ along with the binary operator:

*	a	b	c
a	a	b	c
b	b	b	c
c	c	c	c

Is $\langle S, * \rangle$ a group?

39. Let $S = \{a, b, c\}$ along with the binary operator:

*	2	0	1
2	2	0	1
0	0	1	2
1	1	2	0

Is $\langle S, * \rangle$ a group?

40. Let $S = \{(x, y) | x, y \in \Re\}$. Show that $\langle S, * \rangle$ with $(x, y)*(\bar{x}, \bar{y}) = (x + \bar{x} - 1, y + \bar{y} + 1)$ is a group. Is the group abelian? Cyclic?

41. For $n \geq 0$, let P_n denote the set of polynomials of degree less than or equal to n. Show that

$$\langle P_n, * \rangle \text{ with } \left(\sum_{i=0}^{n} a_i x^i \right) * \left(\sum_{i=0}^{n} b_i x^i \right) = \sum_{i=0}^{n} (a_i + b_i) x^i \text{ is a group. Is the group abelian?}$$

42. Let $S = \{(x, y) | x, y \in \Re\}$. Show that $\langle S, * \rangle$ with $(x, y)*(\bar{x}, \bar{y}) = (x + \bar{x} + 2, y + \bar{y})$ is a group. Is the group abelian?

43. Let Q denote the set of rational numbers. Show that $\langle Q, * \rangle$ with $a*b = a + b + ab$ is not a group.

44. Let $\bar{Q} = \{a \in Q | a \neq -1\}$. Show that $\langle \bar{Q}, * \rangle$ with $a*b = a+b+ab$ is a group. Is the group abelian?

45. Let $G = \{e, a_1, a_2, ..., a_{n-1}\}$. Show that the function $f_{a_i}: G \to G$ given by $f_{a_i}(g) = a_i g$ is a bijection

46. (a) Give an example of a group G in which the exponent law $(ab)^n = a^n b^n$ does not hold in a group G, for $n \in Z'''$.
    (b) Prove that the exponential law $(ab)^n = a^n b^n$ does hold if the group G is abelian.
    (c) Express the property $(ab)^n = a^n b^n$ in sum-notation form.

47. Let G be a group and $a, b, c \in G$. Show that if $ab = ac$, then $b = c$.

48. Show that the group Z_n of Theorem 5.1 is cyclic for any $n \in Z^+$.

49. Let $F(\Re)$ denote the set of all real-valued functions. For f and g in $F(\Re)$, let $f+g$ be given by $(f+g)(x) = f(x) + g(x)$. Show that $\langle F(\Re), + \rangle$ is a group. Is the group abelian?

50. Let $G = \{e, a_1, a_2, ..., a_{n-1}\}$ be a group. Show that the function $f_{a_i}(g) = a_i g$ is a bijection.

51. Prove Theorem 5.2.

52. Let G and H be groups. Let $G \times H = \{(g, h) | g \in G, h \in H\}$ with:
$$(g, h) * (\bar{g}, \bar{h}) = (g\bar{g}, h\bar{h})$$
    (a) Show that $\langle G \times H, * \rangle$ is a group.
    (b) Prove that $\langle G \times H, * \rangle$ is abelian if and only both G and H are abelian.

53. Let X be a set and let $P(X)$ be the set of all subsets of X. Is $\langle P(X), * \rangle$ a group if:
    (a) $A*B = A \cup B$ (b) $A*B = A \cap B$

PROVE OR GIVE A COUNTEREXAMPLE

54. The set $\Re$ of real numbers under multiplication is a group

55. The set $\Re^+ = \{r \in \Re | r > 0\}$ of positive real numbers under multiplication.

56. Let G be a group and $a, b, c \in G$. If $b \neq c$, then $ab \neq ac$.

57. Let G be a group and $a, b \in G$. If $ab = b$, then $ac = c$ for every $c \in G$.

58. Let G be a group and $a, b \in G$. If $ab = ba$, then $ac = ca$ for every $c \in G$.

59. The cyclic group $\langle Z, + \rangle$ has exactly two distinct generators.

§2 ELEMENTARY PROPERTIES OF GROUPS

We begin by recalling the group axioms, featuring both the product and the sum notation:

	Product Form	Sum Form (Typically reserved for abelian groups)
Closure	$ab \in G$	$a + b \in G$
Associative Axiom	1. $a(bc) = (ab)c$	1. $a + (b + c) = (a + b) + c$
Identity Axiom	2. $ae = a$	2. $a + 0 = a$
Inverse Axiom	3. $aa^{-1} = e$	3. $a + (-a) = 0$

For aesthetic reasons, a set of axioms should be independent, in that no axiom or part of an axiom is a consequence of the rest. One should not, for example, replace Axiom 2 in Definition 5.1, page 207:

$\exists e \in G \ni ae = a \forall a \in G$

with:

$\exists e \in G \ni ae = ea = a \forall a \in G$

Actually, as we show below, both the identity element of Axiom 2 and the inverse elements of Axiom 3 work on both sides; but first:

LEMMA 5.1 Let G be a group. If $a \in G$ is such that $a^2 = a$, then $a = e$.

PROOF:
$$aa = a \Rightarrow (aa)a^{-1} = aa^{-1} \underset{\text{Axiom 1}}{\Rightarrow} a(aa^{-1}) = e \underset{\text{Axiom 3}}{\Rightarrow} ae = e \underset{\text{Axiom 2}}{\Rightarrow} a = e$$

THEOREM 5.7 Let G be a group. For $a \in G$:

(a) $aa^{-1} = e \Rightarrow a^{-1}a = e$

(b) $ae = a \Rightarrow ea = a$

PROOF:

(a) $(a^{-1}a)(a^{-1}a) \underset{\text{Axiom 1}}{=} [a^{-1}(aa^{-1})]a \underset{\text{Axiom 3}}{=} (a^{-1}e)a \underset{\text{Axiom 2}}{=} a^{-1}a$

Since $(a^{-1}a)(a^{-1}a) = a^{-1}a$: $a^{-1}a = e$ (see Lemma 5.1).

(b) $ea \underset{\text{Axiom 3}}{=} (aa^{-1})a \underset{\text{Axiom 1}}{=} a(a^{-1}a) \underset{\text{part (a)}}{=} ae \underset{\text{Axiom 3}}{=} a$

Axioms 2 and 3 stipulates the existence of an identity in a group, and of an inverse for each element of the group. Are they necessarily unique? Yes:

THEOREM 5.8 (a) There is but one identity in a group G.

(b) Every element in G has a unique inverse.

PROOF: (a) We assume that e and $\bar{e}$ are identities, and go on to show that $e = \bar{e}$:

$$e = e\bar{e} = \bar{e}$$

(Since $\bar{e}$ is an identity; Since e is an identity)

(b) We assume that a^{-1} and $\bar{a}^{-1}$ are inverses of $a \in G$, and show that $a^{-1} = \bar{a}^{-1}$.

Since aa^{-1} and $a\bar{a}^{-1}$ are both equal to **the** identity they must be equal to each other:

$$aa^{-1} = a\bar{a}^{-1}$$

Multiply both sides by a^{-1}: $a^{-1}(aa^{-1}) = a^{-1}(a\bar{a}^{-1})$

Associativity: $(a^{-1}a)a^{-1} = (a^{-1}a)\bar{a}^{-1}$

$$ea^{-1} = e\bar{a}^{-1}$$
$$a^{-1} = \bar{a}^{-1}$$

CHECK YOUR UNDERSTANDING 5.6

Show that if a, b, c are elements of a group such that $abc = e$, then $bca = e$.

Answer: See page A-28.

The **left and right cancellation laws** hold in groups.

THEOREM 5.9 In any group G:

(a) If $ab = cb$, then $a = c$.

(b) If $ba = bc$, then $a = c$

Sum form:
$a + b = c + b \Rightarrow a = c$
$b + a = b + c \Rightarrow a = c$

PROOF:

(a)
$$ab = cb$$
$$(ab)b^{-1} = (cb)b^{-1}$$
$$a(bb^{-1}) = c(bb^{-1})$$
$$ae = ce$$
$$a = c$$

(b)
$$ba = bc$$
$$b^{-1}(ba) = b^{-1}(bc)$$
$$(b^{-1}b)a = (b^{-1}b)c$$
$$ea = ec$$
$$a = c$$

Just in case you are asking yourself:
What if b is 0 and has no inverse?
Tisk, every element in a group has an inverse.

CHECK YOUR UNDERSTANDING 5.7

PROVE OR GIVE A COUNTEREXAMPLE:

(a) In any group G, if $ab = bc$ then $a = c$.

(b) In any abelian group G, if $a + b = b + c$ then $a = c$.

Answer: See page A-28.

In the real number system, do linear equations $ax = b$ have unique solutions for every $a, b \in \Re$? No — the equation $0x = 5$ has no solution, while the equation $0x = 0$ has infinitely many solutions. This observation assures us that the reals is not a group under multiplication, for:

THEOREM 5.10 Let G be a group. For any $a, b \in G$, the linear equations $ax = b$ and $ya = b$ have unique solutions in G.

PROOF: Existence:

(a) $\quad ax = b$
$\quad a^{-1}(ax) = a^{-1}b$
$\quad (a^{-1}a)x = a^{-1}b$
$\quad ex = a^{-1}b$
$\quad x = a^{-1}b$

(b) $\quad ya = b$
$\quad (ya)a^{-1} = ba^{-1}$
$\quad y(aa^{-1}) = ba^{-1}$
$\quad ye = ba^{-1}$
$\quad y = ba^{-1}$

Uniqueness: We assume (as usual) that there are two solutions, and then proceed to show that they are equal:

(a) $ax_1 = b$ and $ax_2 = b \Rightarrow ax_1 = ax_2 \overset{\text{Theorem 5.9(b)}}{\Rightarrow} x_1 = x_2$

(b) $y_1a = b$ and $y_2a = b \Rightarrow y_1a = y_2a \overset{\text{Theorem 5.9(a)}}{\Rightarrow} y_1 = y_2$

CHECK YOUR UNDERSTANDING 5.8

Since the set of real numbers under addition is a group, Theorem 5.9 applies. Show, **directly**, that any linear equation in $\langle \Re, + \rangle$ has a unique solution.

Answer: See page A-29.

The inverse of a product is the product of the inverses, but in reverse order:

This is another *shoe-sock theorem* (see page 73).

THEOREM 5.11 For every a, b in a group G:
$$(ab)^{-1} = b^{-1}a^{-1}$$

PROOF: To show that $b^{-1}a^{-1}$ is the inverse of ab is to show that $(b^{-1}a^{-1})(ab) = e$. No problem:

$$(b^{-1}a^{-1})(ab) = b^{-1}(a^{-1}a)b = b^{-1}eb = b^{-1}b = e$$

> **CHECK YOUR UNDERSTANDING 5.9**
>
> Give an example of a group G for which $(ab)^{-1} = a^{-1}b^{-1}$ does not hold for every $a, b \in G$.

Answer: See page A-29.

The associative axiom of a group G assures us that an expression such as abc, sans parentheses, is unambiguous [since $(ab)c$ and $a(bc)$ yield the same result]. It is plausible to expect that this nicety extends to any product $a_1 a_2 \cdots a_n$ of elements of G. Plausible, to be sure; but more importantly, True:

THEOREM 5.12 Let $a_1 a_2 \cdots a_n \in G$. The product expression $a_1 a_2 \cdots a_n$ is unambiguous in that its value is independent of the order in which adjacent factors are multiplied.

PROOF: [By induction (page 33)]:

I. The claim holds for $n = 3$ (the associative axiom).

II. Assume the claim holds for $n = k$, with $k > 3$.

III. (Now for the fun part) We show the claim holds for $n = k + 1$:

Let x denote the product $a_1 a_2 \cdots a_{k+1}$ under a certain pairing of its elements, and y the product under another pairing of its elements. We are to show that $x = y$.

Assume that one starts the multiplication process with the following pairing for x and y:

$$x = \underbrace{(a_1 a_2 \cdots a_i)}_{A} \underbrace{(a_{i+1} \cdots a_{k+1})}_{B} \text{ and } y = \underbrace{(a_1 a_2 \cdots a_j)}_{C} \underbrace{(a_{j+1} \cdots a_{k+1})}_{D}$$

Case 1. $i = j$: By the induction hypothesis (II), no matter how the products in A and C are performed, A will equal C. The same can be said concerning B and D. Consequently $x = AB = CD = y$.

Case Assume, without loss of generality, that $i < j$. Breaking the "longer" product B into two pieces M and D we have:

$$x = \underbrace{(a_1 a_2 \cdots a_i)}_{A} \underbrace{(a_{i+1} \cdots a_j)}_{M} \underbrace{(a_{j+1} \cdots a_{k+1})}_{D}$$

By the induction hypothesis, A, M, and D are well defined (independent of the pairing of its elements in their products). Bringing us to:

$$x = AB = A(MD) \stackrel{\downarrow}{=} (AM)D = CD = y$$

I: Claim holds for $n = 3$

CHECK YOUR UNDERSTANDING 5.10

Use the Principle of Mathematical Induction, to show that for any $a_1 a_2 \ldots a_n \in G$:

$$(a_n \ldots a_2 a_1)^{-1} = a_1^{-1} a_2^{-1} \ldots a_n^{-1}$$

Answer: See page A-29.

THEOREM 5.13 For any given element a of a finite group G:
$$a^m = e \text{ for some } m \in Z^+.$$

PROOF: Let G be of order n. Surely not all of the $n+1$ elements $a, a^2, a^3, \ldots, a^{n+1}$ can be distinct. Choose $1 \leq s < t \leq n+1$ such that $a^t = a^s$. Since $a^t a^{-s} = a^{t-s} = e$:

$$a^m = e, \text{ for } m = t-s.$$

In the additive notation, $a^m = e$ translates to $na = 0$; which is to say:
$a + a + \ldots + a = 0$
(sum of n a's)

DEFINITION 5.5

ORDER OF AN ELEMENT OF G

Let G be a finite group, and let $a \in G$

The smallest positive integer m such that $a^m = e$ is called the **order of** a and is denoted by $o(a)$. If no such element exists, then a is said to have infinite order.

EXAMPLE 5.3 (a) Determine the order of the element 4 in the group $\langle Z_6, +_6 \rangle$.

(b) Determine the order of the element
$$\sigma = \begin{pmatrix} 1 & 2 & 3 & 4 & 5 \\ 3 & 2 & 4 & 1 & 5 \end{pmatrix}$$
in the symmetric group S_5.

SOLUTION: (a) Since:
$$1(4) = 4$$
$$2(4) = 4 +_6 4 = 2$$
$$3(4) = 4 +_6 4 +_6 4 = 2 +_6 4 = 0$$

The element 4 has order 3 in Z_6.
(b) Since:

$$\begin{pmatrix} 1 & 2 & 3 & 4 & 5 \\ 3 & 2 & 4 & 1 & 5 \\ 4 & 2 & 1 & 3 & 5 \\ 1 & 2 & 3 & 4 & 5 \end{pmatrix}$$

The element $\sigma = \begin{pmatrix} 1 & 2 & 3 & 4 & 5 \\ 3 & 2 & 4 & 1 & 5 \end{pmatrix}$ has order 3 in S_5.

CHECK YOUR UNDERSTANDING 5.11

(a) Determine the order of the element $\sigma = \begin{pmatrix} 1 & 2 & 3 & 4 \\ 2 & 3 & 4 & 1 \end{pmatrix}$ in S_4.

(b) Determine the order of the element 4 in Z_{24}.

Answer: (a) 4   (b) 6

Note: There is no "subtraction" in a group $\langle G, + \rangle$. For convenience, however, for given $a, b \in G$, we define the symbol $a - b$ as follows:
$$a - b = a + (-b)$$
(add the additive inverse of b *to* a)

There is no "division" in a group $\langle G, \cdot \rangle$. In this setting, however, one does not ever substitute the symbol $\frac{a}{b}$ for ab^{-1}. Why not? Convention.

	EXERCISES	

1. Let G be a group and $a, b, c \in G$. Solve for x, if:
   (a) $axa^{-1} = e$ (b) $axa^{-1} = a$ (c) $axb = c$ (d) $ba^{-1}xab^{-1} = ba$

2. Let G be a group. Prove that $(a^{-1})^{-1} = a$ for every $a \in G$.

3. Prove that for any element a in a group G the functions $f_a: G \to G$ given by $f_a(b) = ab$ and the function $g_a: G \to G$ given by $g_a(b) = ba$ are bijections.

4. Let a be an element of a group G. Show that $G = \{ab | b \in G\}$

5. Let G be a group and let $a \in G$. Show that if there exists one element $x \in G$ for which $ax = x$, then $a = e$.

6. Let a be an element of a group G for which there exists $b \in G$ such that $ab = b$. Prove that $a = e$.

7. Prove that a group G is abelian if and only if $(ab)^{-1} = a^{-1}b^{-1}$ for every $a, b \in G$.

8. Let G be group for which $a^{-1} = a$ for every $a \in G$. Prove that G is abelian.

9. Let G be group for which $(ab)^2 = a^2 b^2$ for every $a, b \in G$. Prove that G is abelian.

10. Let G be a finite group consisting of an even number of elements. Show that there exists $a \in G$, $a \neq e$, such that $a^2 = e$.

11. (a) Let G be a group. Show that if, for any $a, b \in G$, there exist three consecutive integers i such that $(ab)^i = a^i b^i$ then G is abelian.
    (b) Let G be an abelian group. Show that for any $a \in G$ and $n \in Z$: $(-n)a = n(-a)$.

12. Let $*$ be an associative operator on a set S. Assume that for any $a, b \in S$ there exists $c \in S$ such that $a*c = b$, and an element $d \in S$ such that $d*a = b$. Show that $\langle S, * \rangle$ is a group.

13. Let G be a group and $a \in G$. Define a new operation $*$ on G by $b*c = ba^{-1}c$ for all $b, c \in G$. show that $\langle G, * \rangle$ is a group.

14. Let G be a group and $a, b \in G$. Use the Principle of Mathematical Induction to show that for any positive integer n: $(a^{-1}ba)^n = a^{-1}b^n a$.

15. Let $a \in G$ be of order n. Find a^{-1}.

16. List the order of each element in the Symmetric group S_3 of Figure 5.6, page 213.

17. Let $a \in G$ be of order n. Prove that $a^s = a^t$ if and only if n divides $s - t$.

18. Prove that if $a^2 = e$ for every element a in a group G, then G is abelian.

19. Let $*$ be an associative operator on a finite set S. Show that if both the left and right cancellation laws of Theorem 5.9 hold under $*$, then $\langle S, * \rangle$ is a group.

| | **PROVE OR GIVE A COUNTEREXAMPLE** | |

20. If a, b, c are elements of a group such that $abc = e$, then $cba = e$.

21. In any group G there exists exactly one element a such that $a^2 = a$.

22. In any group G: $(ab)^{-2} = b^{-2}a^{-2}$.

23. Let G be a group. If $abc = bac$ then $ab = ba$.

24. Let G be a group. If $abcd = bacd$ then $ab = ba$.

25. Let G be a group. If $(abc)^{-1} = a^{-1}b^{-1}c^{-1}$ then $a = c$.

§3. SUBGROUPS

DEFINITION 5.6
SUBGROUP
A **subgroup** of a group G is a nonempty subset H of G which is itself a group under the imposed binary operation of G.

Apart from closure, to challenge whether or not a non-empty subset of a group is a subgroup you need but challenge Axiom 3:

GROUP AXIOMS
Closure: $ab \in G$
Axiom 1. $a(bc) = (ab)c$
Axiom 2. $ae = a$
Axiom 3. $aa^{-1} = e$

THEOREM 5.14 A nonempty subset S of a group G is a subgroup of G if and only if:

(i) S is closed with respect to the operation in G.

(ii) $a \in S$ implies that $a^{-1} \in S$.

PROOF: If S is a subgroup, then (i) and (ii) must certainly be satisfied.

Conversely, if (i) and (ii) hold in S, then Axioms 1 and 2 hold:

Axiom 1: Since $a(bc) = (ab)c$ holds for every $a, b, c \in G$, that associative property must surely hold for every $a, b, c \in S$.

Axiom 2: Since $ae = a$ for every $a \in G$, then surely $ae = a$ for every $a \in S$. It remains to be shown that $e \in S$. Lets do it:

Choose any $a \in S$. By (ii): $a^{-1} \in S$.

By (i): $aa^{-1} = e \in S$.

> When challenging if $S \subset G$ is a subgroup, we suggest that you first determine if it contains the identity element. For if not, then S is not a subgroup, period. If it does, then $S \neq \emptyset$ and you can then proceed to challenge (i) and (ii) of Theorem 2.14.

For example:
$5Z = \{..., -10, -5, 0, 5, 10, ...\}$

EXAMPLE 5.4 Show that for any fixed $n \in Z$ the subset
$$nZ = \{nm | m \in Z\}$$
is a subgroup of $\langle Z, + \rangle$.

SOLUTION: Since $0 = n \cdot 0 \in nZ$, $nZ \neq \emptyset$.

(i) nZ is closed under addition:
$$nm_1 + nm_2 = n(m_1 + m_2) \in nZ$$

We remind you that, under addition, $-a$ rather than a^{-1} is used to denote the inverse of a.

(ii) For any $nm \in nZ$:
$$-(nm) = n(-m) \in nZ$$

Conclusion: nZ is a subgroup of Z (Theorem 5.14).

> **CHECK YOUR UNDERSTANDING 5.12**
>
> The previous example assures us that $3Z$ is a subgroup of $\langle Z, +\rangle$. As such, it is itself a group. Show that $6Z$ is a subgroup $3Z$.

Answer: See page A-29.

You are invited to show in the exercises that the following result holds for any collection of subgroups of a given group:

THEOREM 5.15 If H and K are subgroups of a group G, then $H \cap K$ is also a subgroup of G.

PROOF: Since H and K are subgroups, each contains the identity element. It follows that $e \in H \cap K$ and that therefore $H \cap K \neq \varnothing$. We now verify that conditions (i) and (ii) of Theorem 5.14 are satisfied:

(i) (Closure) If $a, b \in H \cap K$, then $a, b \in H$ and $a, b \in K$. Since H and K are subgroups, $ab \in H$ and $ab \in K$. It follows that $ab \in H \cap K$.

(ii) (Inverses) If $a \in H \cap K$, then $a \in H$ and $a \in K$. Since H and K are subgroups, $a^{-1} \in H$ and $a^{-1} \in K$. consequently, $a^{-1} \in H \cap K$.

> **CHECK YOUR UNDERSTANDING 5.13**
>
> **PROVE OR GIVE A COUNTEREXAMPLE:**
> If H and K are subgroups of a group G, then $H \cup K$ is also a subgroup of G.

Answer: See page A-30.

We recall the definition of a cyclic group appearing on page 214:

> A group G is **cyclic** if there exists $a \in G$ such that $G = \{a^n | n \in Z\}$.

DEFINITION 5.7 Let G be a group, and $a \in G$. The cyclic group $\langle a \rangle = \{a^n | n \in Z\}$ is called the cyclic **subgroup** of G **generated** by a.
(In sum form: $\langle a \rangle = \{na | n \in Z\}$)

> **CHECK YOUR UNDERSTANDING 5.14**
>
> For $G = Z_8$, determine $\langle 3 \rangle$ and $\langle 4 \rangle$. (Use sum notation.)

Answer: $\langle 3 \rangle = Z_8$
$\langle 4 \rangle = \{0, 4\}$

THEOREM 5.16 Every subgroup of a cyclic group is cyclic.

PROOF: Let G be a cyclic group generated by a: $G = \langle a \rangle$. Let H be a subgroup of G.

Case 1. $H = \{e\}$, then $H = \langle e \rangle$ is cyclic with generator e.

Case 2. $H \neq \{e\}$. Let m be the smallest positive integer such that $a^m \in H$. We show $H = \langle a^m \rangle$ by showing that every $h \in H$ is a power of a^m:

Let $h = a^n \in H$. Employing the Division Algorithm of page 43, we chose integers q and r, with $0 \leq r < m$, such that: $n = mq + r$. And so we have:

$$h = a^n = a^{mq+r} = (a^m)^q a^r \quad (*) \text{ or:}$$

$$a^r = (a^m)^{-q} a^n \quad (**)$$

Since a^n and a^m are both in H, and since H is a group: $(a^m)^{-q} a^n \in H$. Consequently, from (**): $a^r \in H$.

Since $0 \leq r < m$ and since m is the smallest positive integer such that $a^m \in H$: $r = 0$. Consequently, from (*):

$$h = a^n = (a^m)^q a^0 = (a^m)^q \text{ — a power of } a^m.$$

Here is a particularly important result:

Joseph-Louis Lagrange
(1736-1813)

THEOREM 5.17 If G is a finite group and H is a subgroup of
(Lagrange) G, then the order of H divides the order of G:

$$|H| \,\big|\, |G|$$

(see Definition 5.2, page 209)

While the converse of Theorem 5.17 holds for abelian groups, it does not hold in general. In particular, there exists a group of order 12 that does not contain a subgroup of order 6 (The so called alternating group of degree 4).

To illustrate: If a group G contains 35 elements, it cannot contain a subgroup of 8 elements, as 8 does not divide 35.

A proof of Lagrange's Theorem is offered at the end of the section. We now turn to a few of its consequences, beginning with:

THEOREM 5.18 Any group G of prime order is cyclic.

PROOF: Let $|G| = p$, where p is prime. Since $p \geq 2$, we can choose an element $a \in G$ distinct from e. By Lagrange's theorem, the order of the cyclic group $\langle a \rangle = \{a^n | n \in Z\}$ must divide p. But only 1 and p divide p, and since $\langle a \rangle$ contains more than one element, it must contain p elements, and is therefore all of G.

> The symmetric group S_3 is an example of a non-abelian group of order 6.

THEOREM 5.19 Every group of order less than 6 is abelian.

PROOF: We know that Z_4 and the Klein group K are the only groups of order 4, and each is abelian. Clearly the trivial group $\{e\}$ of order 1 is abelian. Any group or order 2 or 3, being of prime order, must be cyclic (Theorem 5.18), and therefore abelian (Theorem 5.6, page 215).

> We remind you that $o(a)$ denotes the order of a. (Definition 5.5, page 223).

THEOREM 5.20 For any element a in a finite group G:
$$o(a) \mid |G|$$

PROOF: If $o(a) = m$, then $\langle a \rangle = \{a, a^2, \ldots, a^{m-1}, a^m = e\}$ is a subgroup of G consisting of m elements. Consequently: $o(a) \mid |G|$.

> If $n = 1$, then $G = \{e\}$.

THEOREM 5.21 If G is a finite group of order n, then $a^n = e$ for every $a \in G$.

PROOF: Let $a \in G$, with $o(a) = m$. Since m divides n (Lagrange's Theorem), $n = tm$ for some $t \in Z^+$. Thus:
$$a^n = a^{tm} = (a^m)^t = e^t = e$$

CHECK YOUR UNDERSTANDING 5.15

PROVE OR GIVE A COUNTEREXAMPLE

Let G be a group with $a, b \in G$. If $o(a) = n$ and $o(b) = m$, then $(ab)^{nm} = e$.

> Answer: See page A-30.

PROOF OF LAGRANGE'S THEOREM

We begin by recalling some material from Chapter 1:

> See Definition 2.20 page 88.

An **equivalence relation** $\sim$ on a set X is a relation which is

Reflexive: $x \sim x$ for every $x \in X$,

Symmetric: If $x \sim y$, then $y \sim x$,

Transitive: If $x \sim y$ and $y \sim z$, then $x \sim z$.

> See Definition 2.21 page 91.

For $x_0 \in X$ the **equivalence class** of x_0 is the set:
$$[x_0] = \{x \in X \mid x \sim x_0\}.$$

LEMMA 5.2 Let H be a subgroup of a group G. The relation $a \sim b$ if $ab^{-1} \in H$ is an equivalence relation on G. Moreover:
$$[a] = \{ha | h \in H\}$$

PROOF:

$\sim$ is reflexive: $x \sim x$ since $xx^{-1} = e \in H$.

$\sim$ is symmetric:

$$a \sim b \Rightarrow ab^{-1} = h \text{ for some } h \in H$$
$$\Rightarrow (ab^{-1})^{-1} = h^{-1}$$
Theorem 5.11, page 221: $\Rightarrow (b^{-1})^{-1}a^{-1} = h^{-1}$
Exercise 2, page 225: $\Rightarrow ba^{-1} = h^{-1} \Rightarrow \boldsymbol{b \sim a}$ since $h^{-1} \in H$

(bracketed: H is a group)

$\sim$ is transitive: If $a \sim b$ and $b \sim c$, then:

$$ab^{-1} \in H \text{ and } bc^{-1} \in H$$
$$\Rightarrow (ab^{-1})(bc^{-1}) \in H$$
$$\Rightarrow a(\boldsymbol{b^{-1}b})c^{-1} \in H$$
$$\Rightarrow aec^{-1} \in H$$
$$\Rightarrow ac^{-1} \in H \Rightarrow \boldsymbol{a \sim c}$$

Having established the equivalence part of the theorem, we now verify that $[a] = \{ha | h \in H\}$:

$$\boldsymbol{b \in [a]} \Leftrightarrow b \sim a \Leftrightarrow ba^{-1} = h \text{ for some } h \in H$$
$$\Leftrightarrow b = ha \text{ for some } h \in H$$
$$\Leftrightarrow \boldsymbol{b \in \{ha | h \in H\}}$$

> **NOTE:** The above set $\{ha | h \in H\}$ will be denoted by Ha:
> $$Ha = \{ha | h \in H\}$$

We are now in a position to offer a proof of Lagrange's Theorem:

If H is a subgroup of a finite group G, then $|H| \,\big|\, |G|$.

PROOF: Theorem 2.15(a), page 93, and Lemma 5.2, tell us that sets $\{Ha | a \in G\}$ partition G. Since G is finite, we can choose $a_1, a_2, \ldots, a_k$ such that $G = \bigcup_{i=1}^{k} Ha_i$ with $Ha_i \cap Ha_j = \emptyset$ if $i \neq j$. We now show that each Ha_i has the same number of elements as H by verifying that the function $f_i: H \to Ha_i$ given by $f_i(h) = ha_i$ is a bijection:

f_i **is one-to-one**:
$$f_i(h_1) = f_i(h_2) \Rightarrow h_1 a = h_2 a$$
$$\Rightarrow (h_1 a)a^{-1} = (h_2 a)a^{-1}$$
$$\Rightarrow h_1(aa^{-1}) = h_2(aa^{-1}) \Rightarrow h_1 = h_2$$

f_i **is onto**:

For any given $ha_i \in Ha_i$, $f_i(h) = ha_i$.

Since G is the disjoint union of the k sets $Ha_1, Ha_2, \ldots, Ha_k$, and since each of those sets contains $|H|$ elements: $|G| = k|H|$, and therefore: $|H| \big| |G|$.

EXERCISES

Exercise 1-5. Determine if the given subset S is a subgroup of $\langle Z, + \rangle$.

1. $S = \{n \mid n \text{ is even}\}$
2. $S = \{n \mid n \neq 1\}$
3. $S = \{n \mid n \text{ is odd}\}$
4. $S = \{n \mid n \text{ is divisible by 2 and 3}\}$
5. $S = \{n \mid n \text{ is divisible by 2 or 3}\}$

Exercise 6-8. Determine if the given subset S is a subgroup of $\langle Z_8, +_8 \rangle$ (see Theorem 5.2, page 208).

6. $S = \{0, 2, 4, 6\}$
7. $S = \{0, 3, 6\}$
8. $S = \{0, 2, 3, 4\}$

Exercise 9-12. Determine if the given subset S is a subgroup of $\langle \mathfrak{R}, + \rangle$.

9. $S = \{x \mid x = 7y \text{ for } y \in \mathfrak{R}\}$
10. $S = \{x \mid x = 7y \text{ for } y \geq 0\}$
11. $S = \{x \mid x = 7 + y \text{ for } y \in \mathfrak{R}\}$
12. $S = \{x \mid x = 7 + y \text{ for } y \geq 0\}$

Exercise 13-18. Determine if the given subset S is a subgroup of $(S_3, \circ)$ where:

$\alpha_0 = \begin{pmatrix} 1 & 2 & 3 \\ 1 & 2 & 3 \end{pmatrix}, \alpha_1 = \begin{pmatrix} 1 & 2 & 3 \\ 2 & 3 & 1 \end{pmatrix}, \alpha_2 = \begin{pmatrix} 1 & 2 & 3 \\ 3 & 1 & 2 \end{pmatrix}, \alpha_3 = \begin{pmatrix} 1 & 2 & 3 \\ 1 & 3 & 2 \end{pmatrix}, \alpha_4 = \begin{pmatrix} 1 & 2 & 3 \\ 3 & 2 & 1 \end{pmatrix}, \alpha_5 = \begin{pmatrix} 1 & 2 & 3 \\ 2 & 1 & 3 \end{pmatrix}$

13. $S = \{\alpha_0, \alpha_1\}$
14. $S = \{\alpha_0, \alpha_2\}$
15. $S = \{\alpha_0, \alpha_3\}$
16. $S = \{\alpha_0, \alpha_1, \alpha_2\}$
17. $S = \{\alpha_0, \alpha_3, \alpha_5\}$
18. $S = \{\alpha_1, \alpha_2, \alpha_3, \alpha_4, \alpha_5\}$

Exercise 19-21. Determine if the given subset S is a subgroup of $\langle R^3, + \rangle$.

19. $S = \{(a, b, 0) \mid a, b \in \mathfrak{R}\}$
20. $S = \{(a, b, 1) \mid a, b \in \mathfrak{R}\}$
21. $S = \{(a, b, c) \mid c = a + b\}$
22. $S = \{(a, b, c) \mid c = ab\}$

Exercise 23-26. Determine if the given subset S is a subgroup of $\langle M_{2 \times 2}, + \rangle$.

23. $S = \left\{ \begin{bmatrix} a & b \\ a+b & 0 \end{bmatrix} \bigg| a, b \in \mathfrak{R} \right\}$
24. $S = \left\{ \begin{bmatrix} a & b \\ a+b & 1 \end{bmatrix} \bigg| a, b \in \mathfrak{R} \right\}$
25. $S = \left\{ \begin{bmatrix} a & b \\ a+b & ab \end{bmatrix} \bigg| a, b \in \mathfrak{R} \right\}$
26. $S = \left\{ \begin{bmatrix} a & b \\ c & 2a+c \end{bmatrix} \bigg| a, b, c \in \mathfrak{R} \right\}$

Exercise 27-30. Determine if the given subset S is a subgroup of $\langle F(\mathfrak{R}), + \rangle$ (see Exercise 49, page 218).

27. $S = \{f \mid f \text{ is continuous}\}$
28. $S = \{f \mid f \text{ is differentiable}\}$
29. $S = \{f \mid f(1) = 1\}$
30. $S = \{f \mid f(1) = 0\}$

Exercise 31-34. Determine if the given subset S is a subgroup of $\langle S_\mathfrak{R}, \circ \rangle$ (see Theorem 5.4, page 212).

31. $S = \{f \mid f \text{ is continuous}\}$
32. $S = \{f \mid f \text{ is differentiable}\}$
33. $S = \{f \mid f(1) = 1\}$
34. $S = \{f \mid f(1) = 0\}$

35. Prove that all subgroups of Z are of the form nZ.

36. Find all subgroups of $\langle Z_6, +_n \rangle$.

37. Prove that if $\{e\}$ and G are the only subgroups of a group G, then G is cyclic of order p, for p prime.

38. Show that a nonempty subset S of a group G is a subgroup of G if and only if
$$s, \bar{s} \in S \Rightarrow s\bar{s}^{-1} \in S$$

39. Show that for any group G the set $Z(G) = \{a \in G | ag = ga \ \forall g \in G\}$ is a subgroup of G.

40. Let G be an abelian group. Show that for any integer n, $\{a \in G | a^n = e\}$ is a subgroup of G.

41. Prove that the subset of elements of finite order in an abelian group G is a subgroup of G (called the **torsion subgroup** of G).

42. Let G be a cyclic group of order n. Show that if m is a positive integer, then G has an element of order m if and only if m divides m.

43. Let a be an element of a group G. The set of all elements of G which commute with a:
$$C(a) = \{b \in G | ab = ba\}$$
is called the **centralizer of a in G**. Prove that $C(a)$ is a subgroup of G.

44. Let H be a subgroup of a group G. The **centralizer** $C(H)$ of H is the set of all elements of G that commute with every element of H: $C(H) = \{a \in G | ah = ha \text{ for all } h \in H\}$. Prove that $C(H)$ is a subgroup of G.

45. The **center** $Z(G)$ of a group G is the set of all elements in G that commute with ever element of G: $Z(G) = \{a \in G | ab = ba \text{ for all } b \in G\}$.

    (a) Prove that $Z(G)$ is a subgroup of G.

    (b) Prove that $a \in Z(G)$ if and only if $C(a) = G$ (see Exercise 43.)

    (c) Prove that $Z(G) = \bigcap_{a \in G} C(a)$.

46. Let H and K be subgroups of an abelian group G. Verify that $HK = \{hk | h \in H \text{ and } k \in K\}$ is a subgroup of G.

47. Let H and K be subgroups of a group G such that $k^{-1}Hk \subseteq H$ for every $k \in K$. Show that $HK = \{hk | h \in H \text{ and } k \in K\}$ is a subgroup of G.

48. Let G be a finite group, and $a, b \in G$.

    (a) Prove that the elements a, a^{-1} and bab^{-1} have the same order.
    (b) Prove that ab and ba have the same order.
    Suggestion: Apply (a) to $ab = a(ba)a^{-1}$

49. Prove that H is a subgroup of a group G if and only if $HH^{-1} = \{ab^{-1} | a, b \in H\} \subseteq H$.

50. Let H and K be subgroups of an abelian group G of orders n and m respectively. Show that if $H \cap K = \{e\}$, then $HK = \{hk | h \in H \text{ and } k \in K\}$ is a subgroup of G of order nm.

51. (a) Prove that the group $\langle Z, + \rangle$ contains an infinite number of subgroups.
    (b) Prove that any infinite group contains an infinite number of subgroups.

52. Let S be a finite subset of a group G. Prove that S is a subgroup of G if and only if $ab \in S$ for every $a, b \in S$.

53. (a) $\{H_i\}_{i=1}^{n}$ be subgroups of a group G. Show that $\bigcap_{i=1}^{n} H_i$ is also a subgroup of G.

    (b) Let $\{H_i\}_{i=1}^{\infty}$ be a collection of subgroups of a group G. Show that $\bigcap_{i=1}^{\infty} H_i$ is also a subgroup of G.

    (c) Let $\{H_\alpha\}_{\alpha \in A}$ be a collection of subgroups of a group G. Show that $\bigcap_{\alpha \in A} H_\alpha$ is also a subgroup of G.

	PROVE OR GIVE A COUNTEREXAMPLE	

54. If H and K are subgroups of a group G, then $H \cup K$ is also a subgroup of G.

55. It is possible for a group G to be the union of two disjoint subgroups of G.

56. In any group G, $\{a \in G | a^n = e \text{ for some } n \in Z\}$ is a subgroup of G.

57. In any abelian group G, $\{a \in G | a^n = e \text{ for some } n \in Z^+\}$ is a subgroup of G.

58. Let G be a group with $a, b \in G$. If $o(a) = n$ and $o(b) = m$, then $(ab)^{nm} = e$.

59. If H and K are subgroups of a group G, then $HK = \{hk | h \in H \text{ and } k \in K\}$ is also a subgroup of G.

60. In any group G, $\{a \in G | a^3 = e\}$ is a subgroup of G.

61. No nontrivial group can be expressed as the union of two disjoint subgroups.

§4. HOMOMORPHISMS AND ISOMORPHISMS

Up until now we have focused our attention exclusively on the internal nature of a given group. The time has come to consider links between groups:

The word homomorphism comes from the Greek homo meaning "same" and morph meaning "shape."

DEFINITION 5.8
HOMOMORPHISM

A function $\phi: G \to G'$ from a group G to a group G' is said to be a **homomorphism** if $\phi(ab) = \phi(a)\phi(b)$ for every $a, b \in G$.

Let's focus a bit on the equation:
$$\phi(ab) = \phi(a)\phi(b)$$

The operation, ab, on the left side of the equation is taking place in the group G while the one on the right, $\phi(a)\phi(b)$, takes place in the group G'. What the statement is saying is that you can perform the product in G and then carry the result over to the group G' (via ϕ), or you can first carry a and b over to G' and then perform the product in that group. Those groups and products, however, need not resemble each other. Consider the following examples:

You can easily verify that $G' = \{-1, 1\}$, under standard multiplication

*	1	-1
1	1	-1
-1	-1	1

is a group.

EXAMPLE 5.5

Let $G = \langle Z, + \rangle$, and let $G' = \{-1, 1\}$ under standard integer multiplication (see margin). Show that $f: G \to G'$ given by:

$$\phi(n) = \begin{cases} 1 & \text{if } n \text{ is even} \\ -1 & \text{if } n \text{ is odd} \end{cases}$$

is a homomorphism.

SOLUTION: We consider three cases:

Case 1. (Both integers are even). If $a = 2n$ and $b = 2m$, then:
$\phi(a+b) = \phi(2n+2m) = 1$ (since $2n+2m$ is even)
And also: $\phi(a)\phi(b) = \phi(2n)\phi(2m) = 1 \cdot 1 = 1$.

Case 2. (Both are odd). If $a = 2n+1$ and $b = 2m+1$, then:
$\phi(a+b) = \phi[(2n+1) + (2m+1)] = \phi(2n+2m+2) = 1$
And also:
$\phi(a)\phi(b) = \phi(2n+1)\phi(2m+1) = (-1)(-1) = 1$.

Since $\langle Z, + \rangle$ is abelian, we need not consider $a = 2n+1$ and $b = 2m$

Case 3. (Even and odd). If $a = 2n$ and $b = 2m+1$, then:
$\phi(a+b) = \phi[(2n) + (2m+1)] = \phi[2(n+m)+1] = -1$
And also: $\phi(a)\phi(b) = \phi(2n)\phi(2m+1) = (1)(-1) = -1$.

5.4 Homomorphisms and Isomorphisms

See page 208 for a discussion of the group $\langle Z_n, +_n \rangle$.

EXAMPLE 5.6 Show that the function $\phi: \langle Z, + \rangle \to \langle Z_n, +_n \rangle$ given by $\phi(m) = r$ where $m = nq + r$ with $0 \leq r < n$ is a homomorphism.

SOLUTION: Let $a = nq_1 + r_1$ with $0 \leq r_1 < n$, $b = nq_2 + r_2$ with $0 \leq r_2 < n$, and $r_1 + r_2 = nq_3 + r_3$ with $0 \leq r_3 < n$. Then:

$$\phi(a+b) = \phi[(nq_1 + r_1) + (nq_2 + r_2)]$$
$$= \phi[n(q_1 + q_2) + (r_1 + r_2)]$$
$$= \phi[n(q_1 + q_2) + (nq_3 + r_3)] \text{ with } 0 \leq r_3 < n$$
$$= \phi[n(q_1 + q_2 + q_3) + r_3] = r_3 \text{ (since } 0 \leq r_3 < n\text{)}$$

And: $\phi(a) +_n \phi(b) = \phi(nq_1 + r_1) +_n \phi(nq_2 + r_2)$
$$= r_1 +_n r_2 = r_3 \text{ (since } r_1 + r_2 = nq_3 + r_3 \text{ with } 0 \leq r_3 < n\text{)}$$

(same)

See page 212 for a discussion on the symmetric group $\langle S_G, \circ \rangle$.

EXAMPLE 5.7 For any fixed element a in a group G, let $f_a: G \to G$ be given by $f_a(g) = ag$. Show that the function $\phi: G \to \langle S_G, \circ \rangle$ given by $\phi(a) = f_a$ is a one-to-one homomorphism.

SOLUTION: ϕ is one-to-one:

$$\phi(a) = \phi(b) \Rightarrow f_a = f_b \Rightarrow f_a(e) = f_b(e) \Rightarrow ae = be \Rightarrow a = b$$

in particular

To show that ϕ is a homomorphism we need to show that $\phi(ab) = \phi(a) \circ \phi(b)$, which is to say, that the function $f_{ab}: G \to G$ is equal to the function $f_a \circ f_b: G \to G$. Let's do it:

For any $x \in G$: $f_{ab}(x) = (ab)x$
and $(f_a \circ f_b)(x) = f_a[f_b(x)] = f_a(bx) = a(bx)$

By associativity, $(ab)x = a(bx)$, and we are done.

CHECK YOUR UNDERSTANDING 5.16

(a) Show that for any two groups G and G' the function $\phi: G \to G'$ given by $\phi(a) = e$ for every $a \in G$ is a homomorphism (called the **trivial homomorphism** from G to G').

(b) Let $\mathfrak{R}^+$ denote the group of positive real numbers under multiplication. Prove that the function $\phi: \langle Z, + \rangle \to \mathfrak{R}^+$, given by $\phi(n) = \pi^n$ is a homomorphism.

Answer: See page A-30.

Homomorphisms preserve identities, inverses, and subgroups:

THEOREM 5.22 Let $\phi: G \to G'$ be a homomorphism. Then:
(a) $\phi(e) = e'$
(b) $\phi(a^{-1}) = [\phi(a)]^{-1}$
(c) If H is a subgroup of G, then $\phi(H)$ is a subgroup of G'.
(d) If H' is a subgroup of G', then $\phi^{-1}(H')$ is a subgroup of G.

PROOF:
(a) Since ϕ is a homomorphism: $\phi(e) = \phi(ee) = \phi(e)(\phi(e))$.
Multiplying both sides by $[\phi(e)]^{-1}$ yields the desired result:
$$[\phi(e)]^{-1}\phi(e) = [\phi(e)]^{-1}[\phi(e)\phi(e)]$$
$$e' = ([\phi(e)]^{-1}[\phi(e)])\phi(e)$$
$$e' = e'\phi(e) = \phi(e)$$

(b) Since $\phi(a^{-1})\phi(a) = \phi(a^{-1}a) = \phi(e) \underset{(a)}{=} e'$:
$$\phi(a^{-1}) = [\phi(a)]^{-1}.$$

(c) We use Theorem 5.14, page 227, to show that the nonempty set $\phi(H)$ is a subgroup of G':

Since $\phi(a)\phi(b) = \phi(ab)$: $\phi(H)$ is closed with respect to the operation in G'.

Since, for any $a \in G$, $\phi(a^{-1}) = [\phi(a)]^{-1}$:
$$[\phi(a)]^{-1} \in \phi(H) \text{ for every } \phi(a) \in \phi(H).$$

(d) We use Theorem 5.14, page 227, to show that the nonempty set $\phi^{-1}(H')$ is a subspace of G:

Let $a, b \in \phi^{-1}(H')$. To say that $ab \in \phi^{-1}(H')$ is to say that $\phi(ab) \in H'$, and it is:

Since $\phi(ab) = \phi(a)\phi(b)$, and since H', being a subgroup of G', is closed with respect to the operations in G': $\phi(ab) \in H'$.

Let $a \in \phi^{-1}(H')$. To say that $a^{-1} \in \phi^{-1}(H')$ is to say that $\phi(a^{-1}) \in H'$, and it is:

Since $\phi(a^{-1}) = [\phi a]^{-1}$, and since H' contains the inverse of each of its elements: $\phi(a^{-1}) \in H'$.

CHECK YOUR UNDERSTANDING 5.17

Let $\phi: G \to G'$ and $\theta: G' \to G''$ be homomorphisms. Prove that the composite function $\theta \circ \phi: G \to G''$ is also a homeomorphism.

Answer: See page A-30.

5.4 Homomorphisms and Isomorphisms

IMAGE AND KERNEL

For any given homomorphism $\phi: G \to G'$, we define the kernel of ϕ to be the set of elements in G which map to the identity $e' \in G'$ [see Figure 5.7(a)]. We define the set of all elements in G' which are "hit" by some $\phi(a)$ to be the image of ϕ [see Figure 5.7(b)].

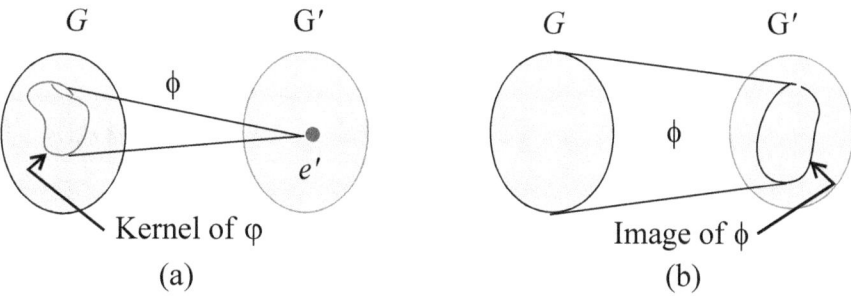

(a)   (b)

Figure 5.7

More formally:

DEFINITION 5.9 Let $\phi: G \to G'$ be a homomorphism.

KERNEL The **kernel** of ϕ, denoted by $\text{Ker}(\phi)$, is given by:
$$\text{Ker}(\phi) = \{a \in G | \phi(a) = e'\}$$

IMAGE The **image** of ϕ, denoted by $\text{Im}(\phi)$, is given by:
$$\text{Im}(\phi) = \{\phi(a) | a \in G\}$$

Utilizing the notation of Definition 2.8, page 63:
$\text{Ker}(\phi) = \phi^{-1}[\{e'\}]$
$\text{Im}(\phi) = \phi[G]$

Both the kernel and image of a homomorphism turn out to be subgroups of their respective groups:

THEOREM 5.23 Let $\phi: G \to G'$ be a homomorphism. Then:

(a) $\text{Ker}(\phi)$ is a subgroup of G.

(b) $\text{Im}(\phi)$ is a subgroup of G'.

PROOF: We appeal to Theorem 5.14, page 227 to establish the desired results.

(a) Nonempty: Since $\phi(e) = e'$ [Theorem 5.22(a)], $e \in \text{Ker}(\phi)$.

Closure: $a, b \in \text{Ker}(\phi) \Rightarrow \phi(ab) \underset{\phi \text{ is a homomorphism}}{=} \phi(a)\phi(b) = e'e' = e'$

Inverses: For $a \in \text{Ker}(\phi)$, we are to show that $a^{-1} \in \text{Ker}(\phi)$. Let's do it: $\phi(a^{-1}) \underset{\text{Theorem 5.22(b)}}{=} [\phi(a)]^{-1} = (e')^{-1} = e'$

(b) Nonempty: Since $\phi(e) = e'$, $e' \in \text{Im}(\phi)$

Closure: For $a', b' \in \text{Im}(\phi)$ choose $a, b \in G$ such that $\phi(a) = a'$ and $\phi(b) = b'$. Then:

$$\phi(ab) \underset{\phi \text{ is a homomorphism}}{=} \phi(a)\phi(b) = a'b' \Rightarrow a'b' \in \text{Im}(\phi).$$

Inverses: For $a' \in \text{Im}(\phi)$ choose $a \in G$ such that $\phi(a) = a'$. Then:

$$\phi(a^{-1}) \underset{\text{Theorem 5.22(b)}}{=} [\phi(a)]^{-1} = (a')^{-1} \Rightarrow (a')^{-1} \in \text{Im}(\phi)$$

CHECK YOUR UNDERSTANDING 5.18

Show that the function $\phi: 2Z \to 4Z$ given by $\phi(2n) = 8n$ is a homomorphism. Determine the kernel and image of ϕ.

Answer: See page A-31.

Definition 5.9 tells us that a homomorphism $\phi: G \to G'$ is onto if and only if $\text{Im}(\phi) = G'$. The following result is a bit more interesting, in that it asserts that in order for a homomorphism to be one-to-one, it need only behave "one-to-one-ish" at e (see margin):

THEOREM 5.24 A homomorphism $\phi: G \to G'$ is one-to-one if and only if $\text{Ker}(\phi) = \{e\}$.

PROOF: Suppose ϕ is one-to-one. If $a \in \text{Ker}(\phi)$, then both $\phi(a) = e'$ and $\phi(e) = e'$ [Theorem 5.22(a)]. Consequently $a = e$ (since ϕ is assumed to be one-to-one). Hence: $\text{Ker}(\phi) = \{e\}$.

Conversely, assume that $\text{Ker}(\phi) = \{e\}$. We need to show that if $\phi(a) = \phi(b)$, then $a = b$. Let's do it:

$$\phi(a) = \phi(b)$$
$$\phi(a)[\phi(b)]^{-1} = e'$$
Theorem 5.22(b): $\phi(a)\phi(b^{-1}) = e'$
ϕ is a homomorphism: $\phi(ab^{-1}) = e'$
$\text{Ker}(\phi) = \{e\}$: $ab^{-1} = e$
$$(ab^{-1})b = eb$$
$$a = b$$

Margin note: A homomorphism $\phi: G \to G'$ must map e to e'. What this theorem is saying is that if e is the only element that goes to e', then no element of G' is going to be hit by more that one element of G. This is certainly not true for arbitrary functions:

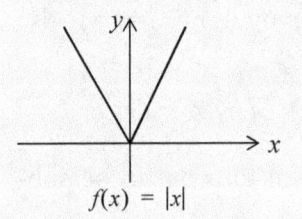

$f(x) = |x|$

CHECK YOUR UNDERSTANDING 5.19

Prove that a homomorphism $\phi: G \to G'$ is one-to-one if and only if there exists an element $a \in G$ (not necessarily the identity e) such that if $\phi(a) = \phi(b)$ then $a = b$.

In other words: for a homomorphism $\phi: G \to G'$ to be one-to-one, it need only behave "one-to-one-ish" at any one-point in G."

Answer: See page A-31.

The word *isomorphism* comes from the Greek *iso* meaning "equal" and *morph* meaning "shape."

ISOMORPHISMS

As previously noted, a homomorphism $\phi: G \to G'$ preserves the algebraic structure in that $\phi(ab) = \phi(a)\phi(b)$. An isomorphism also preserves the set structures in that it pairs up the elements of G with those of G'. More formally:

DEFINITION 5.10

ISOMORPHISM — A homomorphism $\phi: G \to G'$ which is also a bijection is said to be an **isomorphism** from the group G to the group G'.

ISOMORPHIC — Two groups G and G' are **isomorphic**, written $G \cong G'$, if there exists an isomorphism from one of the groups to the other.

EXAMPLE 5.8 Show that the group $\langle \Re, + \rangle$ of real numbers under addition is isomorphic to the group $\langle \Re^+, \cdot \rangle$ of positive real numbers under multiplication.

SOLUTION: We show that the function $\phi: \langle \Re, + \rangle \to \langle \Re^+, \cdot \rangle$ given by $\phi(a) = e^a$ is a homomorphism that is also a bijection:

ϕ is a homomorphism:
$$\phi(a + b) = e^{a+b} = e^a e^b = \phi(a)\phi(b)$$

In this discussion we are not using e to denote the identity element in $\langle \Re^+, \cdot \rangle$ (which is 1). Here, e is the transcendental number $e \approx 2.718$.

ϕ is one-to-one:

$$\phi(a) = \underset{\text{The identity in } \langle \Re^+, \cdot \rangle}{1} \Rightarrow e^a = 1 \Rightarrow a = \underset{\text{The identity in } \langle \Re, + \rangle}{0}$$

ϕ is a onto: For $a \in \langle \Re^+, \cdot \rangle$, we have: $\phi(\ln a) = e^{\ln a} = a$.

CHECK YOUR UNDERSTANDING 5.20

(a) Prove that $\cong$ is an equivalence relation on any set of groups (see Definition 2.19, page 88).
(b) Let $g \in G$. Prove that the map $i_g: G \to G$ given by
$$i_g(x) = gxg^{-1} \; \forall x \in G$$
is an automorphism (called an **inner automorphism**.)

Answer: See page A-31.

A ROSE BY ANY OTHER NAME

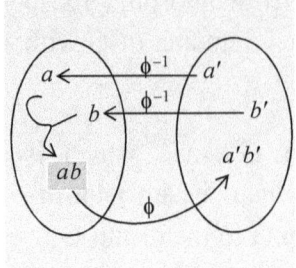

Let $\phi: G \to G'$ be an isomorphism. Being a bijection it links every element in G with a unique element in G' (every element in G has its own G' counterpart, and vice versa). Moreover, if you know how to function algebraically in G, then you can also figure out how to function algebraically in G' (and vice versa). Suppose, for example, that you forgot how to multiply in the group G', but remember how to multiply in G. To figure out $a'b'$ in G' you can take the "ϕ^{-1} bridge" back to G to find the elements a and b such that $\phi(a) = a'$ and $\phi(b) = b'$, perform the product ab in G, and then take the "ϕ bridge" back to G' to find the product $a'b'$: $\phi(ab)$.

Basically, if a group G is isomorphic to G', then the two groups can only differ in appearance, but not "algebraically." Consider, for example, the following two groups which appeared previously in Figure 5.1, page 209:

Z_4:

$+_4$	0	1	2	3
0	0	1	2	3
1	1	2	3	0
2	2	3	0	1
3	3	0	1	2

(a)

K:

*	e	a	b	c
e	e	a	b	c
a	a	e	c	b
b	b	c	e	a
c	c	b	a	e

(b)

Both contain four elements ($\{0,1,2,3\}$ and $\{e,a,b,c\}$); so, as far as sets go, they "are one and the same" (different element-names, that's all). But as far as groups go, they are **not** the same (not isomorphic). Here are two algebraic differences (either one of which would serve to prove that the two groups are not isomorphic):

1. Z_4 is cyclic while the Klein 4-group, K, is not.
2. There exist three elements in K of order 2 (see Definition 5.5, page 223), while Z_4 contains but one (the element 2).

To better substantiate the above claims:

THEOREM 5.25 If $G \cong G'$, then:

(a) G is cyclic if and only if G' is cyclic.

(b) For any given integer n, there exists an element $a \in G$ such that $a^n = e$ if and only if there exists an element $a' \in G'$ such that $(a')^n = e'$.

PROOF: Let $\phi: G \to G'$ be an isomorphism.

(a) Suppose G is cyclic, with $G = \langle a \rangle$. We show $G' = \langle \phi(a) \rangle$ by showing that for any $b' \in G'$, there exists $n \in Z$ such that $b' = [\phi(a)]^n$:

Let $b \in G$ be such that $\phi(b) = b'$. Since $G = \langle a \rangle$, there exists $n \in Z$ such that $b = a^n$. Then:
$$b' = \phi(b) = \phi[a^n] \underset{\text{Exercise 13}}{=} [\phi(a)]^n$$

The "only-if" part follows from the fact that if G is isomorphic to G', then G' is isomorphic to G (see CYU 5.20).

(b) Let $a \in G$ be such that $a^n = e$. Then:
$$[\phi(a)]^n = \phi(a^n) = \phi(e) = e'$$
The "only-if" part follows from CYU 5.20.

CHECK YOUR UNDERSTANDING 5.21

Answer: See page A-31.

Prove that if $G \cong G'$, then G is abelian if and only if G' is abelian.

A property of a group G that is shared by all groups isomorphic to G is said to be a **group invariant** property. For example, abelian and cyclic are group invariant properties. Other group invariant properties are cited in the exercises.

In general, one can show that two groups are not isomorphic by exhibiting a group invariant property that holds in one of the groups but not in the other. For example, the permutation group S_3 is not isomorphic to $\langle Z_6, +_6 \rangle$ as one is abelian while the other is not.

The following results underlines the importance of symmetric groups (see discussion on page 212).

Arthur Cayley (1821-1885)

THEOREM 5.26 (Cayley) Every group is isomorphic to a subgroup of a symmetric group.

PROOF: The function $\phi: G \to S_G$ given by
$$\phi(g) = f_g : G \to G \text{ where } f_g(x) = gx \quad (\forall x \in G)$$
was shown to be a one-too-one homomorphism in Example 5.7. Since ϕ it is onto the subspace $\phi(G)$ of S_G:
$$\phi: G \to \phi(G) \text{ is an isomorphism.}$$

EXERCISES

Exercise 1-14. Determine if the given function $\phi: G \to G'$ is a homomorphism.

1. $G = G' = \langle Z, + \rangle$ and $\phi(n) = 2n$.

2. $G = G' = \langle Z, + \rangle$ and $\phi(n) = n + 1$.

3. $G = \langle \Re, + \rangle$, $G' = \langle Z, + \rangle$ and $\phi(x) = n$ where n is the smallest integer greater than or equal to x.

4. $G = \langle Z, + \rangle$, $G' = \langle \Re, + \rangle$ and $\phi(n) = n$.

5. $G = \langle Z, + \rangle$, $G' = Z_3$ and $\phi(n) = r$ where $n = 3m + r$ with $0 \le r < 3$.

6. $G = \langle Z, + \rangle$, $G' = \langle \{-1, 1\}, \cdot \rangle$ and $\phi(n) = 1$ if n is even and $f(n) = -1$ if n is odd.

7. $G = G'$ and $\phi(a) = a^{-1}$ for $a \in G$.

8. $G = G'$ with G abelian, and $\phi(a) = a^{-1}$ for $a \in G$.

9. $G = G'$ with G abelian, $n \in Z^+$, and $\phi(a) = a^n$ for $a \in G$.

10. $G = Z_5$, $G' = Z_2$ and $\phi(n) = r$ where $n = 2d + r$ with $0 \le r < 2$.

11. $G = Z_6$, $G' = Z_2$ and $\phi(n) = r$ where $n = 2d + r$ with $0 \le r < 2$.

12. $G = G' = \langle M_{2 \times 2}, + \rangle$ and $\phi\left(\begin{bmatrix} a & b \\ c & d \end{bmatrix}\right) = \begin{bmatrix} a+b & d \\ -c & 0 \end{bmatrix}$.

13. $G = \langle M_{2 \times 2}, + \rangle$, $G' = \Re$ and $\phi\left(\begin{bmatrix} a & b \\ c & d \end{bmatrix}\right) = ad - bc$.

14. $G = G' = \langle M_{2 \times 2}, + \rangle$ and $\phi\left(\begin{bmatrix} a & b \\ c & d \end{bmatrix}\right) = \begin{bmatrix} a+b & d \\ -c & 1 \end{bmatrix}$.

15. Let $\langle \Re, + \rangle$ denote the group of all real numbers under addition, and $\langle \Re^+, \cdot \rangle$ the group of all positive real numbers under multiplication. Show that the map $\phi: \Re^+ \to \Re$ given by $\phi(x) = \ln x$ is an isomorphism.

16. Let $\phi: G \to G'$ be a homomorphism and let $a \in G$. Prove that $\phi(a^n) = [\phi(a)]^n$ for every $n \in Z$.

17. Let $\phi: G \to G'$ be a homomorphism. Prove that for all $a, b \in G$:
$$\phi(ab^{-1}) = \phi(a)\phi(b)^{-1} \text{ and } \phi(a^{-1}b) = \phi(a)^{-1}\phi(b)$$

18. Let $\phi: G \to G'$ be a homomorphism, Show that:
    (a) If ϕ is onto and if G is abelian, then G' is abelian.

    (a) If ϕ is one-to-one and if G' is abelian, then G is abelian.

19. Prove that a group G is abelian if and only if the function $f: G \to G$ given by $f(g) = g^{-1}$ is a homomorphism.

20. Let $G = \langle a \rangle$ be cyclic and let G' be any group. Let $\phi: G \to G'$ be a homomorphism. Prove that $\text{Im}(\phi)$ is cyclic.

21. Let $\phi: G \to G'$ be a homomorphism. Show that if $k \in \text{Ker}(\phi)$, then $gkg^{-1} \in \text{Ker}(\phi)$ for every $g \in G$.

22. Let $G = \langle a \rangle$ be cyclic and let H be any group. Prove that for any chosen $h \in H$ there exists a unique homomorphism $\phi: G \to H$ such that $\phi(a) = h$.
    So, a homomorphism on a cyclic group $G = \langle a \rangle$ is completely determined by its action on a.

23. Let $\phi: G \to G'$ be a homomorphism. Prove that, for any given $x \in G$:
    $$\{g \in G | \phi(g) = \phi(x)\} = \{xk | k \in \text{Ker}(\phi)\}$$

24. Let A, B, C, and D be groups. Show that if $A \cong B$ and $C \cong D$, then $A \times C \cong B \times D$ (see Exercise 52, page 218).

25. Let G and G' be groups. Show that $G \times G' \cong G' \times G$ (see Exercise 52, page 218).

26. (a) Show that the set $Z \times Z = \{(a, b) | a, b \in Z\}$, with $(a, b) * (c, d) = (a+c, b+d)$ is a group.
    (b) Verify that the functions $\phi_1: Z \times Z \to Z$ and $\phi_2: Z \times Z \to Z$ given by $\phi_1(a, b) = a$ and $\phi_2(a, b) = b$, respectively, are homomorphisms.
    (c) Show that the function $\phi: Z \times Z \to Z$ given by $\phi(a, b) = 2\phi_1(a, b) + 3\phi_2(a, b)$ is a homomorphism.
    (d) Show that the function $\theta: Z \times Z \to Z \times Z$ given by $\theta(a, b) = [\phi_2(a, b), \phi_1(a, b)]$ is an isomorphism.

27. For $m \in Z$, $m \neq 0$, let $\phi_m: Z \to Z$ be given by $\phi_m(n) = mn$.
    (a) Show that ϕ_m is a one-to-one homomorphism.
    (b) Show that ϕ_m is an isomorphism if and only if $m = \pm 1$.

28. Let $F(\Re)$ denote the additive group of real valued function (see Exercise 49, page 218), and let $\Re$ denote the additive group of real numbers. Prove that for any $c \in \Re$ the function $\phi_c: F(\Re) \to \Re$ given by $\phi_c(f) = f(c)$ for $f \in F(\Re)$ is a homomorphism (called an **evaluation homomorphism**.)

29. Let $D(\Re)$ denote the set of differentiable functions from $\Re$ to $\Re$.
    (a) Show that $\langle D(\Re), +\rangle$ is a group.
    (b) Show that for any $c \in \Re$ the function $\phi_c: D(\Re) \to \Re$ given by $\phi_c(f) = f(c)$ is a homomorphism.
    (c) Is ϕ_c one-to-one for any c?
    (d) Is ϕ_c onto for any c?

30. Let $C(\Re)$ denote the set of continuous real valued functions.
    (a) Show that $\langle C(\Re), +\rangle$ is a group.
    (b) Show that for any closed interval $[a, b]$ in $\Re$ the function $\phi: C(\Re) \to \Re$ given by $\phi(f) = \int_a^b f(x)dx$ is a homomorphism.
    (c) Show that the function $\theta: C(\Re) \to \Re$ given by $\phi(f) = \int_0^1 f(x)dx + 2\int_2^3 f(x)dx$ is a homomorphism.

31. Show that for any $\tau \in S_3$, the function $\phi: S_3 \to S_3$ given by $\phi(\sigma) = \sigma \circ \tau$ is a homomorphism. Is it necessarily an isomorphism?

32. Show that the function $\phi: S_3 \to S_4$ given by $[\phi(\sigma)](i) = \begin{cases} \sigma(i) & \text{if } i < 4 \\ 4 & \text{if } i = 4 \end{cases}$ is a homomorphism. Determine the image and kernel of ϕ

33. Let G be a group. Prove that $\text{Aut}(G) = \langle \{\phi | \phi: G \to G \text{ is an automorphism}\}, \circ\rangle$ is a group.

Exercise 34-40. Show that the give property on a G is an invariant.

34. $|G|$ — the order of a finite group G.

35. G contains a nontrivial cyclic subgroup.

36. G contains an element of order n for given $n \geq 1$.

37. G contains m elements of order n for given $n \geq 1$.

38. G contains a subgroup of order of order n for given $n \geq 1$.

39. The number of elements in $T_n = \{g \in G | o(g) = n\}$ (see Definition 5.5, page 224).

40. The number of elements in $Z(G)$ — the center of a finite group G. (See Exercise 45, page 235.)

| **PROVE OR GIVE A COUNTEREXAMPLE** |

41. The additive group $\Re$ is isomorphic to the additive group Q of rational numbers)

42. The additive group Z is isomorphic to the additive group Q of rational numbers)

43. If ϕ is a homomorphism from a group G to a cyclic group $G' = \langle a \rangle$, then Ker(ϕ) is a cyclic subgroup of G.

44. If ϕ is an isomorphism from a group G to a cyclic group $G' = \langle a \rangle$, then Ker(ϕ) is a cyclic subgroup of G.

45. For $C(\Re)$ the group of continuous real valued functions under addition the function $\phi: C(\Re) \to \Re$ given by $\phi(f) = \left(\int_0^1 f(x)dx\right)\left(\int_2^3 f(x)dx\right)$ is a homomorphism.

46. If $n \neq m$, S_n and S_m are not isomorphic.

CHECK YOUR UNDERSTANDING SOLUTIONS
CHAPTER 1
A LOGICAL BEGINNING

1.1 PROPOSITIONS

CYU 1.1 (a) Since q is True ($3 + 5 = 8$), $p \vee q$ is True, independently of p.

(b) Since p is False ($7 = 5$), $p \wedge q$ is False, independently of q.

CYU 1.2 (a) Since p is True ($5 > 3$), its negation $\sim p$ is False.

(b) Since q is False ($3 = 5$), its negation p is True, so $\sim p$ is False.

CYU 1.3 (a) Since p and q are True, so is $p \wedge q$. It follows that $\sim(p \wedge q)$ is False.

(b) Since p and q are True, so is $p \vee q$. It follows that $\sim(p \vee q)$ is False.

(c) Since q is True, so is $s \vee q$. It follows that $\sim(s \vee q)$ is False.

(d) Since s is False, so is $s \wedge q$. It follows that $\sim(s \wedge q)$ is True.

(e) Since p is True, $\sim p$ is False. It follows that $\sim p \wedge q$ is False.

(f) Since q is True, so is $\sim p \vee q$.

(g) Since s is False, $\sim s$ is True. It follows that $\sim s \vee q$ is True.

(h) Since both $\sim s$ and q are True, so is $\sim s \wedge q$.

(i) Since s is False, so is $(q \wedge s)$. It follows that $(p \vee s) \wedge (q \wedge s)$ is False.

(j) Since p is True, so is $(p \vee s)$. It follows that $(p \vee s) \vee (q \wedge s)$ is True.

(k) Since p is True, so is $(p \vee s)$. It follows that $\sim(p \vee s)$ is False, as is $\sim(p \vee s) \wedge (q \vee s)$.

CYU 1.4 We show that $[(p \to q) \wedge \sim q] \to \sim p$ is a tautology:

p	q	$p \to q$	$\sim q$	$(p \to q) \wedge \sim q$	$\sim p$	$[(p \vee q) \wedge \sim q] \to \sim p$
T	T	T	F	F	F	T
T	F	F	T	F	F	T
F	T	T	F	F	T	T
F	F	T	T	T	T	T

CYU 1.5 (a) Negating *Mary is going to a movie **or** she is going shopping*:

$$\sim[(\text{Mary is going to a movie}) \vee (\text{Mary is going shopping})]$$

Theorem 1.1(b): $\sim(\text{Mary is going to a movie}) \wedge \sim(\text{Mary is going shopping})$

*Mary is neither going to a movie **nor** going shopping.*

(b) Negating *Bill weighs more than 200 pounds **and** is less than 6 feet tall*:

$$\sim[(\text{Bill weighs more that 200 pounds}) \wedge (\text{Bill is less than 6 feet tall})]$$

Theorem 1.1(a): $\sim(\text{Bill weighs more that 200 pounds}) \vee \sim(\text{Bill is less than 6 feet tall})$

*Bill weighs at most 200 pounds **or** is at least 6 feet tall.*

(c) Negating $x > 0$ **or** $x \leq -5$: $\sim(x > 0 \text{ **or** } x \leq -5)$

Theorem 1.1(b): $\sim(x > 0)$ **and** $\sim(x \leq -5)$

$$x \leq 0 \text{ **and** } x > -5$$

$$-5 < x \leq 0$$

CYU 1.6 (a) $\sim(\sim p) \Leftrightarrow p$

p	$\sim p$	$\sim(\sim p)$
T	F	T
F	T	F

(b) $p \rightarrow q \Leftrightarrow \sim p \vee q$

p	q	$p \rightarrow q$	$\sim p$	$\sim p \vee q$
T	T	T	F	T
T	F	F	F	F
F	T	T	T	T
F	F	T	T	T

(c) $\sim(p \rightarrow q) \Leftrightarrow \sim(\sim p \vee q)$

p	q	$p \rightarrow q$	$\sim(p \rightarrow q)$	$\sim p$	$\sim p \vee q$	$\sim(\sim p \vee q)$
T	T	T	F	F	T	F
T	F	F	T	F	F	T
F	T	T	F	T	T	F
F	F	T	F	T	T	F

(d) $p \leftrightarrow q \Leftrightarrow (p \rightarrow q) \wedge (q \rightarrow p)$

p	q	$p \leftrightarrow q$	$p \rightarrow q$	$q \rightarrow p$	$(p \rightarrow q) \wedge (q \rightarrow p)$
T	T	T	T	T	T
T	F	F	F	T	F
F	T	F	T	F	F
F	F	T	T	T	T

CYU 1.7 (a) $[(p \vee q) \to \sim s] \Leftrightarrow [s \to (\sim p \wedge \sim q)]$:

p	q	s	$p \vee q$	$\sim s$	$(p \vee q) \to \sim s$	$\sim p$	$\sim q$	$\sim p \wedge \sim q$	$s \to (\sim p \wedge \sim q)$
T	T	T	T	F	F	F	F	F	F
T	T	F	T	T	T	F	F	F	T
T	F	T	T	F	F	F	T	F	F
T	F	F	T	T	T	F	T	F	T
F	T	T	T	F	F	T	F	F	F
F	T	F	T	T	T	T	F	F	T
F	F	T	F	F	T	T	T	T	T
F	F	F	F	T	T	T	T	T	T

(b) $[(p \vee q) \to \sim s] \underset{\uparrow}{\Leftrightarrow} [\sim(\sim s) \to \sim(p \vee q)] \underset{\uparrow}{\Leftrightarrow} [s \to \sim(p \wedge \sim q)]$
                        Theorem 1.4   CYU 1.6(a) and Theorem 1.1(b)

CYU 1.8 Go with a Truth Table if you wish. For our part: $q \to p \Leftrightarrow \sim p \to \sim q$ by Theorem 1.4.

1.2 QUANTIFIERS

CYU 1.9 (a) All months have at least thirty days is False: February has 28 (or 29) days.

(b) Every month contains (at least) three Sundays is True.

(c) $\forall n, m \in Z^+, n + m \in Z^+$ is True.

(d) $\forall n \in Z^+$ and $\forall m \in Z, n + m \in Z^+$ is False: $3 + (-4) = -1 \notin Z^+$.

CYU 1.10 (a) There exists a month with more than thirty days is True: January has 31 days.

(b) There exists a week with more than seven day is False.

(c) $\exists n, m \in Z^+ \ni n + m = 100$ is True: $50 + 50 = 100$.

(d) $\exists n, m \in Z^+ \ni nm = n + m$ is True: $2 \cdot 2 = 2 + 2$.

CYU 1.11 (a) $\forall n, m \in Z^+, \exists s \in Z^+ \ni s > nm$ is True: Let $s = nm + 1$.

(b) $\forall s \in Z^+, \exists n, m \in Z^+ \ni s > nm$ is False: If $s = 1$, then $s \leq nm$ for every $n, m \in Z^+$.

(c) $\exists x \in \Re \ni \forall n \in Z^+, x^n > x$ is False: $\forall x \in \Re, x^1 \not> x$.

(d) $\exists n \in Z^+ \ni \forall m \in Z^+, n^m = n$ is True: $1^m = 1 \forall m \in Z^+$

CYU 1.12 (a) Negation of *All college students study hard*: Some college student does not study hard.

(b) Negation of *Everyone takes a bath at least once a week*: Someone does not take a bath at least once a week.

(c) Negation of $\forall x \in X, p(x) \wedge q(x)$: $\exists x \in X \ni [\sim p(x) \vee \sim q(x)]$.

(d) Negation of $\forall x \in X, p(x) \vee q(x)$: $\exists x \in X \ni [\sim p(x) \wedge \sim q(x)]$.

CYU 1.13 (a) Negation of *There are days when I don't want to get up*: Every day I want to get up.

(b) Negation of $\exists x \in X \ni [p(x) \wedge q(x)]$: $\forall x \in X, \sim p(x) \vee \sim q(x)$.

(b) Negation of $\exists x \in X \ni [p(x) \vee q(x)]$: $\forall x \in X, \sim p(x) \wedge \sim q(x)$.

CYU 1.14 (a) Negation of *For every $x \in X$ there exists a $y \in Y$ such that y blips at x*: There exists some $x \in X$ such that no $y \in Y$ blips at that x.

(b) Negation of $\forall x \in \mathfrak{R}, \exists n \in Z^+ \ni x = 2n$: $\exists x \in \mathfrak{R} \ni \forall n \in Z^+, x \neq 2n$.

CYU 1.15 (a) Negation of *There is a motorcycle that gets better mileage than any car*: For every motorcycle there is some car that gets the same or better mileage than that motorcycle.

(b) Negation of $\exists n \in Z \ni \forall m \in Z^+ n < m$: $\forall n \in Z \; \exists m \in Z^+ \ni n \geq m$.

1.3 METHODS OF PROOF

CYU 1.16 (a) Let $n = 2k+1$ and $m = 2h+1$. We show that $n + m$ is even:
$$n + m = (2k+1) + (2h+1) = 2(k+h+1).$$

(b) The sum of any even integer with any odd integer is odd. Proof:
For $n = 2k$ and $m = 2h+1$, $n + m = (2k) + (2h+1) = 2(k+h) + 1$.

CYU 1.17 $2m + n$ is even $\Leftrightarrow 2m + n = 2k$
$\Leftrightarrow n = 2k - 2m$
$\Leftrightarrow n = 2(k-m) \Leftrightarrow n$ is even

CYU 1.18 (a) $3n$ is odd if and only if n is odd

n odd $\Rightarrow 3n$ odd: (**a direct proof**): n odd $\Rightarrow n = 2k+1 \Rightarrow 3n = 3(2k+1)$
$$\Rightarrow 3n = 6k+3$$
$$\Rightarrow 3n = (6k+2)+1$$
$$\Rightarrow 3n = 2(3k+1)+1 \Rightarrow 3n \text{ odd}$$

$3n$ odd $\Rightarrow n$ odd (**a contrapositive proof**):

$\boldsymbol{n}$ **not odd** $\Rightarrow n$ even $\Rightarrow n = 2k$
$$\Rightarrow 3n = 6k$$
$$\Rightarrow 3n = 2(3k) \Rightarrow 3n \text{ even} \Rightarrow \boldsymbol{3n} \text{ **not odd**}$$

(b) n^3 is odd if and only if n is odd

n odd $\Rightarrow n^3$ odd: (**a direct proof**): n odd $\Rightarrow n = 2k+1$
$$\Rightarrow n^3 = (2k+1)^3 = 8k^3 + 12k^2 + 6k + 1$$
$$\Rightarrow n^3 = 2(4k^3 + 6k^2 + 3k) + 1 \Rightarrow n^3 \text{ odd}$$

n^3 odd $\Rightarrow n$ odd (**a contrapositive proof**):

$\boldsymbol{n}$ **not odd** $\Rightarrow n$ even $\Rightarrow n = 2k$
$$\Rightarrow n^3 = 8k^3 = 2(4k^3) \Rightarrow n^3 \text{ even} \Rightarrow \boldsymbol{n} \text{ **not odd**}$$

CYU 1.19 A direct proof can be use to show that n odd $\Rightarrow 3n+2$ odd:

n odd $\Rightarrow n = 2k+1$
$$\Rightarrow 3n+2 = 3(2k+1)+2 = 6k+5 = 2(3k+2)+1 \Rightarrow 3n+2 \text{ odd}$$

An attempt to show, directly, that $3n+2$ odd $\Rightarrow n$ odd leads us to a dead end:
$$3n+2 = 2k+1 \Rightarrow 3n = 2k-1 \text{ — Now what?}$$

We try something else, that's what:

$$3n+2 \text{ odd} \Rightarrow n \text{ odd}$$

Contrapositive Proof:	**Proof by Contradiction:**
n even $\Rightarrow 3n+2$ even	Let $\boldsymbol{3n+2}$ **be odd** (given condition)
$\boldsymbol{n}$ **even** $\Rightarrow n = 2k$ (for some k)	**Assume** that n is even, say $n = 2k$. Then:
$\Rightarrow 3n+2 = 3(2k)+2$	$3n+2 = 3(2k)+2 = 2(3k+1)$ even
$ = 2(3k+1)$	— contradicting the stated condition that $\boldsymbol{3n+2}$ **is odd.**
$\Rightarrow \boldsymbol{3n+2}$ **even**	

CYU 1.20 (a-i) If $a|n$ and $a|m$, then $n = ha$ and $m = ka$ for some h and k. Consequently:
$$n + m = ha + ka = (h + k)a \Rightarrow a|(n + m)$$

(a-ii) $n|a \Rightarrow a = nh$ for some h. For $c \in Z$ we then have:
$$ca = c(nh) = (ch)n \Rightarrow n|ca$$

(b-i) If $a|b$ or $a|c$, then $a|(b + c)$ is False:
Counterexample: $a = b = 2$ and $c = 1$. Since $2|2 : 2|2$ or $2|1$ is True. However $2|(2 + 1)$ is False.

(b-ii) If a and b are even and if $a|(b + c)$, then c must be even is True:
Proof: Let k_1, k_2, h be such that $a = 2k_1$, $b = 2k_2$, and $b + c = ha$.
${}_{a|(b+c)}$

Then: $ha \underset{(**)}{=} b + c \underset{(*)}{=} 2k_2 + c \Rightarrow 2k_2 + c = ha$

$\Rightarrow 2k_2 + c \underset{(*)}{=} h(2k_1) \Rightarrow c = 2hk_1 - 2k_2$

$\Rightarrow c = 2(hk_1 - k_2)$ (even)

(b-iii) If $(a + b)|(c + d)$, then there exist k such that $ak + bk + c + d = 0$ is True.
Proof: Let h be such that $c + d = h(a + b)$.
Then: $c + d - ha - hb = 0$, or $ak + bk + c + d = 0$, where $k = -h$.

1.4 PRINCIPLE OF MATHEMATICAL INDUCTION

CYU 1.21 (a) The equation $2 + 4 = 1 + 2 + 3 + 4 - (1 + 3)$ illustrate that the sum of the first two even integers can be expressed as the sum of the first four integers minus the sum of the first two odd integer. Generalizing, we anticipate that the sum of the first n even integers is the sum of the first **2n integers** minus the sum of the first **n odd integers**; leading us to the conjecture that the sum of the first n even integers equals $n^2 + n$:

$$\underbrace{\frac{2n(2n+1)}{2}}_{\text{sum or first 2n integers (Eample 1.16)}} - \underbrace{n^2}_{\text{sum of first } n \text{ odd integers (page 34)}} = 2n^2 + n - n^2 = n^2 + n$$

(b) Let $P(n)$ be the proposition that the sum of the first n even integers equals $n^2 + n$.

I. Since the sum of the first 1 even integers is 2, $P(1) = 1^2 + 1 = 2$ is true.

II. Assume $P(k)$ is true; that is: $\mathbf{2 + 4 + 6 + \cdots + 2k = k^2 + k}$.

III. We complete the proof by verifying that $P(k + 1)$ is true; which is to say, that $2 + 4 + 6 + \cdots + 2k + 2(k + 1) = (k + 1)^2 + (k + 1)$:

$$\underbrace{2 + 4 + 6 + \cdots + 2k}_{\text{by II}} + 2(k + 1) = k^2 + k + 2(k + 1)$$
$$= (k^2 + 2k + 1) + (k + 1) = (k + 1)^2 + (k + 1)$$

CYU 1.22 Let $P(n)$ be the proposition that $6|(n^3 + 5n)$ for all integers $n \geq 1$.

I. True at $n = 1$: $6|(1^3 + 5 \cdot 1)$.

II. Assume $P(k)$ is true; that is: $6|(k^3 + 5k)$.

III. To establish that $6|[(k+1)^3 + 5(k+1)]$, we begin by noting that $(k+1)^3 + 5(k+1) = (k^3 + 3k^2 + 8k) + 6$ and then set our sights on showing that $6|(k^3 + 3k^2 + 8k)$ (for clearly $6|6$).

Wanting to get II into play we rewrite $k^3 + 3k^2 + 8k$ in the form $(k^3 + 5k) + (\mathbf{3k^2 + 3k})$. Our induction hypothesis allows us to assume that $6|(k^3 + 5k)$. If we can show that $6|(\mathbf{3k^2 + 3k})$, then we will be done, by virtue of Theorem 1.6(b), page 28. Let's do it:

Since $3k^2 + 3k = 3k(k+1)$, and since either k or $k+1$ is even:
6 is a factor of $3x^2 + 3k$.

CYU 1.23 Let $P(n)$ be the proposition that $2^n < (n+2)!$ for all integers $n \geq 0$

I. Since $2^0 < (0+2)!$, $P(0)$ is true.

II. Assume $P(k)$ is true; that is: $\mathbf{2^k < (k+2)!}$.

III. We complete the proof by showing that $2^{k+1} < [(k+1)+2]! = (k+3)!$:

$$2^{k+1} = 2 \cdot \underbrace{2^k < (k+3)(k+2)!}_{\text{II}} = (k+3)!$$

CYU 1.24 Let $P(n)$ be the proposition that the n lines, no two of which are parallel and no three of which pass through a common point, will separate the plane into $\dfrac{n^2 + n + 2}{2}$ regions.

I. One line does separate the plane into $\dfrac{1^2 + 1 + 2}{2} = 2$ regions.

II. Assume $P(k)$ is true.

III. Consider $k+1$ lines. The $(k+1)^{\text{th}}$ line will pass through **region**-line-**region**-line-**region**... (one more region than the k lines) and will split each of the encountered regions into two parts, adding $k+1$ regions to the $\dfrac{k^2 + k + 2}{2}$ preceding regions generated by the first k lines. (see above figure for $k = 3$). It follows that the number of regions stemming from the $k+1$ lines is given by:

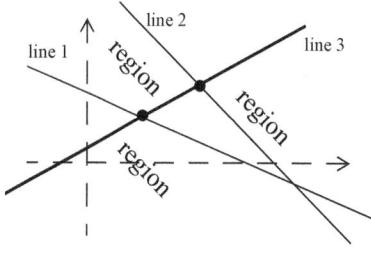

$$\underbrace{\boxed{\dfrac{k^2 + k + 2}{2}}}_{\text{II}} + k + 1 = \dfrac{k^2 + k + 2 + 2k + 2}{2} = \dfrac{(k+1)^2 + (k+1) + 2}{2}$$

CYU 1.25 Let $P(n)$ be a proposition for which $P(1)$ is True and for which the validity at k implies the validity at $k+1$. We are to show, using the Well-Ordering Principle, that $P(n)$ is True for all n. Suppose not (we will arrive at a contradiction):

Let $S = \{n \in Z^+ | P(n) \text{ is False}\}$. Since $P(1)$ is True, $S \neq \emptyset$. The Well-Ordering Principle tells us that S contains a least element, n_0. But since the validity at $n_0 - 1$ implies the validity at n_0, $n_0 - 1$ must be in S — contradicting the minimality of n_0.

1.5 THE DIVISION ALGORITHM AND BEYOND

CYU 1.26 The division algorithm tells us that n must be of the form $3m$, or $3m+1$, or $3m+2$, for some integer m. We show that, in each case, $n^2 = 3q$ or $n^2 = 3q+1$ for some integer q:

If $n = 3m$, then $n^2 = 9m^2 = 3q$ with $q = 3m^2$.

If $n = 3m+1$, then $n^2 = 9m^2 + 6m + 1 = 3(3m^2 + 2m) + 1 = 3q+1$.

If $n = 3m+2$, then $n^2 = 9m^2 + 12m + 4 = 3(3m^2 + 4m + 1) + 1 = 3q+1$.

CYU 1.27 (a) As in Example 1.22: (1) $5605 = 2(1870) + 1865$
(2) $1870 = 1865 + 5$
(3) $1865 = 373(5) + 0 \Rightarrow \gcd(1870, 5605) = 5$

(b) $5 \overset{\text{above line 2}}{=} 1870 - 1865 \overset{\text{above line 1}}{=} 1870 - [5605 - 2(1870)] = 3(1870) + (-1)5605$

(c) We simply show that $c > 0$ divides $n \in Z$ if and only if $c \,|\,|n|$:

$$c = kn \Leftrightarrow |c| = |kn| \Leftrightarrow |c| = |k||n| \underset{\text{since } c > 0}{\Leftrightarrow} c = h|n| \text{ where } h = |k|$$

CYU 1.28 Proof by contradiction: Assume that $\gcd(a, c) = 1$. From Theorem 1.9: if $a|bc$, and if $\gcd(a, c) = 1$, then $a|b$ — contradicting the given condition that $a \nmid b$.

CYU 1.29 Let $P(n)$ be the proposition that if $p|a_1 a_2 \cdots a_n$, then $p|a_i$ for some $1 \leq i \leq n$.

I. $P(1)$ is trivially True.

II. Assume $P(k)$ is True: If $p|a_1 a_2 \cdots a_k$, then $p|a_i$ for some $1 \leq i \leq k$.

III. Suppose $p|a_1 a_2 \cdots a_k a_{k+1}$; or, to write it another way: $p|(a_1 a_2 \cdots a_k) a_{k+1}$.
If $p|a_{k+1}$ then we are done. If not, then by Theorem 1.9: $p|(a_1 a_2 \cdots a_k)$.
Invoking II we conclude that $p|a_i$ for some $1 \leq i \leq k$.

CYU 1.30 CYU 1.20(iii) enables us to restrict our attention to the case where $a > 1$ and $b > 1$.

Let $a = p_1^{r_1} p_2^{r_2} \ldots p_s^{r_s}, b = q_1^{m_1} q_2^{m_2} \ldots q_t^{m_t}$ be the prime decompositions of a and b, with distinct primes $p_1, p_2, \ldots, p_s$, and distinct primes $q_1, q_2, \ldots, q_t$.

Since $a|n$: $n = ak = p_1^{r_1} p_2^{r_2} \ldots p_s^{r_s} \cdot k$. It follows that and each $p_i^{r_i}$ must appear in the prime decomposition of n, for $1 \leq i \leq s$ (with possibly additional p_i's appearing in the prime decomposition of k). Similarly, since $b|n$, each $q_i^{m_i}$ must appear in the prime decomposition of n, for $1 \leq i \leq t$.

Since a and b are relatively prime, none of the p_i's is equal to any of the q_i's. It follows that $p_1^{r_1} p_2^{r_2} \ldots p_s^{r_s} q_1^{m_1} q_2^{m_2} \ldots q_t^{m_t}$ appears in the prime decomposition of n, and that therefore $ab = p_1^{r_1} p_2^{r_2} \ldots p_s^{r_s} q_1^{m_1} q_2^{m_2} \ldots q_t^{m_t}$ divides n.

CHAPTER 2
A TOUCH OF SET THEORY

2.1 BASIC DEFINITIONS

CYU 2.1 (a) (i) $(A \cup B)^c = \{1, 2, 3, 4, 5, 7\}^c = \{6\}$

(ii) $(A \cap B^c) \cup (A \cup B)^c = (\{1, 3, 5\} \cap \{1, 5, 6\}) \cup \{6\} = \{1, 5, 6\}$

(iii) $(A - B)^c \cap C = \{1, 5\}^c \cap \{3, 4, 5\} = \{2, 3, 4, 6, 7\} \cap \{3, 4, 5\} = \{3, 4\}$

(iv) $A - (B \cap C)]^c = [\{1, 3, 5\} - \{3, 4\}]^c = \{1, 5\}^c = \{2, 3, 4, 6, 7\}$

(v) $\{x \in U | x = y + 2, y \in B\} = \{4, 5, 6\}$

(b-i) True. (b-ii) True. (b-iii) True. (b-iv) True: Every element in $\varnothing$ is contained in $\{1, 2\}$, by default, since $\varnothing$ contains no element.

(b-v) False: The empty set is not an element of $\{1, 2\}$ — it is not contained in $\{1, 2\}$.
(b-vi) False: $\varnothing$ is not an element of $\{\{\varnothing\}\}$. (b-vii) True.

CYU 2.2 (a) The proposition $A - (B \cap C) = (A - B) \cap (A - C)$ is False. Here is a counterexample: For $A = \{1, 2, 3\}$, $B = \{3, 4\}$, and $C = \{2, 3\}$:

$$A - (B \cap C) = \{1, 2, 3\} - (\{3, 4\} \cap \{2, 3\})$$
$$= \{1, 2, 3\} - \{3\} = \{1, 2\}$$

While:
$$(A - B) \cap (A - C) = (\{1, 2, 3\} - \{3, 4\}) \cap (\{1, 2, 3\} - \{2, 3\})$$
$$= \{1, 2\} \cap \{1\} = \{1\}$$

(b) The proposition $(A \cap B^c)^c \cup B = A^c \cup B$ is True:

A	B	B^c	$A \cap B^c$	$(A \cap B^c) \cup B$	A^c	$A^c \cup B$
1	1	0	0	0	0	0
1	0	1	1	1	0	1
0	1	1	0	1	1	1
0	0	0	0	0	1	1

↑ same ↑

CYU 2.3 (a)

A	B	A^c	B^c	$A \cap B$	$(A \cap B)^c$	$A^c \cup B^c$
1	1	0	0	1	0	0
1	0	0	1	0	1	1
0	1	1	0	0	1	1
0	0	1	1	0	1	1

(b) $x \in (A \cup B)^c \Leftrightarrow x \notin A \cup B \Leftrightarrow x \notin A$ and $x \notin B \Leftrightarrow x \in A^c$ and $x \in B^c \Leftrightarrow x \in A^c \cap B^c$

CYU 2.4 $x \in \left(\bigcap_{\alpha \in A} S_\alpha \right)^c \Leftrightarrow x \notin \bigcap_{\alpha \in A} S_\alpha \Leftrightarrow x \notin S_{\alpha_0}$ for some $\alpha_0 \in A$

$\Leftrightarrow x \in S_{\alpha_0}^c$ for some $\alpha_0 \in A \Leftrightarrow x \in \bigcup_{\alpha \in A} S_\alpha^c$

CYU 2.5 (a) $(-2, 2) \cap [0, 5] = [0, 2)$ (b) $[-1, 3]^c \cap (5, \infty) = (5, \infty)$

(c) $(-2, 0) \cup [-1, 2] \cup [3, 5] = (-2, 2] \cup [3, 5]$

2.2 FUNCTIONS

CYU 2.6 (a) No: $(2, 3) \in f$ and $(2, 6) \in f$ (b) Yes. Range: $\{2\}$

(c) No: $3 \in X$ but there is no $(3, \text{--}) \in f$ (d) No: $4 \notin X$

CYU 2.7 (a-i) $(g \circ f)(3) = g[f(3)] = g(5) = \dfrac{15}{6} = \dfrac{5}{2}$

(a-ii) $(f \circ g)(3) = f[g(3)] = f\left(\dfrac{9}{4}\right) = \dfrac{9}{4} + 2 = \dfrac{17}{4}$

(a-iii) $(g \circ f)(x) = g[f(x)] = g(x + 2) = \dfrac{3(x + 2)}{(x + 2) + 1} = \dfrac{3x + 6}{x + 3}$

(a-iv) $(f \circ g)(x) = f[g(x)] = f\left(\dfrac{3x}{x + 1}\right) = \dfrac{3x}{x + 1} + 2 = \dfrac{5x + 2}{x + 1}$

(b) $g \circ f = \{(1, a), (c, t), (x, 4)\}$. $f \circ g$ is not defined. [$g(a)$, for example, is not in the domain of f].

CYU 2.8 (a) $f(a) = f(b) \Rightarrow \dfrac{a}{a+1} = \dfrac{b}{b+1} \Rightarrow a(b+1) = b(a+1) \Rightarrow ab + a = ba + b \Rightarrow a = b$

(b) The function $g(x) = x^5 - x = x(x^4 - 1)$ is easily seen not to be one-to-one, since $g(0) = g(1) = 0$. It follows that the function $f(x) = x^5 - x + 777$ is also not one-to-one, since $f(0) = f(1) = 777$.

CYU 2.9 One-to-one: $f\left(\begin{bmatrix} a & b \\ c & d \end{bmatrix}\right) = f\left(\begin{bmatrix} \bar{a} & \bar{b} \\ \bar{c} & \bar{d} \end{bmatrix}\right) \Rightarrow (d, -c, 3a, b) = (\bar{d}, -\bar{c}, 3\bar{a}, \bar{b})$

$$\Rightarrow \begin{array}{c} d = \bar{d} \\ -c = -\bar{c} \\ 3a = 3\bar{a} \\ b = \bar{b} \end{array} \Rightarrow \begin{bmatrix} a & b \\ c & d \end{bmatrix} = \begin{bmatrix} \bar{a} & \bar{b} \\ \bar{c} & \bar{d} \end{bmatrix}$$

Onto: For given (x, y, z, w), we find $\begin{bmatrix} a & b \\ c & d \end{bmatrix}$ such that $f\left(\begin{bmatrix} a & b \\ c & d \end{bmatrix}\right) = (x, y, z, w)$:

$f\left(\begin{bmatrix} a & b \\ c & d \end{bmatrix}\right) = (x, y, z, w) \Rightarrow (d, -c, 3a, b) = (x, y, z, w) \Rightarrow \begin{array}{c} d = x \\ -c = y \\ 3a = z \\ b = w \end{array} \Rightarrow \begin{array}{c} d = x \\ c = -y \\ a = z/3 \\ b = w \end{array}$

Hence: $f\left(\begin{bmatrix} z/3 & w \\ -y & x \end{bmatrix}\right) = (x, y, z, w)$

CYU 2.10 Let $y \in Y$. Since $[y, f^{-1}(y)] \in f^{-1}$, $[f^{-1}(y), y] \in f$, which is to say: $f[f^{-1}(y)] = y$.

CYU 2.11 The function $f: M_{2 \times 2} \to R^4$ given by $f\left(\begin{bmatrix} a & b \\ c & d \end{bmatrix}\right) = (d, -c, 3a, b)$ is a bijection [see CYU 2.9]. To find its inverse we determine $\begin{bmatrix} a & b \\ c & d \end{bmatrix}$ for which $f\left(\begin{bmatrix} a & b \\ c & d \end{bmatrix}\right) = (x, y, z, w)$:

$f\left(\begin{bmatrix} a & b \\ c & d \end{bmatrix}\right) = (x, y, z, w) \Rightarrow (d, -c, 3a, b) = (x, y, z, w) \Rightarrow \begin{array}{c} d = x \\ -c = y \\ 3a = z \\ b = w \end{array} \Rightarrow \begin{array}{c} a = z/3 \\ b = w \\ c = -y \\ d = x \end{array}$

Conclusion: $f^{-1}(x, y, z, w) = \begin{bmatrix} z/3 & w \\ -y & x \end{bmatrix}$

Moreover: $f[f^{-1}(x,y,z,w)] = f\left(\begin{bmatrix} z/3 & w \\ -y & x \end{bmatrix}\right) = \left(x, -(-y), 3\left(\frac{z}{3}\right), w\right) = (x,y,z,w)$

and: $f^{-1}\left[f\left(\begin{bmatrix} a & b \\ c & d \end{bmatrix}\right)\right] = f^{-1}(d, -c, 3a, b) = \begin{bmatrix} 3(a/3) & b \\ -(-c) & d \end{bmatrix} = \begin{bmatrix} a & b \\ c & d \end{bmatrix}$

2.3 INFINITE COUNTING

CYU 2.12 The function $f: Z^+ \to \{0, 1, 2, 3, \ldots\}$ given by $f(n) = n - 1$ is a bijection:

On-to-one: $f(a) = f(b) \Rightarrow a - 1 = b - 1 \Rightarrow a = b$

Onto: For any $a \in \{0, 1, 2, 3, \ldots\}$, $a + 1 \in Z^+$ and $f(a + 1) = a$.

CYU 2.13

$$\begin{array}{cccc} a_{11} & a_{12} & a_{13} & a_{14} & \cdots \\ a_{21} & a_{22} & a_{23} & a_{24} & \cdots \\ a_{31} & a_{32} & a_{33} & a_{34} & \cdots \\ a_{41} & a_{42} & a_{43} & a_{44} & \cdots \\ \cdots & \cdots & \cdots & \cdots & \cdots \end{array}$$

CYU 2.14 We show that the function $f: [a, b] \to [c, d]$ given by $f(x) = \left(\frac{d-c}{b-a}\right)x + i$ (inspired by the adjacent figure) is a bijection.
$\underset{y\text{-intercept}}{\uparrow}$

One-to-one: $f(x_1) = f(x_2)$

$\left(\frac{d-c}{b-a}\right)x_1 + i = \left(\frac{d-c}{b-a}\right)x_2 + i$

$(d-c)x_1 = (d-c)x_2$

$x_1 = x_2$

ONTO: For $y_0 \in [c, d]$, we are to find $x \in [a, b]$ such that $f(x) = y_0$. Let's do it:

$f(x) = y_0 \Leftrightarrow \left(\frac{d-c}{b-a}\right)x + i = y_0$

$\Leftrightarrow (d-c)x + i(b-a) = y_0(b-a)$

$\Leftrightarrow x = \frac{y_0(b-a) - i(b-a)}{d-c} = (y_0 - i)\left(\frac{b-a}{c-d}\right)$

We complete the proof by showing that $a \leq (y_0 - i)\left(\left(\frac{b-a}{c-d}\right) \leq b\right)$:

$$f(a) \le y_0 \le f(b) \Rightarrow \left(\frac{d-c}{b-a}\right)a + i \le y_0 \le \left(\frac{d-c}{b-a}\right)b + i \Rightarrow \left(\frac{d-c}{b-a}\right)a \le y_0 - i \le \left(\frac{d-c}{b-a}\right)b$$

$\uparrow \quad \uparrow$
$c \quad \ d$

$$\Rightarrow a \le (y_0 - i)\left(\frac{b-a}{d-c}\right) \le b$$

CYU 2.15 For any given finite interval F, choose closed intervals $L = [a, b]$, $M = [c, d]$ such that $[a, b] \subset F \subset [c, d]$. CYU 2.15 tells us that $Card(L) = Card(M)$. It follows, from Theorem 2.9, that the finite interval F is of the same cardinality as that of (any) finite closed interval. That being the case, any two finite intervals must be of the same cardinality (Theorem 2.7).

CYU 2.16 (a) Since $(-\infty, 3) \subseteq (-\infty, 3) \cup (5, 6] \cup [9, \infty) \subseteq \Re$ and since
$Card[(-\infty, 3)] = Card(\Re) = c$, $Card[(-\infty, 3) \cup (5, 6] \cup [9, \infty)] = c$

(b) Since $Card(Q) \ne Card(\Re)$, there cannot exist a bijection from $\Re$ to Q.

CYU 2.17 Assume the set of all sets exists. Let's call it S. Consider the set $T = \bigcup_{A \in S} A$, and its power set $P(T)$. Since S contains all sets, $P(T) \in S$. As such, $P(T) \subseteq T$. But then: $Card[P(T)] \le Card(T)$ [Theorem 2.10(a)] — a contradiction [Theorem 2.13].

2.4 EQUIVALENCE RELATIONS

CYU 2.18 (a-i) Not symmetric: $1 \sim 2$ but $2 \not\sim 1$. (a-ii) Yes.
(a-iii) Not transitive: $2 \sim 1$ and $1 \sim 3$ but $2 \not\sim 3$.
(b-i) Yes.   (b-ii) Yes.   (b-iii) Not reflexive nor symmetric.

CYU 2.19 Reflexive: Let $S \in P(X)$. Since $I: S \to S$, given by $I(s) = s \ \forall s \in S$ is a bijection, $S \sim S$.

Symmetric: If $S \sim T$ for $S, T \in P(X)$, then there exists a bijection $f: S \to T$. Theorem 2.4(a), page 70, tells us that $f^{-1}: T \to S$ is a bijection. Hence, $T \sim S$

Transitive: If $S \sim T$ and $T \sim W$ with $S, T, W \in P(X)$, then there exists bijections $f: S \to T$ and $g: T \to W$. Theorem 2.5(c), page 72, tells us that $g \circ f: S \to W$ is a bijection. Hence, $S \sim W$.

CYU 2.20 (a) No: $[1, 2] \cap [2, 3] \ne \varnothing$.

(b) Yes: Every element of $\Re$ is either an integer or is contained in some $(i, i+1)$ for some integer $i \ge 0$ or in some $(-i, -i-1)$ for some $i \ge 1$. Moreover the sets in $\{\{n\} | n \in Z\} \cup \{(i, i+1)\}_{i=0}^{\infty} \cup \{(-i, -i-1)\}_{i=1}^{\infty}$ are mutually disjoint.

CYU 2.21 Let's find the equivalence classes, starting with $[0]$:

$$[0] = \{b \mid 2 \mid (3(0) - 7b)\} = \{b \mid 7b \text{ is even}\} = \{\text{even integers}\}$$

Moving on to $[1]$:

$$1 = \{b \mid 2 \mid (3 - 7b)\} = \{b \mid 7b \text{ is odd}\} = \{\text{odd integers}\}$$

Conclusion: The equivalence relation partitions the integers into two disjoint pieces: the class of even integers and the class of odd integers.

CHAPTER 3
A TOUCH OF ANALYSIS

CYU 3.1 (a) $\text{lub}\{(3, 5) \cup [4, 7]\} = 7$, $\text{glb}\{(3, 5) \cup [4, 7]\} = 3$. Maximum: 7. No minimum.

(b) $\text{lub}[(-\infty, 0) \cup [1, 3] \cup \{9\}] = 9$. Maximum: 9. Greatest lower bound and minimum do not exist.

(c) $\text{lub}[\{x < 0 \mid x^2 < 2\} \cup \{x \geq 0 \mid x^2 \leq 2\}] = \sqrt{2}$

$\text{glb}[\{x < 0 \mid x^2 < 2\} \cup \{x \geq 0 \mid x^2 \leq 2\}] = -\sqrt{2}$. Maximum: $\sqrt{2}$. No minimum.

CYU 3.2 If β is the greatest lower bound, then it is a lower bound. To establish (b-ii), we consider a given $\varepsilon > 0$. Since $\beta + \varepsilon > \beta$, and since nothing greater than β can be a lower bound of S, there must exist some $s \in S$ to the left of $\beta + \varepsilon$.

Conversely, suppose β satisfies (b-i) and (b-ii). To show that β is the greatest lower bound of S we consider an arbitrary lower bound b of S, and show that $\beta \geq b$: Assume, to the contrary, that $\beta < b$. Let $\varepsilon = b - \beta > 0$. By (b-ii)

$$\exists s \in S \ni s < \beta + \varepsilon = \beta + (b - \beta) = b$$

Contradicting the assumption that b is a lower bound of S.

CYU 3.3 As observed in Example 3.1: $\dfrac{n}{n+7} > 1 - \varepsilon \Leftrightarrow n > \dfrac{7 - 7\varepsilon}{\varepsilon}$. In particular:

(a) $\dfrac{n}{n+7} > \dfrac{99}{100} = 1 - \dfrac{1}{100} \Leftrightarrow \dfrac{7 - \frac{7}{100}}{\frac{1}{100}} = 693$. It follows that $\dfrac{694}{694 + 7}$ is the smallest element of S which lies to the right of $\dfrac{99}{100}$.

(b) $\dfrac{n}{n+7} > \dfrac{99.9}{100} \Leftrightarrow n > \dfrac{7 - \frac{7}{1000}}{\frac{1}{1000}} = 6993$. It follows that $\dfrac{6994}{6949 + 7}$ is the smallest element of the set S which lies to the right of $\dfrac{99.9}{100}$.

CYU 3.4 (a) Theorem 3.2, with $a = 1$, assures us that such an integer n exists.

(b) There exists a positive integer n such that $\dfrac{1}{n} < \varepsilon$ if and only if there exists a positive integer n such that $1 < n\varepsilon$. Theorem 3.2, with 1 playing the role of b and ε that of a, assures us that such an integer n exists.

CYU 3.5 Let $x \in (0, 1]$ be given. CYU 3.4(b) assures us of the existence of a positive integer n_x such that $\frac{1}{n_x} < x$ (let x play the role of ε). In particular, x is not contained in $J_{n_x} = \left(0, \frac{1}{n_x}\right]$. It follows that no $x \in (0, 1]$ is contained in every $J_n = \left(0, \frac{1}{n}\right]$; or, to put it another way: $\bigcap_{n=1}^{\infty} J_n = \emptyset$.

CYU 3.6 (a) Let x be irrational, and let $\frac{a}{b}$ be a rational number with $a \neq 0$. Assume that $\frac{a}{b}x$ is rational, say $\frac{a}{b}x = \frac{c}{d}$. Then: $x = \frac{c}{d} \cdot \frac{b}{a} = \frac{cb}{da}$ — contradicting the given condition that x is irrational.

(b) The sum of two irrational numbers can be rational: $\sqrt{2} - \sqrt{2} = \sqrt{2} + [(-1)\sqrt{2}] = 0$.

CYU 3.7 (a) If S is a finite subset of $\Re$, then $S \subseteq [-N, N]$ for some $N \in Z^+$. Since no element of S is contained in $(N, N+1)$, S is not dense in $\Re$.

(b) Suppose there exists an interval (a, b) containing only finitely many elements of S. Let c be the smallest of those finitely many elements of S. It follows that the interval (a, c) contains no element of S — contradicting the given condition that S is dense in $\Re$.

CYU 3.8 (a-i) Let $\varepsilon > 0$ be given. We want N such that: $n > N \Rightarrow \left|7 - \frac{101}{n} - 7\right| < \varepsilon$

$$\Leftrightarrow \frac{101}{n} < \varepsilon$$

$$\Leftrightarrow n > \frac{101}{\varepsilon}$$

From the above, we see that if N is any integer greater than $\frac{101}{\varepsilon}$, then $n > N \Rightarrow \left|7 - \frac{101}{n} - 7\right| < \varepsilon$.

(a-ii) If $\varepsilon = \frac{1}{100}$, then $\frac{101}{\varepsilon} = \frac{101}{\frac{1}{100}} = 10,100$. It follows, from (i) that $N = 10,101$ is the smallest integer for which $n > N \Rightarrow \left|7 - \frac{101}{n} - 7\right| < \varepsilon$.

(b) For any given $r \in \Re$ we observe that: $\frac{n-5}{333} > r + 1 \Leftrightarrow n - 5 > 333r + 333$
$$\Leftrightarrow n > 333r + 338$$

It follows that the sequence $\frac{n-5}{333}$ will be larger than $r + 1$ for any $n > 333r + 338$, and that, consequently, the sequence does not converge to r. Since r was arbitrary, the sequence does not converge, period.

(Why not start the above argument with $\frac{n-5}{333} > r$ rather than $\frac{n-5}{333} > r + 1$?

CYU 3.9 Let $a_n \to \alpha$. Taking $\varepsilon = 1$ we choose N such that $n > N \Rightarrow |a_n - \alpha| < 1$. Noting that only the elements $a_1, a_2, \ldots, a_N$ can be more than 1 unit from α, we let $K = \max\{|a_i - \alpha|\}_{i=1}^{N}$. Taking M to be the larger of the two numbers, 1 and K, we see that $|a_n - \alpha| \leq M$ for all n. That is, the sequence is bounded below by $\alpha - M$ and above by $\alpha + M$.

CYU 3.10 (a) Suppose, to the contrary, that $\alpha > \beta$. For $\varepsilon = \dfrac{\alpha - \beta}{2}$, let N be such that $n > N \Rightarrow |a_n - \alpha| < \varepsilon$ and $n > N \Rightarrow |b_n - \beta| < \varepsilon$. It follows that $a_{N+1} > b_{N+1}$:

$$\underset{\beta\uparrow\ \ \ \ \alpha\uparrow}{(\ \ \bullet\overset{|\leftarrow\varepsilon\rightarrow|\leftarrow\varepsilon\rightarrow|}{}\bullet\ \ )}\ :\ \text{a contradiction.}$$

(b) One possible answer: For $(a_n) = (0, 0, 0, \ldots)$ and $(b_n) = \left(1, \dfrac{1}{2}, \dfrac{1}{3}, \ldots\right)$: $0 \leq a_n < b_n$ and $\lim a_n = \lim b_n = 0$.

CYU 3.11 The sequence $g: Z^+ \to \Re$ is a subsequence of the sequence $f: Z^+ \to \Re$ if $g = f \circ h$, where $h: Z^+ \to Z^+$ is a strictly increasing function.

CYU 3.12 Let $(a_n)_{n=1}^{\infty}$ be Cauchy. Choose N such that $n, m > N \Rightarrow |a_n - a_m| < 1$. In particular: $|a_n - a_{N+1}| < 1$ for every $n > N$. It follows that $|a_n| < 1 + |a_{N+1}|$ for every $n > N$. Letting $K = \max\{|a_i|\}_{i=1}^{N}$ and $M = \max\{K, 1 + |a_{N+1}|\}$, we find that $-M \leq a_n \leq M$ for every a_n.

CYU 3.13 The sequence $\left(\dfrac{1}{n}\right)_{n=1}^{\infty}$ is Cauchy but does not converge in $X = (0, \infty)$, since the number 0 is not in X.

CYU 3.14 $|x| = |(x-y) + y| \overset{\text{triangle inequality}}{\leq} |x-y| + |y| \Rightarrow |x| \leq |x-y| + |y| \Rightarrow |x-y| \geq |x| - |y|$.

CYU 3.15 (a) For $x \in (a, b)$ let $\varepsilon = \min(x - a, b - x)$. Then $S_\varepsilon(x) \subseteq (a, b)$.

(b) For $x \in (-\infty, a)$ let $\varepsilon = a - x$. Then $S_\varepsilon(x) \subseteq (-\infty, a)$.

For $x \in (a, \infty)$ let $\varepsilon = x - a$. Then $S_\varepsilon(x) \subseteq (a, \infty)$.

(c) Let $S = \{x_1, x_2, \ldots, x_n\}$ and let $x_i \in S$. For any given $\varepsilon > 0$, since $S_\varepsilon(x)$ is infinite and S is finite $S_\varepsilon(x) - S \neq \emptyset$. It follows that $S_\varepsilon(x)$ is not contained in S.

CYU 3.16 For $O_i = \left(-\frac{1}{i}, \frac{1}{i}\right)$, $\bigcap_{i=1}^{\infty} O_i = \{0\}$ (not open).

CYU 3.17 (a) $[a, b]^c = (-\infty, a) \cup (b, \infty)$. Since both $(-\infty, a)$ and (b, ∞) are open [CYU 3.15(b)] and since unions of open sets are open [Theorem 3.13(ii)], $[a, b]^c$ is open. Consequently, $[a, b]$ is closed.

(b) Since $(1, 3]^c = (-\infty, 1] \cup (3, \infty)$ is not open [no $S_\varepsilon(1)$ is contained in $(-\infty, 1] \cup (3, \infty)$], $(1, 3]$ is not closed.

CYU 3.18 If $H_i = \left[\frac{1}{i}, 1\right]$, then $\bigcup_{i=1}^{\infty} H_i = (0, 1]$ — not closed since:

$(0, 1]^c = (-\infty, 0] \cup (1, \infty)$ is not open [no $S_\varepsilon(0)$ is contained in $(-\infty, 0] \cup (1, \infty)$].

CYU 3.19 (a) $B = \left\{\frac{1}{n}\right\}_{n=1}^{\infty}$ is not compact: The open cover $\left\{S_{\frac{1}{2n}}\left(\frac{1}{n}\right)\right\}_{n=1}^{\infty}$ of B has no finite subcover (note that $S_{\frac{1}{2n}}\left(\frac{1}{n}\right) \cap B = \left\{\frac{1}{n}\right\}$).

(b) $B = \{0\} \cup \left\{\frac{1}{n}\right\}_{n=1}^{\infty}$ is compact: For $U = \{O_\alpha\}_{\alpha \in A}$ an open cover of B, let O_{α_0} be an element of U which contains 0, and let $\varepsilon > 0$ be such that $0 \in S_\varepsilon(0) \subseteq O_{\alpha_0}$. Choose N such that $\frac{1}{N} < \varepsilon$ [CYU 3.4(b), page 115]. For each $n \leq N$ let O_{α_n} be an element of U containing $\frac{1}{n}$. It follows that $\{O_{\alpha_i}\}_{i=0}^{N}$ covers B.

CYU 3.20 A consequence of CYU 3.12 (page 131), Theorem 3.10 (page 129), and Theorem 3.19.

CYU 3.21 For a given $\varepsilon > 0$ we are to find $\delta > 0$ such that:

$$|x - 2| < \delta \Rightarrow |(5x + 1) - 11| < \varepsilon$$

i.e: $|x - 2| < \delta \Rightarrow |5x - 10| < \varepsilon$

$|x - 2| < \delta \Rightarrow 5|x - 2| < \varepsilon$

$|x - 2| < \delta \Rightarrow |x - 2| < \frac{\varepsilon}{5}$ Let $\delta = \frac{\varepsilon}{5}$

CYU 3.22 For given $\varepsilon > 0$ we are to find $\delta > 0$ such that:

$$|x - 2| < \delta \Rightarrow |(x^2 + 1) - 5| < \varepsilon$$
$$|x - 2| < \delta \Rightarrow |(x + 2)(x - 2)| < \varepsilon$$
$$|x - 2| < \delta \Rightarrow |x + 2||x - 2| < \varepsilon \quad (*)$$

Since we are interested in what happens near $x = 2$, we decide to focus on the interval: $(1, 3) = \{x | |x - 2| < 1\}$. **Within that interval** $|x + 2| < 5$. Consequently, within that interval: $|x + 2||x - 2| < 5|x - 2|$.

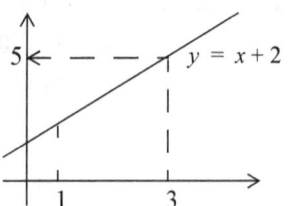

We observe that (*) is satisfied for $\delta = \min\left(1, \frac{\varepsilon}{5}\right)$:

$$|x - 2| < \delta \Rightarrow |x + 2||x - 2| < 5\delta < 5\left(\frac{\varepsilon}{5}\right) = \varepsilon$$

CYU 3.23 (a) Let $\varepsilon > 0$ be given. We are to find $\delta > 0$ such that:

$$|x - 0| < \delta \Rightarrow |\sqrt{x} - \sqrt{0}| < \varepsilon$$

$x \geq 0$ and $\sqrt{x} \geq 0$: $\quad x < \delta \Rightarrow \sqrt{x} < \varepsilon$

$$x < \delta \Rightarrow x < \varepsilon^2 \quad \text{Let } \delta = \varepsilon^2$$

(b) Let x_0 be a fixed but arbitrary number. We are to find $\delta > 0$ such that:

$$|x - x_0| < \delta \Rightarrow |x^2 - x_0^2| < \varepsilon$$

i.e: $|x - x_0| < \delta \Rightarrow |x - x_0||x + x_0| < \varepsilon$

We observe (see adjacent figure) that in the chosen interval $(x_0 - 1, x_0 + 1)$:

$$|x + x_0| < M = \max[|2x_0 + 1|, |2x_0 - 1|]$$

It follows that for $\delta = \min\left(1, \frac{\varepsilon}{M}\right)$:

$$|x - x_0| < \delta \Rightarrow |x - x_0||x + x_0| < \frac{\varepsilon}{M} \cdot M = \varepsilon$$

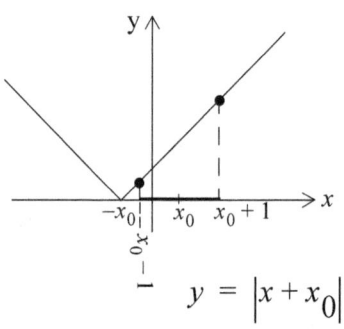

CYU 3.24 We show that $f(x) = \begin{cases} x & \text{if } x \neq 1 \\ \dfrac{999}{1000} & \text{if } x = 1 \end{cases}$ is not continuous at $x = 1$:

Let $\varepsilon = 1 - \dfrac{999}{1000} = \dfrac{1}{1000}$ (the "jump" at 1), and let δ be **ANY** positive number whatsoever.

Since, for every $x \in (\delta, 1)$, $|f(x) - f(1)| = x - \dfrac{999}{1000} > \dfrac{1}{1000} = \varepsilon$, f is not continuous at 1.

CYU 3.25 For given $\varepsilon > 0$ we are to find $\delta > 0$ such that:

$$|x - c| < \delta \Rightarrow |(f - g)(x) - (f - g)(c)| < \varepsilon$$
$$|x - c| < \delta \Rightarrow |f(x) - g(x) - f(c) + g(c)| < \varepsilon$$
$$|x - c| < \delta \Rightarrow |[f(x) - f(c)] + [g(c) - g(x)]| < \varepsilon \quad (*)$$

Let $\delta_1 > 0$ be such that $|x - c| < \delta_1 \Rightarrow |f(x) - f(c)| < \frac{\varepsilon}{2}$, and let $\delta_2 > 0$ such that $|x - c| < \delta_2 \Rightarrow |g(c) - g(x)| < \frac{\varepsilon}{2}$. Letting $\delta = \min(\delta_1, \delta_2)$, we see that if $|x - c| < \delta$, then (*) holds: $|[f(x) - f(c)] + [g(c) - g(x)]| \underset{\text{triangle inequality}}{\leq} |f(x) - f(c)| + |g(x) - g(c)| < \frac{\varepsilon}{2} + \frac{\varepsilon}{2} = \varepsilon$

CYU 3.26 Let $f: \Re \to \Re$ and $g: \Re \to \Re$ be continuous. We show that $g \circ f: \Re \to \Re$ is also continuous at an arbitrarily chosen $x_0 \in \Re$:

For given $\varepsilon > 0$ we are to find $\delta > 0$ such that:

$$|x - x_0| < \delta \Rightarrow |(g \circ f)x - (g \circ f)x_0| < \varepsilon$$

i.e: $|x - x_0| < \delta \Rightarrow |g[f(x)] - g[f(x_0)]| < \varepsilon$

Since g is continuous at $f(x_0)$, we can find $\bar{\delta} > 0$ such that:

$$|y - f(x_0)| < \bar{\delta} \Rightarrow |g(y) - g[f(x_0)]| < \varepsilon \quad (*)$$

Since f is continuous at x_0, we can find $\delta > 0$ such that:

$$|x - x_0| < \delta \Rightarrow |f(x) - f(x_0)| < \bar{\delta} \quad (**)$$

Consequently: $|x - x_0| < \delta \overset{(**)}{\Rightarrow} |f(x) - f(x_0)| < \bar{\delta} \overset{(*)}{\Rightarrow} |g[f(x)] - g[f(x_0)]| < \varepsilon$

CHAPTER 4
A TOUCH OF TOPOLOGY

CYU 4.1 $|x| = |(x - y) + y| \underset{\text{triangle inequality}}{\leq} |x - y| + |y| \Rightarrow |x| \leq |x - y| + |y| \Rightarrow |x - y| \geq |x| - |y|$

CYU 4.2 For: $d(x, y) = \begin{cases} 1 & \text{if } x \neq y \\ 0 & \text{if } x = y \end{cases}$:

(i) $d(x, y) = 0$ if and only if $x = y$ (see above definition of d).

(ii) $d(x, y) = d(y, x)$ (see above definition of d).

(iii) Let $x, y, z \in X$. If $x = y = z$, then: $d(x, z) = d(x, y) + d(y, z) = 0$.

If no two of the elements are equal, then: $d(x, z) = 1 < 1 + 1 = d(x, y) + d(y, z)$

If two of the elements are equal but differ from the third element, say $x = y$ and $y \neq z$, then: $d(x, z) = 1 = 0 + 1 = d(x, y) + d(y, z)$

CYU 4.3 Since $S_1(5) = \{n \in Z^+ | d((n, 5) < 1)\}$, and since 5 is the only element of Z^+ which is within one unit of 5: $S_1(5) = \{5\}$.

Since every element of Z^+ falls within five units of 1: $S_5(1) = Z^+$.

CYU 4.4 (a) Since for all $\varepsilon > 0: S_\varepsilon(5) \nsubseteq (1, 5]$, $(1, 5]$ is not open.

Since $(1, 5]^c = (-\infty, 1] \cup (5, \infty)$, and since for all $\varepsilon > 0: S_\varepsilon(1) \nsubseteq (1, 5]^c$, $(1, 5]^c$ is not open. It follows that $(1, 5]$ is not closed.

(b). Since every subset of X is open [Example 4.2(c)], and since a set of n element has 2^n subsets (Theorem 2.12, page 84), X has 2^n open subsets.

CYU 4.5 (i) Since both $X^c = \emptyset$ and $\emptyset^c = X$ are open, both X and $\emptyset$ are closed.

(ii) Let $\{H_\alpha\}_{\alpha \in A}$ be a collection of closed sets. Since: $\left(\bigcap_{\alpha \in A} H_\alpha\right)^c = \bigcup_{\alpha \in A} (H_\alpha)^c$

[Theorem 2.3(b), page 58], and since each $(H_\alpha)^c$ is open, $\bigcap_{\alpha \in A} H_\alpha$ is closed.

(iii) Let $\{H_i\}_{i=1}^n$ be a collection of closed sets. Since $\left(\bigcup_{i=1}^n H_i\right)^c = \bigcap_{i=1}^n (H_i)^c$ [Theorem 2.2(a), page 57] and since $\bigcap_{i=1}^n (H_i)^c$ is open, $\bigcup_{i=1}^n H_i$ is closed.

CYU 4.6 (a) True. Let A and B be bounded subsets of (X, d), bounded by M_A and M_B, respectively. If either A or B is empty, then $A \cup B$ is bounded by the bound of the other. If neither A nor B is empty, then choose elements $a \in A$ and $b \in B$. We show that $M_A + d(a, b) + M_B$ is a bound for $A \cup B$:

Let $x, y \in A \cup B$. If both x and y are in either A or B, then clearly $d(x, y) < M_A + d(a, b) + M_B$. For $x \in A$ and $y \in B$ we have:

$$d(x, y) \underset{\uparrow}{\leq} d(x, a) + d(a, b) + d(b, y) \leq M_A + d(a, b) + M_B$$
triangle inequality

(b) False. Each of the sets $\{[i, i+1]\}_{i=0}^\infty$ in the Euclidean space $\Re$ is bounded by 1, but the set $\bigcup_{i=0}^\infty [i, i+1] = [0, \infty)$ is not bounded.

(c) True. Let $\{S_\alpha\}_{\alpha \in A}$ be a collection of bounded sets in a space X. Since $\bigcap_{\alpha \in A} S_\alpha$ is contained in each S_α, the bound M_α of any chosen S_α will be a bound for $\bigcap_{\alpha \in A} S_\alpha$.

CYU 4.7 (One possible answer): Consider the discrete metric d on $Z^+ = \{1, 2, 3, 4, \ldots\}$; namely: $d(n, m) = \begin{cases} 1 & \text{if } n \neq m \\ 0 & \text{if } n = m \end{cases}$ (see CYU 4.2). (Z^+, d) is a subset of itself, that is closed and bounded by 1. Since the open cover $\{S_1(n)\}_{n=1}^{\infty}$ contains no finite subcover, it is not compact.

CYU 4.8 (a-i) $y \in f(A \cup B) \Leftrightarrow y = f(x)$ for $x \in A \cup B$
$\Leftrightarrow y = f(x)$ for $x \in A$ or $x \in B$
$\Leftrightarrow y \in f(A)$ or $y \in f(B) \Leftrightarrow y \in f(A) \cup f(B)$

(a-ii) $x \in f^{-1}(A \cup B) \Leftrightarrow f(x) \in A \cup B \Leftrightarrow f(x) \in A$ or $f(x) \in B$
$\Leftrightarrow x \in f^{-1}(A)$ or $x \in f^{-1}(B) \Leftrightarrow x \in f^{-1}(A) \cup f^{-1}(B)$

(a-iii) $x \in f^{-1}(A \cap B) \Leftrightarrow f(x) \in A \cap B \Leftrightarrow f(x) \in A$ and $f(x) \in B$
$\Leftrightarrow x \in f^{-1}(A)$ and $x \in f^{-1}(B) \Leftrightarrow x \in f^{-1}(A) \cap f^{-1}(B)$

(a-iv) $x \in f^{-1}(A^c) \Leftrightarrow f(x) \in A^c \Leftrightarrow f(x) \notin A \Leftrightarrow x \notin f^{-1}(A) \Leftrightarrow x \in [f^{-1}(A)]^c$

(b) $f(A) \subseteq [f(A^c)]^c$:

$y \in f(A) \Rightarrow y = f(x)$ for $x \in A \underset{\text{since } f \text{ is one-to-one}}{\Rightarrow} y \neq f(x)$ for $x \in A^c \Rightarrow y \notin f(A^c) \Rightarrow y \in [f(A^c)]^c$

$[f(A^c)]^c \subseteq f(A)$: $y \in [f(A^c)]^c \Rightarrow y \notin f(A^c) \underset{\text{since } f \text{ is onto}}{\Rightarrow} y = f(x)$ for $x \in A \Rightarrow y \in f(A)$

CYU 4.9 To show that $I: (\mathfrak{R}^2, \bar{d}) \to (\mathfrak{R}^2, d)$ is continuous, we take an arbitrary open set O in $(\mathfrak{R}^2, d)$ and go on to show that $I^{-1}(O) = O$ is open in $(\mathfrak{R}^2, \bar{d})$:

For given $(x_0, y_0) \in O$ let $\varepsilon > 0$ be such that $\{(x, y) | d[(x, y), (x_0, y_0)] < \varepsilon\}$ is contained in the open subset O of $(\mathfrak{R}^2, d)$. Since $\bar{d}[(x_0, y_0), (x, y)] = |x - x_0| + |y - y_0| < \varepsilon + \varepsilon = 2\varepsilon$ the open sphere of radius 2ε centered at (x_0, y_0) in the space $(\mathfrak{R}^2, \bar{d})$ is contained in O.

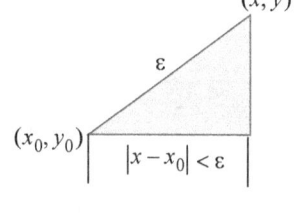

CYU 4.10 No: Let X be the set of real numbers with the discrete metric and let X and Y be the Euclidean space $\mathfrak{R}$. Let $f: X \to Y$ be the identity map $f(x) = x$ for all $x \in X$ and let $g: Y \to Z$ be given by $g(x) = \begin{cases} 1 & \text{if } x = 0 \\ 0 & \text{if } x \neq 0 \end{cases}$. The composite function $g \circ f: X \to Z$ is continuous, since every subset of the discrete space X [in particular, $(g \circ f)^{-1}O$ is open for every O open in Z]. The function g is not continuous since $g^{-1}\left(\frac{1}{2}, \frac{3}{2}\right) = \{0\}$ is not open.

CYU 4.11 Let S be a collection of metric spaces. We show that the relation $X \sim Y$, if X is isometric to Y, is an equivalence relation on S.

(i) For every $X \in S$, $X \sim X$: The identity function $I: X \to X$ is a bijection which satisfies the condition that $d(x_1, x_2) = d[I(x_1), I(x_2)]$.

(ii) For (X, d) and $(Y, \bar{d})$ in S, if $X \sim Y$ then $Y \sim X$: Let $f: X \to Y$ be a bijection such that $d(x_1, x_2) = \bar{d}[f(x_1), f(x_2)]$ (*). The function $f^{-1}: Y \to X$ is a bijection (Theorem 2.4, page 70). Moreover: $\bar{d}(y_1, y_2) = \bar{d}(f[f^{-1}(y_1)], f[f^{-1}(y_2)]) = d[f^{-1}(y_1), f^{-1}(y_2)]$ [read (*) from right to left].

(iii) For (X, d), $(Y, \bar{d})$, $(Z, \hat{d})$ in S, if $X \sim Y$ and $Y \sim Z$ then $X \sim Z$: Let $f: X \to Y$ and $g: Y \to Z$ be bijections such that $d(x_1, x_2) = \bar{d}[f(x_1), f(x_2)]$ (*) and $\bar{d}(y_1, y_2) = \hat{d}[g(y_1), g(y_2)]$ (**). The function $g \circ f: X \to Z$ is a bijection (Theorem 2.5, page 72). Moreover:

$$d(x_1, x_2) \underset{(*)}{=} \bar{d}[f(x_1), f(x_2)] \underset{(**)}{=} \hat{d}(g[f(x_1)], g[f(x_1)]) = \hat{d}([g \circ f](x_1), [g \circ f](x_1))$$

CYU 4.12 (a) For $\tau_0 = \{X, \emptyset\}$. (i) X and $\emptyset$ are certainly in τ_0.

(ii) $X \cup X = X \in \tau_0$, $X \cup \emptyset = X \in \tau_0$, $\emptyset \cup \emptyset = \emptyset \in \tau_0$.

(iii) $X \cap X = X \in \tau_0$, $X \cap \emptyset = \emptyset \in \tau_0$, $\emptyset \cap \emptyset = \emptyset \in \tau_0$.

(b) For $\tau_1 = \{S | S \subseteq X\}$:(i) Since τ_1 contains all subsets of X, it certainly contains X and $\emptyset$. (ii) and (iii) are a consequence that unions and intersections of subsets of X are again subsets of X.

CYU 4.13 We are given that $\emptyset$ and X are contained in τ. Since τ contains but three elements, one can easily check directly that it is closed under unions and intersection.

CYU 4.14 Since any metric space containing at least two points must contain at least four distinct open sets (Exercise 13, page 168), the Sierpinski space is not metrizable.

CYU 4.15 (a) Since every subset of a discrete space is open, the complement of every subset must be open. Consequently, every subset of a discrete space is closed.

(b) Since the only open subsets of an indiscrete space X are X and $\emptyset$, the only closed subsets of X are $X^c = \emptyset$ and $\emptyset^c = X$.

(c) The Surpinski space is $X = \{a, b\}$ with topology $\tau = \{\emptyset, X, \{a\}\}$. Taking complements of the elements of τ we obtain the set of closed subsets of X: $\{\emptyset^c, X^c, \{a\}^c\} = \{X, \emptyset, \{b\}\}$.

CYU 4.16 Let $\{U_i\}_{i=1}^{n}$ be a family of open sets in $(S, \mathcal{T}_S)$. For each i choose $O_i \in \mathcal{T}$ such that $O_i \cap S = U_i$. Since $\mathcal{T}$ is a topology, $\bigcap_{i=1}^{n} O_i \in \mathcal{T}$. The desired result now follows

from the fact that: $\bigcap_{i=1}^{n} U_i = \bigcap_{i=1}^{n} (O_i \cap S) \underset{\text{Exercise 82(b), page 61}}{=} \left(\bigcap_{i=1}^{n} O_i\right) \cap S$.

CYU 4.17 Let O be open in X, and $x \in O$. We already know that $\{S_r(x) | x \in X, r > 0\}$ is a basis for X [Example 4.6(a)]. Since the rationals are dense in $\Re$ (Theorem 3.6, page 118), there exists $\bar{r} \in Q^+$ such that $0 < \bar{r} < r$. It follows that $x \in S_{\bar{r}}(x) \subseteq O$.

CYU 4.18 Let $O \in \mathcal{T}$ be open in the Euclidean topology. For given $x \in O$ choose $r > 0$ such that $x \in S_r(x) \subseteq O$; which is to say: $(x - r, x + r) \subseteq O$. Since $x \in \left[x - \frac{r}{2}, x + \frac{r}{2}\right) \subseteq O$, $O \in \mathcal{T}_S$. At this point we know that $\mathcal{T} \subseteq \mathcal{T}_S$. Since $[0, 1) \in \mathcal{T}_S$ and $[0, 1) \notin \mathcal{T}$, $\mathcal{T} \subset \mathcal{T}_S$.

CYU 4.19 If $\{O_\alpha\}_{\alpha \in A}$ is an open cover of a closed subset H of a compact space X, then $\{O_\alpha\}_{\alpha \in A} \cup H^c$ is an open cover of X. Since X is compact, that cover has a finite subcover $\{O_{\alpha_i}\}_{i=1}^{n} \cup H^c$. It follows that $H \subseteq \{O_{\alpha_i}\}_{i=1}^{n}$.

CYU 4.20 Let d be a metric on X $(X, \mathcal{T})$. For any two distinct point x and y in X, let $r = \frac{d(x, y)}{2}$. Then $S_r(x) \cap S_r(y) = \varnothing$.

CYU 4.21 For each $x \in K$ choose disjoint open sets O_x, $O_{x_0, x}$ containing x and x_0, respectively. Since K is compact the open cover $\{O_x\}_{x \in K}$ of K has a finite subcover $\{O_{x_i}\}_{i=1}^{n}$. It follows that the open neighborhood $\bigcap_{i=1}^{n} O_{x_0, x_i}$ of x_0 is disjoint from the open set $O_K = \bigcup_{i=1}^{n} O_{x_i}$ (note that $K \subseteq O_K$).

CYU 4.22 For $X = \{a, b\}$ with topology $\{\emptyset, X, \{a\}\}$, the function $f: X \to \Re$ given by $f(a) = f(b) = 0$ is continuous [Example 4.8(b)]. The function $g: X \to \Re$ given by $g(a) = 0, g(b) = 2$ is not continuous, since $g^{-1}(1, 3) = \{b\}$ is not open in X.

CYU 4.23 Let f be any function from a topological space X to an indiscrete space Y. Since $f^{-1}(\emptyset) = \emptyset$ and $f^{-1}(Y) = X$ are open in X, and since $\emptyset$ and Y are the only open sets in Y, f is continuous.

CYU 4.24 (a) Let $X = Y = \{a, b\}$ with indiscrete topology, and let $Z = \{a, b\}$ with discrete topology (see CYU 4.12, page 171). Let $f: X \to Y$ and $g: Y \to Z$ be identity maps. $f: X \to Y$ is continuous but $g \circ f: X \to Z$ is not, since $\{a\}$ is open in Z but $(g \circ f)^{-1}(\{a\}) = \{a\}$ is not open in X.

(b) Let $X = \{a, b\}$ with indiscrete topology, and let $Y = Z = \{a, b\}$ with discrete topology. Let $f: X \to Y$ and $g: Y \to Z$ be identity maps. $g: Y \to Z$ is continuous but $g \circ f: X \to Z$ is not, since $\{a\}$ is open in Z but $(g \circ f)^{-1}(\{a\}) = \{a\}$ is not open in X.

(c) Let $X = Y = Z = \{a, b\}$ with topology $\{\emptyset, X, \{a\}\}$ (three copies of the Sierpinski space). Let both $f: X \to Y$ and $g: Y \to Z$ be the function which maps a to b and b to a. Neither of these function is continuous since the inverse of the open set $\{a\}$ is the set $\{b\}$ which is not open in the Sierpinski space. On the other hand $g \circ f: X \to Z$ is the identity map, which is continuous.

CYU 4.25 False. Let $f: \Re \to \Re$ be the constant function $f(x) = 0 \; \forall x \in \Re$. $f(\Re) = \{0\}$ is compact, but $\Re$ is not.

CYU 4.26 Let $f: \Re \to Z^+$ be given by $f([n, n+1)) = n \; \forall x \in [n, n+1)$. Since the subspace Z^+ of $\Re$ is discrete, f is open and closed. It is, however, not continuous since $f^{-1}(\{0\}) = [0, 1)$ is not open in $\Re$.

CYU 4.27 Let $f: X \to Y$ be a homeomorphism.

f is continuous: If U is open in Y, then $U \in \tau_Y$. As such, there exists $O \in \tau_X$ with $f(O) = U$ [recall that $f(\tau_X) = \{f(O) | O \in \tau_X\} = \tau_Y$]. We then have: $f^{-1}(U) = f^{-1}[f(O)] = O$ (recall that f is a bijection).

f^{-1} is continuous: If O is open in X, then $O \in \tau_X$. As such $f(O) = U \in \tau_Y$. We then have: $f^{-1}(U) = f^{-1}[f(O)] = O$.

Conversely, assume that $f: X \to Y$ is a continuous bijection such that $f^{-1}: Y \to X$ is also continuous. We show $f(\tau_X) = \{f(O) | O \in \tau_X\} = \tau_Y$:

Let $U \in \tau_Y$. Is there an $O \in \tau_X$ such that $f(O) = U$? Yes, since $U = f[f^{-1}(U)]$ and $O = f^{-1}(U) \in \tau_X$ by virtue of the continuity of f. So $f(\tau_X) \supseteq \tau_Y$.

Let $O \in \tau_X$. Is $f(O) \in \tau_Y$? Yes, since $[f^{-1}]^{-1}(O) = f(O)$, and f^{-1} is given to be continuous. So, $f(\tau_X) \subseteq \tau_Y$.

CYU 4.28 (a) Let $f: X \to Y$ be a continuous open bijection. We show that $f^{-1}: Y \to X$ is also continuous:

For any O, open in X. Since the inverse of the inverse function f^{-1} is the function f, we have $[f^{-1}]^{-1}(O) = f(O)$ which is open in Y (since f is an open map).

(b) Let $X = \{a, b\}$ with discrete topology, and $Y = \{a, b\}$ with indiscrete topology. The identity function $I: X \to Y$ is a continuous bijection which is not a homomorphism since $\{a\}$ is open in X but $I^{-1}(\{a\}) = \{a\}$ is not open in Y.

CYU 4.29 Let $X = \{a, b\}$ with indiscrete topology, and let $Y = \{a, b\}$ with discrete topology. X is compact and Y is Hausdorff. The identity function $I: X \to Y$ is an open and closed bijection which is not a homomorphism.

CYU 4.30 (a) Let $X \cong Y$ with X compact. We show that Y is also compact:

Let $f: X \to Y$ be a homeomorphism. If $\{O_\alpha\}_{\alpha \in A}$ is an open cover of Y then, since f is continuous, $\{f^{-1}(O_\alpha)\}_{\alpha \in A}$ is an open cover of X. Since X is compact, that cover contains a finite subcover $\{f^{-1}(O_{\alpha_i})\}_{i=1}^{n}$. It follows that $\{f[f^{-1}(O_{\alpha_i})]\}_{i=1}^{n} = \{O_{\alpha_i}\}_{i=1}^{n}$ covers Y.

(b) Since $[0, 1]$ is compact and $(0, 1)$ is not, the two spaces are not homeomorphic.

CYU 4.31 For topological spaces X, Y, and Z:

(i) Since the identity map $I: X \to X$ is a homeomorphism: $X \cong X$.

(ii) If $X \cong Y$, then there exist a homomorphism $f: X \to Y$. CYU 4.25 assures us that $f^{-1}: Y \to X$ is also a homeomorphism, and that consequently: $Y \cong X$.

(iii) If $X \cong Y$ and $Y \cong Z$, then there exist homeomorphisms $f: X \to Y$ and $g: Y \to Z$. Since both $g \circ f: X \to Z$ and $(g \circ f)^{-1} = f^{-1} \circ g^{-1}: Z \to X$ are continuous bijections: $X \cong Z$.

CYU 4.32 (a) We show that every open set in the basis $\beta = \{U \times V | U \in \tau_X, V \in \tau_Y\}$ is the union of elements of $\gamma = \{D \times E | A \in \beta_X \text{ and } B \in \beta_Y\}$:

For $(x, y) \in U \times V$, with $U \in \tau_X, V \in \tau_Y$, choose $D \in \beta_X$ and $E \in \beta_Y$ with $x \in D \subseteq U$ and $y \in E \subseteq V$. Then $(x, y) \in D \times E \subseteq U \times V$.

(b) The open spheres $\gamma = \{S_r(x, y) | r \in \Re^+\}_{(x, y) \in \Re^2}$ is a basis for the topology on $\Re^2$, and the open rectangles $\beta = \{U \times V | U \text{ and } V \text{ are open in } \Re\}$ is a basis for $\Re \times \Re$. We establish the fact that the product topology on $\Re \times \Re$ coincides with the Euclidean topology on $\Re^2$, by showing that: (i) every $S_r(x, y)$ is a union of the elements in β and that: (ii) every $U \times V$ is a union of elements in γ (in other words, that you can build open spheres using open rectangles, and vice versa):

(i) For $(a, b) \in S_r(x, y)$, choose $\delta > 0$ such that $S_\delta(a, b) \subseteq S_r(x, y)$.
Then: $(a, b) \in U \times V \subseteq S_r(x, y)$ for $U = \left(a - \dfrac{\delta}{\sqrt{2}}, a + \dfrac{\delta}{\sqrt{2}}\right)$ and $V = \left(b - \dfrac{\delta}{\sqrt{2}}, b + \dfrac{\delta}{\sqrt{2}}\right)$.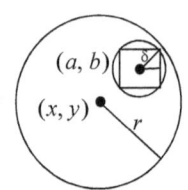

(ii) For $(a, b) \in U \times V$ let r denote the shortest distance between (a, b) and the boundary of the rectangle $U \times V$.
Then $(a, b) \in S_r(a, b) \subseteq U \times V$.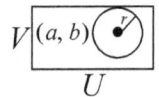

CYU 4.33 Suppose that $X \times Y$ is compact and that either X or Y is not compact (we will arrive at a contradiction). For definiteness, assume that X is not compact, and let $\{O_\alpha\}_{\alpha \in A}$ be an open cover of X containing no finite subcover. This implies that $\{O_\alpha \times Y\}_{\alpha \in A}$ is an open cover of $X \times Y$ containing no finite subcover — contradicting the given condition that $X \times Y$ is compact.

CYU 4.34 We show that $\pi_1 : X \times Y \to X$ is both open and closed. A similar argument can be use dot show that $\pi_2 : X \times Y \to Y$ is also open and closed.

$\pi_1 : X \times Y \to X$ is open: Let O be open in $X \times Y$, and let $x \in \pi_1(O)$. Choose $y \in Y$ such that $(x, y) \in O$. Let $U \in \tau_X, V \in \tau_Y$ be such that $(x, y) \in U \times V \subseteq O$. Then, $x \in U \subseteq \pi_1(O)$.

$\pi_1 : X \times Y \to X$ is closed: Let H be closed in $X \times Y$. We show $[\pi_1(H)]^c$ is open in X:
For any $x \notin \pi_1(H)$, $[\{x\} \times Y] \cap H = \emptyset$. Choose $y \in Y$ such that $(x, y) \notin H$. Since H is closed there exist $U \in \tau_X, V \in \tau_Y$ such that $(x, y) \in U \times V \subseteq H^c$. It follows that $x \in U \subseteq [\pi_1(H)]^c$.

CYU 4.35 Since $X_\alpha = \pi_\alpha(X)$, and since projection maps are continuous, X_α is compact (Theorem 4.12, page 187).

CYU 4.36 Start with the space $X = [0, 2\pi] \times [0, 1]$ and let $\sim$ be the equivalence relation represented by the partition $[(x, y)] = \{(x, y)\}$ for $x \notin (0, 2\pi)$ and $y \notin (0, 1)$, and $[(0, y)] = \{(0, y), (1, y)\}$ along with $[(x, 1)] = \{(x, 1), (x, 0)\}$.

Chapter 5
A Touch of Group Theory

CYU 5.1 (a) $(Q, +)$ is a group, with identity 0 and $-\frac{n}{m}$ the inverse of $\frac{n}{m} \in Q$.

(b) $(\Re, +)$ is a group, with identity 0 and $-r$ the inverse of $r \in \Re$.

(c) $(\Re, \cdot)$ is not a group. It does have a multiplicative identity; namely 1: $r \cdot 1 = 1 \cdot r \ \forall r \in \Re$. However, $0 \in \Re$ does not have a multiplicative inverse: there does not exist $r \in \Re$ such that $0 \cdot r = 1$.

(d) $(\Re^+, \cdot)$ is a group. Unlike the situation in (c) every element in $\Re^+ = \{r \in \Re | r > 0\}$ does have an inverse; namely $\frac{1}{r}$: $r \cdot \frac{1}{r} = 1$ and $\frac{1}{r} \cdot r = 1$.

CYU 5.2 The values in column a follow from the observation that $0 +_n n = n$ for $0 \le n \le 3$.
As for column b, row 3: $3 +_4 1 = 0$, since $3 + 1 = 4 = 1 \cdot 4 + 0$
As for column c, rows 2 and 3: $2 +_4 2 = 0$ and $3 +_4 2 = 1$, since:
$2 + 2 = 4 = 1 \cdot 4 + 0$ and $3 + 2 = 5 = 1 \cdot 4 + 1$.
As for column d, rows 1, 2, and 3: $1 +_4 3 = 0$, $2 +_4 3 = 1$,
and $3 +_4 3 = 2$, since: $1 + 3 = 4 = 1 \cdot 4 + 0$,
$2 + 3 = 5 = 1 \cdot 4 + 1$, $3 + 3 = 6 = 1 \cdot 4 + 2$.

	a	b	c	d
$+_4$	0	1	2	3
0	0	1	2	3
1	1	2	3	0
2	2	3	0	1
3	3	0	1	2

CYU 5.3 Let $G = \{e, a_1, a_2, ..., a_{n-1}\}$. By construction, the ith column of G's group table is precisely $ea_i, a_1 a_i, a_2 a_i, ..., a_{n-1} a_i$. The fact that every element of G appears exactly one time in that row is a consequence of Exercise 37, which asserts that the function $k_{a_i}: G \to G$ given by $k_{a_i}(g) = ga_i$ is a bijection.

CYU 5.4 From $\sigma = \begin{pmatrix} 1 & 2 & 3 & 4 & 5 \\ 1 & 5 & 2 & 3 & 4 \end{pmatrix}$ and $\tau = \begin{pmatrix} 1 & 2 & 3 & 4 & 5 \\ 5 & 3 & 2 & 1 & 4 \end{pmatrix}$ we have:

$$\tau \circ \sigma \begin{matrix} \sigma \\ \downarrow \tau \end{matrix} \begin{pmatrix} 1 & 2 & 3 & 4 & 5 \\ 1 & 5 & 2 & 3 & 4 \\ 5 & 4 & 3 & 2 & 1 \end{pmatrix} \Rightarrow \tau \circ \sigma : \begin{pmatrix} 1 & 2 & 3 & 4 & 5 \\ 5 & 4 & 3 & 2 & 1 \end{pmatrix}$$

$$\sigma \circ \tau \begin{matrix} \tau \\ \downarrow \sigma \end{matrix} \begin{pmatrix} 1 & 2 & 3 & 4 & 5 \\ 5 & 4 & 3 & 2 & 1 \\ 4 & 3 & 2 & 5 & 1 \end{pmatrix} \Rightarrow \sigma \circ \tau : \begin{pmatrix} 1 & 2 & 3 & 4 & 5 \\ 4 & 3 & 2 & 5 & 1 \end{pmatrix}$$

CYU 5.5 (a) We know that 1 and 5 are generators of Z_6 [Example 5.2(a)]. The remaining 4 elements in Z_6 are not:

$$0 +_6 0 = 0 \qquad 2 +_6 2 = 4 \qquad 3 +_6 3 = 0 \qquad 4 +_6 4 = 2$$
$$ \qquad 2 +_6 2 +_6 2 = 0 \qquad \qquad 4 +_6 4 +_6 4 = 0$$

(b) S_2: $\begin{cases} \overset{\alpha_0}{1 \to 1} & \overset{\alpha_1}{1 \to 2} \\ 2 \to 2 & 2 \to 1 \end{cases}$ is cyclic, with generator α_1: $\alpha_1 \circ \alpha_1 = \begin{matrix} 1 \to 2 \to 1 \\ 2 \to 1 \to 2 \end{matrix} = \alpha_0$.

(c) For $n > 2$, consider the following bijections $\beta, \gamma \in S_n$:

$\beta(1) = 2, \beta(2) = 1$, and $\beta(i) = i$ for $3 \le i \le n$

$\gamma(2) = 3, \gamma(3) = 2$, and $\beta(i) = i$ for i not equal to 2 or 3

Since $(\gamma \circ \beta)(1) = \gamma[\beta(1)] = \gamma(2) = 3$ and $(\beta \circ \gamma)(1) = \beta[\gamma(1)] = \beta(1) = 2$, S_n is not abelian, and therefore not cyclic.

CYU 5.6 Since $(bca)(bca) = (bc)(abc)a = (bc)e(a) = bca$: $bca = e$ (Theorem 5.6).

CYU 5.7 (a) False: For $\beta, \gamma, \delta \in S_3$ given by $\beta: \begin{matrix} 1 \to 2 \\ 2 \to 3 \\ 3 \to 1 \end{matrix}$, $\gamma: \begin{matrix} 1 \to 2 \\ 2 \to 1 \\ 3 \to 3 \end{matrix}$ and $\delta: \begin{matrix} 1 \to 3 \\ 2 \to 2 \\ 3 \to 1 \end{matrix}$ we have:

$$\begin{matrix} \beta & \gamma \\ 1 \to 2 \to 1 \end{matrix} \qquad \begin{matrix} \delta & \beta \\ 1 \to 3 \to 1 \end{matrix}$$

$\gamma \circ \beta: 2 \to 3 \to 3$ and $\beta \circ \delta: 2 \to 2 \to 3$ as well.
$ 3 \to 1 \to 2 \phantom{\text{ and } \beta \circ \delta:} 3 \to 1 \to 2$

(b) True: $a + b = b + c \underset{\text{commutativity}}{\Rightarrow} b + a = b + c \underset{\text{Theorem 5.9}}{\Rightarrow} a = c$

CYU 5.8 We show that the equation $a + x = b$ has a unique solution in $\langle \Re, + \rangle$:

Existence: $a + x = b \Rightarrow -a + (a + x) = -a + b \Rightarrow (-a + a) + x = -a + b$
$$\Rightarrow 0 + x = -a + b \Rightarrow x = -a + b$$

Uniqueness: If x and $\bar{x}$ then: $a + x = b$ and $a + \bar{x} = b \Rightarrow a + x = a + \bar{x} \underset{\text{Theorem 5.8}}{\Rightarrow} x = \bar{x}$

CYU 5.9 We know that we have to consider a non-abelian group, and turn to our friend S_3. Specifically for $\alpha, \beta \in S_3$ given by $\alpha: \begin{matrix} 1 \to 3 \\ 2 \to 2 \\ 3 \to 1 \end{matrix}$ and $\beta = \begin{matrix} 1 \to 3 \\ 2 \to 1 \\ 3 \to 2 \end{matrix}$ we have: $\alpha^{-1} = \begin{matrix} 1 \to 3 \\ 2 \to 2 \\ 3 \to 1 \end{matrix}$,

$\beta^{-1} = \begin{matrix} 1 \to 2 \\ 2 \to 3 \\ 3 \to 1 \end{matrix}$, and $\beta \circ \alpha = \begin{matrix} \alpha & \beta \\ 1 \to 3 \to 2 \\ 2 \to 2 \to 1 \\ 3 \to 1 \to 3 \end{matrix} = \begin{matrix} 1 \to 2 \\ 2 \to 1 \\ 3 \to 3 \end{matrix}$, so that:

$(\beta \circ \alpha)^{-1} = \begin{matrix} 1 \to 2 \\ 2 \to 1 \\ 3 \to 3 \end{matrix}$ while $\beta^{-1} \circ \alpha^{-1} = \begin{matrix} \alpha^{-1} & \beta^{-1} \\ 1 \to 3 \to 1 \\ 2 \to 2 \to 3 \\ 3 \to 1 \to 2 \end{matrix} = \begin{matrix} 1 \to 1 \\ 2 \to 3 \\ 3 \to 2 \end{matrix}$

CYU 5.10 I. $(a_n \ldots a_2 a_1)^{-1} = a_1^{-1} a_2^{-1} \ldots a_n^{-1}$ clearly holds for $n = 1$.

II. Assume $(a_k \cdots a_2 a_1)^{-1} = a_1^{-1} a_2^{-1} \ldots a_k^{-1}$. Then:

III. $(a_{k+1} \cdot a_k \cdots a_2 a_1)^{-1} = [a_{k+1} \cdot (a_k \cdots a_2 a_1)]^{-1}$

$\underset{\text{Theorem 2.10:}}{=} (a_k \cdots a_2 a_1)^{-1} a_{k+1}^{-1} \underset{\text{II}}{=} a_1^{-1} a_2^{-1} \ldots a_k^{-1} \cdot a_{k+1}^{-1}$

CYU 5.11 (a) $\begin{pmatrix} 1 & 2 & 3 & 4 \\ 2 & 3 & 4 & 1 \\ 3 & 4 & 1 & 3 \\ 4 & 1 & 2 & 3 \\ 1 & 2 & 3 & 4 \end{pmatrix}$ with arrows $\sigma, \sigma^2, \sigma^3, \sigma^4 = e$

σ has order $\boxed{4}$

(b) $1(4) = 4$
$2(4) = 4 +_{24} 4 = 8$
$3(4) = 8 +_{24} 4 = 12$
$4(4) = 12 +_{24} 4 = 16$
$5(4) = 16 +_{24} 4 = 20$
$6(4) = 20 +_{24} 4 = 0$: 4 has order $\boxed{6}$

CYU 5.12 We already know that $6Z$ is a subgroup of Z. To show that it is a subgroup of $3Z$ we need but observe that $6Z \subseteq 3Z$: $n \in 6Z \Rightarrow n = 6m$ for $m \in Z$
$$\Rightarrow n = 3(2m) \Rightarrow n \in 3Z$$

CYU 5.13 False: $2Z$ and $3Z$ are groups, but $2Z \cup 3Z$ is not:
$$2, 3 \in 2Z \cup 3Z \text{ while } 2 + 3 = 5 \notin 2Z \cup 3Z.$$

CYU 5.14 We show that $\langle 3 \rangle = Z_8 = \{0, 1, 2, 3, 4, 5, 6, 7\}$ by demonstrating that every element of Z_8 is a multiple of 3:

$$1 \cdot 3 = 3, \quad 2 \cdot 3 = 3 +_8 3 = 6, \quad 3 \cdot 3 = 3 +_8 3 +_8 3 = 1, \quad 4 \cdot 3 = 3 +_8 3 +_8 3 +_8 3 = 4$$
$$5 \cdot 3 = 3 +_8 3 +_8 3 +_8 3 +_8 3 = 7, \quad 6 \cdot 3 = 2, \quad 7 \cdot 3 = 5, \quad 8 \cdot 3 = 0$$

Claim: $\langle 4 \rangle = \{0, 4\}$: $1 \cdot 4 = 4$, $2 \cdot 4 = 4 +_8 4 = 0$. Fine, but can we pick up other elements of Z_8 by taking additional multiples of 4? No:

The division algorithm assures us that $n = q4 + r$ for any $n \in Z$, with $0 \leq r < 4$. From the above we know that $1 \cdot 4$ and $2 \cdot 4$ are in $\{0, 4\}$, and surely $0 \cdot 4 \in \{0, 4\}$. The only possible loose end is $3 \cdot 4$. Let's tie it up:

$$3 \cdot 4 = (4 +_8 4) +_8 4 = 0 +_8 4 = 4$$

CYU 5.15 False. In S_3 both the permutations $\alpha = \begin{pmatrix} 1 & 2 & 3 \\ 3 & 2 & 1 \end{pmatrix}$ and $\beta = \begin{pmatrix} 1 & 2 & 3 \\ 2 & 1 & 3 \end{pmatrix}$ have order 2.

Does $\sigma = \alpha \circ \beta$ have order 4? No, it has order 3:

$$\alpha \circ \beta \begin{pmatrix} 1 & 2 & 3 \\ 2 & 1 & 3 \\ 2 & 3 & 1 \end{pmatrix} \Rightarrow \sigma = \begin{pmatrix} 1 & 2 & 3 \\ 2 & 3 & 1 \end{pmatrix} \quad \text{Then:} \quad \begin{pmatrix} 1 & 2 & 3 \\ 2 & 3 & 1 \\ 3 & 1 & 2 \\ 1 & 2 & 3 \end{pmatrix} e$$

(By the way, as you can easily verify, **if G is abelian**, then the assertion in CYU 5.15 does hold)

CYU 5.16 (a) Let $\phi: G \to G'$ be given by $\phi(a) = e$. Since for every $a, b \in G$, $\phi(ab) = e = ee = \phi(a)\phi(b)$, ϕ is a homomorphism.

(b) Let $\phi: \langle Z, + \rangle \to \Re^+$ be given by $\phi(n) = \pi^n$. Since for every $n, m \in Z$, $\phi(n + m) = \pi^{n+m} = \pi^n \pi^m = \phi(n)\phi(m)$, ϕ is a homomorphism.

CYU 5.17 For $a, b \in G$ we have:

$$(\theta \circ \phi)(a + b) = \theta[\phi(a + b)] = \theta[\phi(a) + \phi(b)] = \theta[\phi(a)] + \theta[\phi(b)]$$
$$\text{Definition of composition} \to = (\theta \circ \phi)(a) + (\theta \circ \phi)(b)$$

CYU 5.18 Homomorphism:
$$\phi(2n_1 + 2n_2) = \phi[2(n_1 + n_2)] = 8(n_1 + n_2) = 8n_1 + 8n_2 = \phi(2n_1) + \phi(2n_1)$$
$$\text{Ker}(\phi) = \{2n | \phi(2n) = 0\} = \{2n | 8n = 0\} = \{0\}.$$
$$\text{Im}(\phi) = \{\phi(2n)\} = \{8n\} = 8Z$$

CYU 5.19 Let $a \in G$ be such that $\phi(b) = \phi(a) \Rightarrow b = a$ (*). We establish that ϕ is one-to-one by showing that $\text{Ker}(\phi) = e$ (Theorem 5.24):

Assume that $\phi(c) = e$ (we want to show that $c = e$). Consider the element ca.

Since ϕ is a homomorphism: $\phi(ca) = \phi(c)\phi(a) = e\phi(a) = \phi(a)$. Letting ca play the role of b in (*) we have:
$$ca = a \Rightarrow (ca)a^{-1} = e \Rightarrow c = e$$

CYU 5.20 (a) We show that the relation $\cong$ given by $G \cong G'$ if G is isomorphic to G' is an equivalence relation (see Definition 2.20, page 88):

Reflexive $G \cong G$ since the identity map $I(g) = g$ is clearly an isomorphism.

Symmetric $G \cong G' \Rightarrow G' \cong G$: Let $\phi: G \to G'$ be an isomorphism. Theorem 1.1(a), page 5, assures us that the map $\phi^{-1}: G' \to G$ is a bijection. We show that it is also a homomorphism:

For $a', b' \in G'$ let $a, b \in G$ be such that $\phi(a) = a'$ and $\phi(b) = b'$. Since $\phi(ab) = a'b'$, We then have: $\phi^{-1}(a'b') = ab = \phi^{-1}(a')(\phi^{-1}(b'))$.

Transitive $G_1 \cong G_2$ and $G_2 \cong G_3 \Rightarrow G_1 \cong G_3$: Follows from Theorem 1.2(c), page 7, and CYU 2.19.

(b) Let $g \in G$. The map $i_g: G \to G$ is a bijection:

One-to one: $i_g(a) = i_g(b) \Rightarrow gag^{-1} = gbg^{-1} \Rightarrow g^{-1}gag^{-1}g = g^{-1}gbg^{-1}g \Rightarrow a = b$

Onto: For $a \in G$, $i_g(g^{-1}ag) = g[g^{-1}ag]g^{-1} = a$

$i_g: G \to G$ is a homomorphism: $i_g(ab) = gabg^{-1} = (gag^{-1})(gbg^{-1}) = i_g(a)i_g(b)$

CYU 5.21 Let $\phi: G \to G'$ be an isomorphism. Assume that G is abelian. For $a', b' \in G'$ let $a, b \in G$ be **the** elements in G such that $\phi(a) = a'$ and $\phi(b) = b'$. Then:
$$a'b' = \phi(a)\phi(b) = \phi(ab) = \phi(ba) = \phi(b)\phi(a) = b'a'$$
The same argument can be used to show that if G' is abelian, then so is G.

Appendix B
Answers to Selected Exercises
Chapter 1
A logical Beginning

1.1 PROPOSITIONS

1. True **3.** True **5.** True **7.** False **9.** False **11.** False **13.** False **15.** True **17.** False
19. False **21.** True **23.** True **25.** True **27.** False **29.** False **31.** False **33.** Yes **35.** No
37. Yes **39.** Yes **45.** No **47.** No **49.** Yes **51.** No **53.** No **55.** No **57.** No **59.** No
63. True **65.** False **67.** Joe or Mary is not a math major. **69.** Joe is a math or biology major.
71. $3x + 5 \neq 5$ or x and y are not both integers.
73. x is divisible by 2 and 3 or it is not divisible by 7.

Statement: $p \to q$	contrapositive: $\sim q \to \sim p$	converse $q \to p$	inverse $\sim p \to \sim q$
75. If it rains, then I will stay home.	If it dose not rain, then I will not stay home.	If I stay home, then it will rain.	If it does not rain then I will not stay home.
77. If Nina feels better, then she will either go to the library or go shopping.	if Nina does not go to the library or go shopping, then she will tot feel better.	If Nina goes to the library or shopping, then Nina feels better.	If Nina does not feel better then she will not go to the library nor go shopping.
79. If $X = Z$ then $M > N$.	If $M \leq N$ then $X \neq Z$.	If $M > N$ then $X = Z$.	If $X \neq Z$ then $M \leq N$.
81. If $X \neq Z$ then M or N is a solution of the equation.	If M and N are not solutions of the equation, then $X = Z$.	M or N is a solution of the equation, then $X \neq Z$.	If $X = Z$ then neither nor N is a solution of the equation.
83. If $X < Z$ then the equation has no solution.	If the equation has a solution, then $X \geq Z$.	If the equation has no solution, then $X < Z$.	If $X \geq Z$ then the equation has a solution

1.2 QUANTIFIERS

1. False **3.** True **5.** False **7.** True **9.** False **11.** True **13.** True **15.** False **17.** False
19. False **21.** False **23.** True **25.** True **27.** True **29.** False **31.** True **33.** True
35. False **37.** True **39.** False **41.** True **43.** False **45.** False **47.** False **49.** False **51.** False
53. There is a road that does not lead to Rome.
59. Sometimes it rains but not pennies from heaven.   **57.** $\exists n \in Z \ni n \geq a$
59. $\exists n \in Z \ni n \geq a$ or $n \geq b$ **61.** $\exists x \in X \ni \sim p(x) \vee \sim q(x) \vee \sim s(x)$
63. $\exists x \in X, \sim p((x) \vee [\sim q((x) \wedge \sim s(x))])$ **65.** For every boy there is not some girl.
67. $\forall n \in Z, n \geq a$ **69.** $\forall n \in Z, n \geq a$ or $n < b$ **71.** $\forall x \in X, \sim p(x) \vee \sim q(x) \vee \sim s(x)$

73. $\forall x \in X, \sim p(x) \vee [\sim q(x) \wedge \sim s(x)]$ 75. There is somebody that loves nobody.
77. There is someday with no special moment. 79. $\exists x \in X \ni \forall y \in Y, x + y \neq 0$
81. $\exists a, b \in S \ni \forall m, n \in Z, a + b \neq mn$ 83. $\exists x \in X \ni \forall a, b \in Y, a \neq x + b$ and $b \neq a + x$
85. For every opera there exists a symphony not longer than that opera.
87. For every person there is someone that is greater than that person.
89. $\forall x \in X, \exists y \in Y \ni x + y \neq 0$ 91. $\forall a, b \in S, \exists m, n \in Z \ni a + b \neq mn$
93. $\forall x \in X, \exists a, b \in Y \ni a \neq x + b$ and $b \neq a + x$

1.3 Methods of Proof
Each exercise calls for a verification or proof.

1.4 Principle of Mathematical Induction
Each exercise calls for a verification or proof.

1.5 The Division Algorithm and Beyond
1. $q = r = 0$ 3. $q = -27, r = 0$ 5. 1

Chapter 2
A Touch of Set Theory

2.1 Basic Definitions
1. U 3. $\{15n | n \in U\}$ 5. B 7. D 9. U 11. C 13. F 15. D
17. $\{1, 2, 4, 5, 7, 8, 10, 11, 13, 14\}$ 19. $\{16, 17, 18, \ldots\}$ 21. $\{\emptyset, \{1\}, \{2\}, \{1, 2\}\}$
23. $\{\emptyset, \{\emptyset\}\}$

2.2 Functions
1. Not a function   3. Range: $\{A, B, D\}$ 5. Range: $\{A, B, C, D\}$
7. Not one-to-one, not onto.   9. Both one-to-one and onto.   11. One-to-one and not onto.
13. Onto and not one-to-one.   15. Both one-to-one and onto.   17. One-to-one and not onto.
19. Not one-to-one, not onto.   21. One-to-one and not onto.   23. One-to-one and not onto.
25. Both one-to-one and onto.   27. Not one-to-one, not onto.   29. Not one-to-one, not onto.
31. Not one-to-one, not onto.   33. $f^{-1}(y) = \dfrac{y+2}{3}$ 35. $f^{-1}(y) = \dfrac{y}{2-y}$

37. $f^{-1}(a, b) = \left(\dfrac{a}{5}, b - 3\right)$ 39. $f^{-1}\begin{bmatrix} a & b \\ c & d \end{bmatrix} = \begin{bmatrix} c & -d \\ a & \dfrac{b}{2} \end{bmatrix}$ 41. $f^{-1}(a, b, c) = \begin{bmatrix} \dfrac{a}{2} \\ \dfrac{1}{2}(s - 2b) \\ \dfrac{1}{2}(-a + 2b - 2c) \end{bmatrix}$

2.3 INFINITE COUNTING
Each exercise calls for a verification or proof.

2.4 EQUIVALENCE RELATIONS
1. No **3.** Yes **5.** No **13.** Yes **15.** No **17.** No **19.** No **31.** Yes **33.** Yes
35. No **43.** Yes **45.** Yes **57.** Yes **59.** Yes **61.** Yes **71.** Yes **73.** No
75. Yes **77.** Yes **79.** Yes **81.** No **93.** Yes **95.** No **97.** Yes **99.** No

101. $[a] = \{a, -a-2\}$ **103.** $\left[\dfrac{a}{b}\right] = \left\{\dfrac{a}{d}\,\middle|\, \gcd(a,d) = 1\right\}$ **105.** $[r] = \{-r, r\}$

107. $[(x,y)] = \{(x, \bar{y}) \mid \bar{y} \in \Re\}$ **113.** $[n] = \{n + 10k \mid n + 10k \leq 100\}$

115. $[\varnothing] = \{\varnothing\}$, $[\{1,2,3\}] = \{\{1,2,3\}\}$, for $n = 1, 2$, or 3: $[n] = \{n\}$,
$[\{1,2\}] = [\{1,3\}] = [\{2,3\}] = \{\{1,2\},\{1,3\},\{2,3\}\}$

Chapter 3
A Touch of Analysis

3.1 THE REAL NUMBER SYSTEM
1. Least upper bound: 7, no max or min. **3.** Least upper bound: 10, no max or min.
5. No Least upper bound: 7, no max, min: 7. **7.** 1 **9.** 2 **11.** $\dfrac{124}{25}$

3.2 SEQUENCES

1. $\dfrac{n}{n+1}$ **3.** $\begin{cases} 1 & \text{if } n \text{ is even} \\ 2 & \text{if } n \text{ is odd} \end{cases}$ **5.** $\begin{cases} \left(\dfrac{n}{2}\right)^2 & \text{if } n \text{ is even} \\ \dfrac{n+1}{2} & \text{if } n \text{ is odd} \end{cases}$ **7.** 0 **9.** −1 **11.** 1

13. 2 **15.** 1 **23.** 0 **25.** Does not exist **27.** 1 **29.** 0 **31.** $\dfrac{1}{2}$

3.3 METRIC SPACE STRUCTURE OF $\Re$
1. Neither open nor closed, bounded below and above, not compact.
3. Open, bounded below and above, not compact.
5. Open, bounded below and above, not compact.
7. Closed, bounded below and above, compact.

3.4 CONTINUITY
Each exercise calls for a verification or proof.

Chapter 4
A Touch of Topology

4.1 Metric Spaces

7. Fails property (i) of Definition 4.1.   **9.** Fails Properties (i) and (iii) of Definition 4.1.
9. Fails Properties (i), (ii), and (iii) of Definition 4.1.

4.2 Topological Spaces

1(a). Yes **1(b).** Yes **1(c).** No **1d).** No **1(e).** Yes **1(f).** No **1(g).** No **1(h).** Yes

17(b) No **17(c)** No   **27.** $\overline{\{a\}} = \{a,b,c,d\}, \overline{\{b\}} = \{b,e\}, \overline{\{c\}} = \{c,d\}, \overline{\{d\}} = \{d,c\},$
$\overline{\{e\}} = \{e\}, \overline{\{c,e\}} = \{c,e,d\}, \overline{\{b,e\}} = \{b,e\}$

33. (a) $[1,3]$ (b) Z^+ (c) $\Re$ (d) $[1,3] \cup \{5\}$ (e) $\{0\} \cup \left\{\dfrac{1}{n} \middle| n \in Z^+\right\}$ **37.** Z^+

4.3 Continuous Functions and Homomorphisms

Each exercise calls for a verification or proof.

4.4 Product and Quotient Spaces

1(a). $\tau_1 \times \tau_1$ **1(b).** $\tau_1 \times \tau_2$ **1(c).** $\tau_2 \times \tau_2$

Chapter 5
A Touch of Group Theory

5.1 Definitions and Examples

1. A cyclic group with generator 2.   **2.** Not a group. It does not contain an identity.
3. Not a group. It does not contain an identity.   **5.** Not a group. It does not contain an identity.
7. Not a group. 1 is the identity, but 2 has no inverse.   **9.** Abelian group. Not cyclic.
11. Abelian group. Not cyclic.

13. $\alpha_3^2 = e, \alpha_3^3 = \alpha_3$ **15.** $\alpha_1^n = \begin{cases} e & \text{if } n \equiv 0 \bmod 3 \\ \alpha_1 & \text{if } n \equiv 1 \bmod 3 \\ \alpha_2 & \text{if } n \equiv 2 \bmod 3 \end{cases}$ **17.** $\alpha_3^{-n} = \begin{cases} e & \text{if } n \text{ is even} \\ \alpha_3 & \text{if } n \text{ is odd} \end{cases}$

19. $\alpha_2^2 = \alpha_1, \alpha_2^3 = e$ **21.** $\alpha_2^n = \begin{cases} e & \text{if } n \equiv 0 \bmod 3 \\ \alpha_2 & \text{if } n \equiv 1 \bmod 3 \\ \alpha_1 & \text{if } n \equiv 2 \bmod 3 \end{cases}$ **23.** $\alpha_2^{-n} = \begin{cases} e & \text{if } n \equiv 0 \bmod 3 \\ \alpha_1 & \text{if } n \equiv 1 \bmod 3 \\ \alpha_2 & \text{if } n \equiv 2 \bmod 3 \end{cases}$

25. $\beta\alpha = \begin{pmatrix} 1 & 2 & 3 & 4 & 5 & 6 \\ 1 & 3 & 3 & 6 & 5 & 2 \end{pmatrix}$ **27.** $\gamma\beta = \begin{pmatrix} 1 & 2 & 3 & 4 & 5 & 6 \\ 5 & 6 & 3 & 4 & 1 & 2 \end{pmatrix}$ **29.** $\alpha^5 = \alpha^{-1} = \begin{pmatrix} 1 & 2 & 3 & 4 & 5 & 6 \\ 6 & 1 & 2 & 3 & 4 & 5 \end{pmatrix}$

31. $\alpha^{101} = \alpha^5 = \alpha^{-1} = \begin{pmatrix} 1 & 2 & 3 & 4 & 5 & 6 \\ 6 & 1 & 2 & 3 & 4 & 5 \end{pmatrix}$ **33.** $\beta^{101} = \beta^5 = \beta^{-1} = \begin{pmatrix} 1 & 2 & 3 & 4 & 5 & 6 \\ 2 & 1 & 4 & 3 & 6 & 5 \end{pmatrix}$

35. Abelian   **37.** Abelian   **45.** Not Abelian

5.2 Elementary Properties of Groups

1. (a) e (b) a (c) $a^{-1}cb^{-1}$ (d) aba^{-1}

5.3 Subgroups

1. Yes   **3.** No   **5.** Yes   **7.** Yes   **9.** Yes   **11.** Yes   **13.** No   **15.** Yes   **17.** No

5.4 Homomorphisms and Isomorphisms

1. Yes   **3.** No   **5.** Yes   **7.** Yes   **9.** Yes   **11.** Yes

A

Abelian Group, 209
Addition Modulo n, 208
Alexander's Subbase Theorem, 177
Algebra of Functions, 150
Algebra of Sequences, 126
Alternate Principle of Induction, 38
Archimedian Principle, 114
Automorphism, 242
Axiom of Choice, 110

B

Base, 175
Biconditional Statement, 4
Bijection, 70
Binary Operator, 2
Bolzano-Weierstrass Theorem, 129
Bound, 112
    Greatest Lower, 112
    Least Upper, 112
    Lower Bound, 112
    Upper Bound, 112
Bounded Sequence, 125
Bounded Set, 164

C

Cardinality, 77
Cartesian Product, 63
Cayley's Theorem, 244
Cauchy Sequence, 130
Closed Function, 188
Closed Set, 137, 162, 176
Compact, 139, 164, 181
Complete Ordered Field, 111
Completion Axiom, 112
Composition, 64
Compound Proposition, 2
Conditional Statement, 3
    Converse, 9
    Inverse, 9
    Negation, 8
Congruence Modulo n, 90
Conjunction, 2
Cantor Theorem, 85
Cantor-Bernstein-Schroder Theorem, 82
Continuity at a point, 147

Continuous Function, 149, 166, 185
Contrapositive, 8
Converse, 9
Countable Set, 78
Convergent Sequence, 123
Cyclic Group, 214
    Generator, 214

D

Decreasing Sequence, 125
De Morgan's Laws, 6, 57, 58
Dense Subset, 118
Discrete Metric Space, 161
Discrete Topological Space, 171
Disjunction, 2
Division Algorithm, 43
Divisibility, 27
Domain, 63

E

Equivalence Class, 91
Equivalence Relation, 88
Euclidean Metric, 135
Even Integer, 23
Existential Proposition, 16

F

Finite Complement Space, 171
Finite Subcover, 139, 164
Function, 63
    Bijection, 70
    Closed, 188
    Continuous, 149, 166, 170
    Domain, 63
    Composition, 64
    Homeomorphism, 199
    Homomorphism, 237
    Inverse, 70
    Isomorphism, 241
    One-to-One, 65
    Onto, 66
    Open, 188
    Range, 63
Fundamental Counting Principle, 7
Fundamental Theorem of Arithmetic, 48

G

Generator, 214
Greatest Common Divisor, 44
Greatest Lower Bound, 112
Group, 207
- Abelian, 209
- Cyclic, 214
  - Generator, 215, 230
- Elementary Properties, 220
- Invariant Property, 244
- Klein, 209
- Order, 209
- Subgroup, 228
- Symmetric, 213
- Table, 209

H

Hausdorff Space, 178
Heine-Borell Theorem, 140
Homeomorphic Spaces, 190
Homeomorphism, 190
Homomorphism, 237
- Image, 239
- Kernel, 239
Hypothesis, 3

I

Image, 152, 165
Incompleteness Theorem, 110
Increasing Sequence, 125
Indiscrete Topological Spaces, 171
Induction 33
Integer, 23
- Even, 23
- Odd, 23
- Prime, 47
- Relatively Prime, 46
Inverse Function, 70
Isometry, 167
Isometric Spaces, 167
Isomorphism, 241
Isomorphic Spaces, 241
Interval Notation, 59
Invariant Property, 191, 244
Inverse Function, 70
Inner Automorphism, 242

K

Kernel, 240
Klein Group, 209

L

Lagrange's Theorem, 230
Least Upper Bound, 112
Limit of a Sequence, 123
Logically Equivalent, 5
Lower Bound, 112

M

Matrix, 68
Maximum, 112
Metric, 160
Metric Space, 159
- Compact, 16a
- Discrete, 161
- Euclidean, 135
Metrizable Space, 172
Minimum, 112
Monotone Sequence, 125

N

n-tuple, 68
Negation of a Proposition, 2
- of a Conditional Proposition, 8
- of a Quantified Proposition, 18
Nested Closed Interval Property, 115

O

Odd Integer, 23
One-to-One, 65
Onto, 66
Open Cover, 139, 164
Open Sphere, 161
Open Function, 188
Open Set, 135, 162
Order of a Group, 209
Order of an Element, 224

P

Partition, 92
Path Connected Spaces, 191
Permutation, 212
Power Set, 83
Pre-Image, 152, 165

Proposition, 1
    Compound, 2
Power Set, 83
Prime, 47
Principle of Mathematical Induction, 33
Product Space, 2196, 198, 199
Projection Function, 199, 200
Proof, 23
    by Contradiction, 26
    Contrapositive, 24
    Direct, 23
Proposition, 1
    Compound, 2
    Containing Multiple Quantifiers, 16
    Existential, 15
    Universal, 14

Q

Quantifiers, 14
    Existential, 15
    Universal, 14
Quotient Space, 201
Quotient Topology, 201

R

Range, 63
Relation, 88
    Equivalence, 88
    Reflexive, 88
    Symmetric, 88
Relatively Prime, 46

S

Sequence, 122, 166
    Algebra of, 126
    Bounded, 125
    Cauchy, 130, 167
    Convergent, 123, 166
    Decreasing, 125
    Divergent, 123
    Increasing, 125
    Monotone, 125
    Subsequence, 129

Set, 10
    Bounded, 164
    Closed, 137, 162, 173
    Compact, 139, 164, 176
    Complement, 54
    Dense, 118
    Disjoint, 54
    Equality, 53
    Intersection, 53
    Notation, 10
    Open, 139, 169, 178
    Subset, 53
        Proper, 53
    Union, 53
Sierpinski Space, 172
Sphere of Radius of r $[S_r(x)]$, 123, 161
Statement, 1
Subbase, 175
Subgroup, 227
    Generated by an Element, 228
Subsequence, 129
Subspace, 174
Subset, 55
Successor Set, 103
Symmetric Group, 211

T

Tautology, 4
Topology, 171
Topological Space, 171
    Base, 175
    Compact, 139
    Discrete, 164, 171
    Finite-Complement, 171
    Hausdorff, 178
    Half-Open, 176
    Indiscrete, 171
    Metrizable, 172
    Sierpinski, 172
    Subbase, 175
    Subspace 174
Topologically Invariant Property, 191
Tychonoff's Product Theorem, 200

U
Uncountable Set, 81
Union, 53
Unary Operation, 2
Universal Proposition, 14
Upper Bound, 112

W
Well Ordering Principle, 39

www.ingramcontent.com/pod-product-compliance
Lightning Source LLC
Chambersburg PA
CBHW080651190526
45169CB00006B/2064